# Essential Concepts in Mathematics

# Essential Concepts in Mathematics

Editor: Calanthia Wright

NY Research Press
New York

Published by NY Research Press
118-35 Queens Blvd., Suite 400,
Forest Hills, NY 11375, USA
www.nyresearchpress.com

Essential Concepts in Mathematics
Edited by Calanthia Wright



International Standard Book Number: 978-1-63238-631-1 (Hardback)



**Cataloging-in-Publication Data**

Essential concepts in mathematics / edited by Calanthia Wright.
p. cm.
Includes bibliographical references and index.
ISBN 978-1-63238-631-1
1. Mathematics. 2. Mathematics--Philosophy. I. Wright, Calanthia.
QA36 .E87 2019
510--dc23

# Contents

# Preface

Mathematics is a vast field of study with applications in all major fields of science, engineering, business, economics, computer science and many more. Mathematics uses the tools of measurement, calculations, algebra, etc. to study concepts such as shape, quantity, structure and change. Some of the major branches of mathematics are algebra, calculus, geometry, topology, number theory, probability and statistics. The chapters included herein explore all the important aspects of mathematics in the present day scenario. It strives to provide a fair idea about this discipline and to help develop a better understanding of the latest advances within this field. Coherent flow of topics, student-friendly language and extensive use of examples make this book an invaluable source of knowledge.

The information shared in this book is based on empirical researches made by veterans in this field of study. The elaborative information provided in this book will help the readers further their scope of knowledge leading to advancements in this field.

Finally, I would like to thank my fellow researchers who gave constructive feedback and my family members who supported me at every step of my research.

**Editor**

# 1

# Remarks on an elliptic problem arising in weighted energy estimates for wave equations with space-dependent damping term in an exterior domain

**Motohiro Sobajima**[1,*] **and Yuta Wakasugi**[2]

[1] Department of Mathematics, Faculty of Science and Technology, Tokyo University of Science, 2641 Yamazaki, Noda-shi, Chiba-ken 278-8510, Japan

[2] Graduate School of Mathematics, Nagoya University, Furocho, Chikusaku, Nagoya 464-8602 Japan

[*] **Correspondence:** Email: msobajima1984@gmail.com

**Abstract:** This paper is concerned with weighted energy estimates and diffusion phenomena for the initial-boundary problem of the wave equation with space-dependent damping term in an exterior domain. In this analysis, an elliptic problem was introduced by Todorova and Yordanov. This attempt was quite useful when the coefficient of the damping term is radially symmetric. In this paper, by modifying their elliptic problem, we establish weighted energy estimates and diffusion phenomena even when the coefficient of the damping term is not radially symmetric.

**Keywords:** Damped wave equation; elliptic problem; exterior domain; weighted energy estimates; diffusion phenomena

---

## 1. Introduction

Let $N \geq 2$. We consider the wave equation with space-dependent damping term in an exterior domain $\Omega \subset \mathbb{R}^N$ with a smooth boundary:

$$\begin{cases} u_{tt} - \Delta u + a(x)u_t = 0, & x \in \Omega,\ t > 0, \\ u(x,t) = 0, & x \in \partial\Omega,\ t > 0, \\ (u, u_t)(x,0) = (u_0, u_1)(x), & x \in \Omega, \end{cases} \tag{1.1}$$

where we denote by $\Delta$ the usual Laplacian in $\mathbb{R}^N$ and by $u_t$ and $u_{tt}$ the first and second derivative of $u$ with respect to the variable $t$, and $u = u(x,t)$ is a real-valued unknown function. The coefficient of the damping term $a(x)$ satisfies $a \in C^2(\overline{\Omega})$, $a(x) > 0$ on $\overline{\Omega}$ and

$$\lim_{|x|\to\infty} \left(\langle x\rangle^{\alpha} a(x)\right) = a_0 \tag{1.2}$$

with some constants $\alpha \in [0, 1)$ and $a_0 \in (0, \infty)$, where $\langle y \rangle = (1 + |y|^2)^{\frac{1}{2}}$ for $y \in \mathbb{R}^N$. In this moment, the initial data $(u_0, u_1)$ are assumed to have compact supports in $\Omega$ and to satisfy the compatibility condition of order $k \geq 1$:

$$(u_{\ell-1}, u_\ell) \in (H^2 \cap H_0^1(\Omega)) \times H_0^1(\Omega), \quad \text{for all } \ell = 1, \ldots, k \tag{1.3}$$

where $u_\ell$ is successively defined by $u_\ell = \Delta u_{\ell-2} - a(x) u_{\ell-1}$ $(\ell = 2, \ldots, k)$. We note that existence and uniqueness of solution to the problem (1.1) have been discussed (see e.g., Ikawa [2, Theorem 2]).

It is proved in Matsumura [4] that if $\Omega = \mathbb{R}^N$ and $a(x) \equiv 1$, then the solution $u$ of (1.1) satisfies the energy decay estimate

$$\int_{\mathbb{R}^N} (|\nabla u(x,t)|^2 + |u_t(x,t)|^2)\, dx \leq C(1+t)^{-\frac{N}{2}-1} \|(u_0, u_1)\|^2_{H^1 \times L^2},$$

where the constant $C$ depends on the size of the supprot of initial data. Moreover, it is shown in Nishihara [7] that $u$ has the same asymptotic behavior as the one of the problem

$$\begin{cases} v_t - \Delta v = 0, & x \in \mathbb{R}^N,\ t > 0, \\ v(x, 0) = u_0(x) + u_1(x), & x \in \mathbb{R}^N. \end{cases} \tag{1.4}$$

In particular, we have

$$\|u(\cdot, t) - v(\cdot, t)\|_{L^2} = o(t^{-\frac{N}{4}})$$

as $t \to \infty$. Energy decay properties of solutions to (1.1) for general cases with $a(x) \geq \langle x \rangle^{-\alpha}$ $(0 \leq \alpha \leq 1)$ have been dealt with by Matsumura [5]. On the other hand, Mochizuki [6] proved that if $0 \leq a(x) \leq C\langle x \rangle^{-\alpha}$ for some $\alpha > 1$, then the energy of the solution to (1.1) does not vanish as $t \to \infty$ for suitable initial data. (The solution has an asymptotic behavior similar to the solution of the usual wave equation without damping). Therefore one can expect that diffusion phenomena occur only when $a(x) \geq C\langle x \rangle^{-\alpha}$ for $\alpha \leq 1$.

In this paper, we discuss precise decay rates of the weighted energy

$$\int_{\mathbb{R}^N} (|\nabla u(x,t)|^2 + |u_t(x,t)|^2) \Phi(x,t)\, dx$$

with a special weight function

$$\Phi(x, t) = \exp\left(\beta \frac{A(x)}{1+t}\right)$$

(for some $A \in C^2(\mathbb{R}^N)$ and $\beta > 0$) which is introduced by Todorova and Yordanov [12] based on the ideas in [11] and in [3]. They proved weighted energy estimates

$$\int_{\mathbb{R}^N} a(x) |u(x,t)|^2 \Phi(x,t)\, dx \leq C(1+t)^{-\frac{N-\alpha}{2-\alpha}+\varepsilon},$$

$$\int_{\mathbb{R}^N} (|\nabla u(x,t)|^2 + |u_t(x,t)|^2) \Phi(x,t)\, dx \leq C(1+t)^{-\frac{N-\alpha}{2-\alpha}-1+\varepsilon}$$

when $a(x)$ is radially symmetric and satisfies (1.2). After that, Radu, Todorova and Yordanov [8] extended it to higher-order derivatives. In [13], the second author proved diffusion phenomena for

(1.1) with $\Omega = \mathbb{R}^N$ and $a(x) = \langle x\rangle^{-\alpha}$ ($\alpha \in [0,1)$) by comparing the solution of the following problem

$$\begin{cases} a(x)v_t - \Delta v = 0, & x \in \mathbb{R}^N,\ t > 0, \\ v(x,0) = u_0(x) + \dfrac{1}{a(x)}u_1(x), & x \in \mathbb{R}^N. \end{cases} \tag{1.5}$$

In [10], diffusion phenomena for (1.1) with an exterior domain and for general radially symmetric damping term are obtained. However, the weighted energy estimates and diffusion phenomena for (1.1) with **non-radially symmetric damping** are still remaining open. The difficulty seems to come from the choice of auxiliary function $A$ in the weighted energy, which strongly depends on the existence of positive solution to the Poisson equation $\Delta A(x) = a(x)$. In fact, an example of non-existence of positive solution to $\Delta A = a$ for non-radial $a(x)$ is shown in [10]. Radu, Todorova and Yordanov [9] considered the case $\Omega = \mathbb{R}^N$ and used a solution $A_*(x)$ of $\Delta A_* = a_1(1+|x|)^{-\alpha}$ with $a_1 > 0$ satisfying $a_1(1+|x|)^{-\alpha} \geq a(x)$ for $x \in \mathbb{R}^N$, that is, $A_*(x)$ is a subsolution of the equation $\Delta A = a$. In general one cannot obtain the optimal decay estimate via this choice because of the luck of the precise behavior of $a(x)$ at the spatial infinity which can be expected to determine the precise decay late of weighted energy estimates. Our main idea to overcome this difficulty is to weaken the equality $\Delta A = a$ and consider the inequality $(1-\varepsilon)a \leq \Delta A \leq (1+\varepsilon)a$, and to construct a solution having appropriate behavior, we employ a cut-off argument.

The aim of this paper is to give a proof of Ikehata–Todorova–Yordanov type weighted energy estimates for (1.1) with non-radially symmetric damping and to obtain diffusion phenomena for (1.1) under the compatibility condition of order 1 and the condition (1.2) (without any restriction).

This paper is originated as follows. In Section 2, we discuss related elliptic and parabolic problems. The weighted energy estimates for (1.1) are established in Section 3 (Proposition 3.5). Section 4 is devoted to show diffusion phenomena (Proposition 4.1).

## 2. Related elliptic and parabolic problems

### *2.1. An elliptic problem for weighted energy estimates*

As we mentioned above, in general, existence of positive solutions to the Poisson equation $\Delta A(x) = a(x)$ is false for non-radial $a(x)$. Thus, we weaken this equation and consider the following inequality

$$(1-\varepsilon)a(x) \leq \Delta A(x) \leq (1+\varepsilon)a(x), \quad x \in \Omega, \tag{2.1}$$

where $\varepsilon \in (0,1)$ is a parameter. Here we construct a positive solution $A$ of (2.1) satisfying

$$A_{1\varepsilon}\langle x\rangle^{2-\alpha} \leq A(x) \leq A_{2\varepsilon}\langle x\rangle^{2-\alpha}, \tag{2.2}$$

$$\frac{|\nabla A(x)|^2}{a(x)A(x)} \leq \frac{2-\alpha}{N-\alpha} + \varepsilon \tag{2.3}$$

for some constants $A_{1\varepsilon}, A_{2\varepsilon} > 0$.

**Lemma 2.1.** *For every $\varepsilon \in (0,1)$, there exists $A_\varepsilon \in C^2(\overline{\Omega})$ such that $A_\varepsilon$ satisfies* (2.1)–(2.3).

*Proof.* Firstly, we extend $a(x)$ as a positive function in $C^2(\mathbb{R}^N)$; note that this is possible by virtue of the smoothness of $\partial\Omega$. To simplify the notation, we use the same symbol $a(x)$ as a function defined on

$\mathbb{R}^N$. We construct a solution of approximated equation

$$\Delta A_\varepsilon(x) = a_\varepsilon(x), \quad x \in \mathbb{R}^N$$

for some $a_\varepsilon \in C^2(\mathbb{R}^N)$ satisfying

$$(1-\varepsilon)a(x) \le a_\varepsilon(x) \le (1+\varepsilon)a(x), \quad x \in \mathbb{R}^N. \tag{2.4}$$

Noting (1.2), we divide $a(x)$ as $a(x) = b_1(x) + b_2(x)$ with

$$b_1(x) = \Delta\left(\frac{a_0}{(N-\alpha)(2-\alpha)}\langle x\rangle^{2-\alpha}\right) = a_0\langle x\rangle^{-\alpha} + \frac{a_0\alpha}{N-\alpha}\langle x\rangle^{-\alpha-2},$$
$$b_2(x) = a(x) - a_0\langle x\rangle^{-\alpha} - \frac{a_0\alpha}{N-\alpha}\langle x\rangle^{-\alpha-2}.$$

Then we have

$$\lim_{|x|\to\infty}\left(\frac{b_2(x)}{a(x)}\right) = \lim_{|x|\to\infty}\left[\frac{1}{\langle x\rangle^\alpha a(x)}\left(\langle x\rangle^\alpha a(x) - a_0 - \frac{a_0\alpha}{N-\alpha}\langle x\rangle^{-2}\right)\right] = 0. \tag{2.5}$$

Let $\varepsilon \in (0,1)$ be fixed. Then by (2.5) there exists a constant $R_\varepsilon > 0$ such that $|b_2(x)| \le \varepsilon a(x)$ for $x \in \mathbb{R}^N \setminus B(0,R_\varepsilon)$. Here we introduce a cut-off function $\eta_\varepsilon \in C_c^\infty(\mathbb{R}^N,[0,1])$ such that $\eta_\varepsilon \equiv 1$ on $B(0,R_\varepsilon)$. Define

$$a_\varepsilon(x) := b_1(x) + \eta_\varepsilon(x)b_2(x) = a(x) - (1-\eta_\varepsilon(x))b_2(x), \quad x \in \mathbb{R}^N.$$

Then $a_\varepsilon(x) = a(x)$ on $B(0,R_\varepsilon)$ and for $x \in \mathbb{R}^N \setminus B(0,R_\varepsilon)$,

$$\left|\frac{a_\varepsilon(x)}{a(x)} - 1\right| = (1-\eta_\varepsilon(x))\frac{|b_2(x)|}{a(x)} \le \varepsilon$$

and therefore (2.4) is verified.

Next we define

$$B_{1\varepsilon}(x) := \frac{a_0}{(N-\alpha)(2-\alpha)}\langle x\rangle^{2-\alpha}, \quad x \in \mathbb{R}^N,$$
$$B_{2\varepsilon}(x) := -\int_{\mathbb{R}^N} \mathcal{N}(x-y)\eta_\varepsilon(y)b_2(y)\,dy, \quad x \in \mathbb{R}^N,$$

where $\mathcal{N}$ is the Newton potential given by

$$\mathcal{N}(x) = \begin{cases} \dfrac{1}{2\pi}\log\dfrac{1}{|x|} & \text{if } N = 2, \\ \dfrac{\Gamma(\frac{N}{2}+1)}{N(N-2)\pi^{\frac{N}{2}}}|x|^{2-N} & \text{if } N \ge 3. \end{cases}$$

Then we easily see that $\Delta B_{1\varepsilon}(x) = b_1(x)$ and $\Delta B_{2\varepsilon} = \eta_\varepsilon(x)b_2(x)$. Moreover, noting that $\operatorname{supp}(\eta_\varepsilon b_2)$ is compact, we see from a direct calculation that there exist a constant $M_\varepsilon > 0$ such that

$$|B_{2\varepsilon}(x)| \le \begin{cases} M_\varepsilon(1+\log\langle x\rangle) & \text{if } N = 2, \\ M_\varepsilon\langle x\rangle^{2-N} & \text{if } N \ge 3, \end{cases} \qquad |\nabla B_{2\varepsilon}(x)| \le M_\varepsilon\langle x\rangle^{1-N}, \quad x \in \mathbb{R}^N.$$

This yields that $B_\varepsilon := B_{1\varepsilon} + B_{2\varepsilon}$ is bounded from below and positive for $x \in \mathbb{R}^N$ with sufficiently large $|x|$. Moreover, we have

$$\lim_{|x|\to\infty} \left(\langle x\rangle^{\alpha-2} B_\varepsilon(x)\right) = \frac{a_0}{(N-\alpha)(2-\alpha)}$$

and

$$\begin{aligned}\lim_{|x|\to\infty} \left(\frac{|\nabla B_\varepsilon(x)|^2}{a(x)B_\varepsilon(x)}\right) &= \lim_{|x|\to\infty} \left(\frac{1}{\langle x\rangle^{\alpha} a(x)} \cdot \frac{1}{\langle x\rangle^{\alpha-2} B_\varepsilon(x)} \left|\frac{a_0}{N-\alpha}\langle x\rangle^{-1} x + \langle x\rangle^{\alpha-1}\nabla B_{2\varepsilon}(x)\right|^2\right) \\ &= \frac{2-\alpha}{N-\alpha}.\end{aligned}$$

Using the same argument as in the proof of [10, Lemma 3.1], we can see that there exists a constant $\lambda_\varepsilon \geq 0$ such that $A_\varepsilon(x) := \lambda_\varepsilon + B_\varepsilon(x)$ satisfies (2.1)-(2.3). □

## 2.2. *A parabolic problem for diffusion phenomena*

Here we consider $L^p$-$L^q$ type estimates for solutions to the initial-boundary value problem of the following parabolic equation

$$\begin{cases} a(x)w_t - \Delta w = 0, & x \in \Omega,\ t > 0, \\ w(x,t) = 0, & x \in \partial\Omega,\ t > 0, \\ w(x,0) = f(x), & x \in \Omega. \end{cases} \tag{2.6}$$

Here we introduce a weighted $L^p$-spaces

$$L^p_{d\mu} := \left\{ f \in L^p_{\mathrm{loc}}(\Omega)\ ;\ \|f\|_{L^p_{d\mu}} := \left(\int_\Omega |f(x)|^p a(x)\,dx\right)^{\frac{1}{p}} < \infty \right\}, \quad 1 \leq p < \infty$$

which is quite reasonable because the corresponding elliptic operator $a(x)^{-1}\Delta$ can be regarded as a symmetric operator in $L^2_{d\mu}$.

The $L^p$-$L^q$ type estimates for the semigroup associated with the Friedrichs' extension $-L_*$ (in $L^2_{d\mu}$) of $-a(x)^{-1}\Delta$ are stated in [10]. The proof is based on Beurling–Deny's criterion and Gagliardo–Nirenberg inequality.

**Proposition 2.2** ([10, Proposition 2.6]). *Let $e^{tL_*}$ be a semigroup generated by $L_*$. For every $f \in L^1_{d\mu} \cap L^2_{d\mu}$, we have*

$$\|e^{tL_*} f\|_{L^2_{d\mu}} \leq C t^{-\frac{N-\alpha}{2(2-\alpha)}} \|f\|_{L^1_{d\mu}} \tag{2.7}$$

*and*

$$\|L_* e^{tL_*} f\|_{L^2_{d\mu}} \leq C t^{-\frac{N-\alpha}{2(2-\alpha)}-1} \|f\|_{L^1_{d\mu}}. \tag{2.8}$$

# 3. Weighted energy estimates

In this section we establish weighted energy estimates for solutions of (1.1) by introducing Ikehata–Todorova–Yordanov type weight function with an auxiliary function $A_\varepsilon$ constructed in Subsection 2.1.

To begin with, let us recall the finite speed propagation property of the wave equation (see [2]).

**Lemma 3.1** (Finite speed of propagation). *Let u be the solution of* (1.1) *with the initial data* $(u_0, u_1)$ *satisfying* $\text{supp}\,(u_0, u_1) \subset \overline{B}(0, R_0) = \{x \in \Omega; |x| \le R_0\}$. *Then, one has*

$$\text{supp}\, u(\cdot, t) \subset \{x \in \Omega\ ;\ |x| \le R_0 + t\}$$

*and therefore* $|x|/(R_0 + 1 + t) \le 1$ *for* $t \ge 0$ *and* $x \in \text{supp}\, u(\cdot, t)$.

Before introducing a weight function, we also recall two identities for partial energy functionals proved in [10].

**Lemma 3.2** ([10, Lemma 3.7]). *Let* $\Phi \in C^2(\overline{\Omega} \times [0, \infty))$ *satisfy* $\Phi > 0$ *and* $\partial_t \Phi < 0$ *and let u be a solution of* (1.1). *Then*

$$\frac{d}{dt}\left[\int_\Omega \left(|\nabla u|^2 + |u_t|^2\right)\Phi\, dx\right] = \int_\Omega (\partial_t \Phi)^{-1} \left|\partial_t \Phi \nabla u - u_t \nabla \Phi\right|^2 dx$$
$$+ \int_\Omega \left( - 2a(x)\Phi + \partial_t \Phi - (\partial_t \Phi)^{-1} |\nabla \Phi|^2\right)|u_t|^2\, dx.$$

**Lemma 3.3** ([10, Lemma 3.9]). *Let* $\Phi \in C^2(\overline{\Omega} \times [0, \infty))$ *satisfy* $\Phi > 0$ *and* $\partial_t \Phi < 0$ *and let u be a solution to* (1.1). *Then, we have*

$$\frac{d}{dt}\left[\int_\Omega \left(2uu_t + a(x)|u|^2\right)\Phi\, dx\right] = 2\int_\Omega uu_t(\partial_t \Phi)\, dx + 2\int_\Omega |u_t|^2 \Phi\, dx - 2\int_\Omega |\nabla u|^2 \Phi\, dx$$
$$+ \int_\Omega (a(x)\partial_t \Phi + \Delta \Phi)|u|^2\, dx.$$

Here we introduce a weight function for weighted energy estimates, which is a modification of the one in Todorova-Yordanov [12].

**Definition 3.4.** *Define* $h := \frac{2-\alpha}{N-\alpha}$ *and for* $\varepsilon \in (0, 1)$,

$$\Phi_\varepsilon(x, t) = \exp\left(\frac{1}{h + 2\varepsilon}\frac{A_\varepsilon(x)}{1 + t}\right), \tag{3.1}$$

*where* $A_\varepsilon$ *is given in Lemma 2.1. And define for* $t \ge 0$,

$$E_{\partial x}(t; u) := \int_\Omega |\nabla u|^2 \Phi_\varepsilon\, dx, \quad E_{\partial t}(t; u) := \int_\Omega |u_t|^2 \Phi_\varepsilon\, dx, \tag{3.2}$$

$$E_a(t; u) := \int_\Omega a(x)|u|^2 \Phi_\varepsilon\, dx, \quad E_*(t; u) := 2\int_\Omega uu_t \Phi_\varepsilon\, dx, \tag{3.3}$$

*and also define* $E_1(t; u) := E_{\partial x}(t; u) + E_{\partial t}(t; u)$ *and* $E_2(t; u) := E_*(t; u) + E_a(t; u)$.

Now we are in a position to state our main result for weighted energy estimates for solutions of (1.1).

**Proposition 3.5.** *Assume that* $(u_0, u_1)$ *satisfies* $\text{supp}\,(u_0, u_1) \subset \overline{B}(0, R_0)$ *and the compatibility condition of order* $k_0 \ge 1$. *Let u be a solution of the problem* (1.1). *For every* $\delta > 0$ *and* $0 \le k \le k_0 - 1$, *there exist* $\varepsilon > 0$ *and* $M_{\delta,k,R_0} > 0$ *such that for every* $t \ge 0$,

$$(1 + t)^{\frac{N-\alpha}{2-\alpha} + 2k + 1 - \delta}\left(E_{\partial x}(t; \partial_t^k u) + E_{\partial t}(t; \partial_t^k u)\right) + (1 + t)^{\frac{N-\alpha}{2-\alpha} + 2k - \delta} E_a(t; \partial_t^k u) \le M_{\delta,k,R_0} \|(u_0, u_1)\|^2_{H^{k+1} \times H^k(\Omega)}.$$

To prove, this, we prepare the following two lemmas.

**Lemma 3.6.** *For $t \geq 0$, we have*

$$\frac{1-\varepsilon}{h+2\varepsilon}\frac{1}{1+t}E_a(t;u) \leq E_{\partial x}(t;u). \tag{3.4}$$

*Proof.* As in the proof of [10, Lemma 3.6], by integration by parts we have

$$\int_\Omega \Delta(\log\Phi_\varepsilon)|u|^2\Phi_\varepsilon\,dx = \int_\Omega\left(\Delta\Phi_\varepsilon - \frac{|\nabla\Phi_\varepsilon|^2}{\Phi_\varepsilon}\right)|u|^2\,dx \leq \int_\Omega |\nabla u|^2\,\Phi_\varepsilon\,dx.$$

Noting that

$$\Delta(\log\Phi_\varepsilon(x)) = \frac{1}{h+2\varepsilon}\frac{\Delta A_\varepsilon(x)}{1+t} \geq \frac{1-\varepsilon}{h+2\varepsilon}\frac{a(x)}{1+t},$$

we have (3.4). □

In order to clarify the effect of the finite propagation property, we now put

$$a_1 := \inf_{x\in\Omega}\left(\langle x\rangle^\alpha a(x)\right).$$

Then

**Lemma 3.7.** *For $t \geq 0$, we have*

$$E_{\partial t}(t;u) \leq \frac{1}{a_1}(R_0+1+t)^\alpha E_a(t;\partial_t u), \tag{3.5}$$

$$\int_\Omega \frac{A_\varepsilon(x)}{a(x)}|u_t|^2\Phi_\varepsilon\,dx \leq \frac{A_{2\varepsilon}}{a_1}(R_0+1+t)^2 E_{\partial t}(t;u), \tag{3.6}$$

$$|E_*(t;u)| \leq \frac{2}{\sqrt{a_1}}(R_0+1+t)^{\frac{\alpha}{2}}\sqrt{E_a(t;u)E_{\partial t}(t;u)}. \tag{3.7}$$

*Proof.* By $a(x)^{-1} \leq a_1^{-1}\langle x\rangle^\alpha \leq a_1^{-1}(1+|x|)^\alpha$ and the finite propagation property we have

$$\int_\Omega |u_t|^2\Phi_\varepsilon\,dx = \int_\Omega \frac{a(x)}{a(x)}|u_t|^2\Phi_\varepsilon\,dx \leq \frac{1}{a_1}(R_0+1+t)^\alpha E_a(t;\partial_t u).$$

Using the Cauchy-Schwarz inequality and the above inequality yields (3.6):

$$\begin{aligned}
\left|\int_\Omega uu_t\Phi_\varepsilon\,dx\right|^2 &\leq \left(\int_\Omega |u|^2\Phi_\varepsilon\,dx\right)\left(\int_\Omega |u_t|^2\Phi_\varepsilon\,dx\right)\\
&\leq \frac{(R_0+1+t)^\alpha}{a_1}\left(\int_\Omega a(x)|u|^2\Phi_\varepsilon\,dx\right)E_{\partial t}(t;u)\\
&\leq \frac{(R_0+1+t)^\alpha}{a_1}E_a(t;u)E_{\partial t}(t;u).
\end{aligned}$$

We can prove (3.7) in a similar way. □

**Lemma 3.8.** **(i)** *For every $t \geq 0$, we have*

$$\frac{d}{dt}E_1(t;u) \leq -E_a(t;\partial_t u). \tag{3.8}$$

**(ii)** *For every $\varepsilon \in (0, \frac{1}{3})$ and $t \geq 0$,*

$$\frac{d}{dt}E_2(t;u) \leq -\frac{1-3\varepsilon}{1-\varepsilon}E_{\partial x}(t;u) + \left(\frac{2}{a_1} + \frac{A_{2\varepsilon}(R_0+1)^2}{\varepsilon a_1^2}\right)(R_0+1+t)^{\alpha}E_a(t;\partial_t u). \tag{3.9}$$

*Proof.* Noting (2.3), we have

$$\begin{aligned}
-2a(x)\Phi_\varepsilon + \partial_t\Phi_\varepsilon - (\partial_t\Phi_\varepsilon)^{-1}|\nabla\Phi_\varepsilon|^2 &= \left(-2a(x) - \frac{A_\varepsilon(x)}{(h+2\varepsilon)(1+t)^2} + \frac{1}{h+2\varepsilon}\frac{|\nabla A_\varepsilon(x)|^2}{A_\varepsilon(x)}\right)\Phi_\varepsilon \\
&\leq \left(-2a(x) + \frac{h+\varepsilon}{h+2\varepsilon}a(x)\right)\Phi_\varepsilon \\
&\leq -a(x)\Phi_\varepsilon.
\end{aligned}$$

This implies (3.8). On the other hand, from (2.3) and (2.1) we see

$$\begin{aligned}
a(x)\partial_t\Phi_\varepsilon + \Delta\Phi_\varepsilon &= \frac{1}{h+2\varepsilon}\left(-\frac{a(x)A_\varepsilon(x)}{(1+t)^2} + \frac{|\nabla A_\varepsilon(x)|^2}{(h+2\varepsilon)(1+t)^2} + \frac{\Delta A_\varepsilon(x)}{1+t}\right)\Phi_\varepsilon \\
&\leq \frac{1}{h+2\varepsilon}\left(-\frac{a(x)A_\varepsilon(x)}{(1+t)^2} + \frac{(h+\varepsilon)a(x)A_\varepsilon(x)}{(h+2\varepsilon)(1+t)^2} + \frac{(1+\varepsilon)a(x)}{1+t}\right)\Phi_\varepsilon \\
&\leq \left(-\frac{\varepsilon}{(h+2\varepsilon)^2}\frac{a(x)A_\varepsilon(x)}{(1+t)^2} + \frac{1+\varepsilon}{h+2\varepsilon}\frac{a(x)}{1+t}\right)\Phi_\varepsilon.
\end{aligned}$$

Therefore combining it with Lemma 3.6, we have

$$\int_\Omega (a(x)\partial_t\Phi_\varepsilon + \Delta\Phi_\varepsilon)|u|^2\,dx \leq \frac{1+\varepsilon}{1-\varepsilon}\int_\Omega |\nabla u|^2\Phi_\varepsilon\,dx - \frac{\varepsilon}{(h+2\varepsilon)^2}\frac{1}{(1+t)^2}\int_\Omega a(x)A_\varepsilon(x)|u|^2\Phi_\varepsilon\,dx.$$

Using (3.6), we have

$$\begin{aligned}
2\int_\Omega uu_t(\partial_t\Phi_\varepsilon)\,dx &= -\frac{2}{h+2\varepsilon}\frac{1}{(1+t)^2}\int_\Omega uu_tA_\varepsilon(x)\Phi_\varepsilon\,dx \\
&\leq \frac{2}{h+2\varepsilon}\frac{1}{(1+t)^2}\left(\int_\Omega a(x)A_\varepsilon(x)|u|^2\Phi_\varepsilon\,dx\right)^{\frac{1}{2}}\left(\int_\Omega \frac{A_\varepsilon(x)}{a(x)}|u_t|^2\Phi_\varepsilon\,dx\right)^{\frac{1}{2}} \\
&\leq \frac{2(R_0+1)}{h+2\varepsilon}\frac{1}{1+t}\left(\int_\Omega a(x)A_\varepsilon(x)|u|^2\Phi_\varepsilon\,dx\right)^{\frac{1}{2}}\left(\frac{A_{2\varepsilon}}{a_1}E_{\partial t}(t;u)\right)^{\frac{1}{2}} \\
&\leq \frac{\varepsilon}{(h+2\varepsilon)^2}\frac{1}{(1+t)^2}\int_\Omega a(x)A_\varepsilon(x)|u|^2\Phi_\varepsilon\,dx + \frac{A_{2\varepsilon}(R_0+1)^2}{\varepsilon a_1}E_{\partial t}(t;u).
\end{aligned}$$

Applying (3.5), we obtain (3.9). □

**Lemma 3.9.** *The following assertions hold:*

**(i)** *Set* $t_*(R_0,\alpha,m) := \max\left\{\left(\frac{2m}{a_1}\right)^{\frac{1}{1-\alpha}}, R_0+1\right\}$. *Then for every* $t, m \ge 0$ *and* $t_1 \ge t_*(R_0,\alpha,m)$,

$$\frac{d}{dt}\Big((t_1+t)^m E_1(t;u)\Big) \le m(t_1+t)^{m-1}E_{\partial x}(t;u) - \frac{1}{2}(t_1+t)^m E_a(t;\partial_t u). \tag{3.10}$$

**(ii)** *for every* $t, \lambda \ge 0$ *and* $t_2 \ge R_0+1$,

$$\begin{aligned}\frac{d}{dt}\Big((t_2+t)^\lambda E_2(t;u)\Big) &\le \lambda(1+\varepsilon)(t_2+t)^{\lambda-1}E_a(t;u) - \frac{1-3\varepsilon}{1-\varepsilon}(t_2+t)^\lambda E_{\partial x}(t;u)\\ &\quad + \left(\frac{2}{a_1} + \frac{A_{2\varepsilon}(R_0+1)^2}{\varepsilon a_1^2} + \frac{\lambda}{2\varepsilon a_1^2 t_2^{1-\alpha}}\right)(t_2+t)^{\lambda+\alpha}E_a(t;\partial_t u).\end{aligned} \tag{3.11}$$

**(iii)** *In particular, setting*

$$\nu := \frac{4}{a_1} + \frac{2A_{2\varepsilon}(R_0+1)^2}{\varepsilon a_1^2} + \frac{1}{4\varepsilon a_1},$$

$$t_{**}(\varepsilon,R_0,\alpha,\lambda) := \max\left\{\left(\frac{(1-\varepsilon)(\lambda+\alpha)\nu}{\varepsilon}\right)^{\frac{1}{1-\alpha}}, \left(\frac{2(\lambda+\alpha)}{a_1}\right)^{\frac{1}{1-\alpha}}, R_0+1\right\},$$

*one has that for* $t, \lambda \ge 0$ *and* $t_3 \ge t_{**}(\varepsilon,R_0,\alpha,\lambda)$,

$$\begin{aligned}&\frac{d}{dt}\Big(\nu(t_3+t)^{\lambda+\alpha}E_1(t;u) + (t_3+t)^\lambda E_2(t;u)\Big)\\ &\le -\frac{1-4\varepsilon}{1-\varepsilon}(t_3+t)^\lambda E_{\partial x}(t;u) + \lambda(1+\varepsilon)(t_3+t)^{\lambda-1}E_a(t;u).\end{aligned} \tag{3.12}$$

*Proof.* **(i)** Let $m \ge 0$ be fixed and let $t_1 \ge t_*(R_0,\alpha,m)$. Using (3.8) and (3.5), we have

$$\begin{aligned}(t_1+t)^{-m}\frac{d}{dt}\Big((t_1+t)^m E_1(t;u)\Big) &\le \frac{m}{t_1+t}E_{\partial x}(t;u) + \frac{m}{t_1+t}E_{\partial t}(t;u) + \frac{d}{dt}E_1(t;u)\\ &\le \frac{m}{t_1+t}E_{\partial x}(t;u) + \frac{m}{t_1+t}E_{\partial t}(t;u) - E_a(t;\partial_t u)\\ &\le \frac{m}{t_1+t}E_{\partial x}(t;u) + \left(\frac{m(R_0+1+t)^\alpha}{a_1(t_1+t)} - 1\right)E_a(t;\partial_t u).\end{aligned}$$

Therefore we obtain (3.10).

**(ii)** For $t \ge 0$, and $t \ge R_0+1$,

$$\begin{aligned}&(t_2+t)^{-\lambda}\frac{d}{dt}\Big((t_2+t)^\lambda E_2(t;u)\Big)\\ &\le \frac{\lambda}{t_2+t}E_*(t;u) + \frac{\lambda}{t_2+t}E_a(t;u) + \frac{d}{dt}E_2(t;u)\\ &\le \frac{\lambda}{t_2+t}E_*(t;u) + \frac{\lambda}{t_2+t}E_a(t;u) - \frac{1-3\varepsilon}{1-\varepsilon}E_{\partial x}(t;u) + \left(\frac{2}{a_1} + \frac{A_{2\varepsilon}(R_0+1)^2}{\varepsilon a_1^2}\right)(R_0+1+t)^\alpha E_a(t;\partial_t u).\end{aligned}$$

Noting that by (3.7) and (3.5),

$$\begin{aligned}\frac{\lambda}{t_2+t}E_*(t;u) &\le \frac{2\lambda(R_0+1+t)^{\alpha}}{a_1(t_2+t)}\sqrt{E_a(t;u)E_a(t;\partial_t u)}\\ &\le \frac{\lambda\varepsilon}{t_2+t}E_a(t;u)+\frac{\lambda}{\varepsilon a_1^2}\frac{(R_0+1+t)^{2\alpha}}{t_2+t}E_a(t;\partial_t u)\\ &\le \frac{\lambda\varepsilon}{t_2+t}E_a(t;u)+\frac{\lambda}{\varepsilon a_1^2 t_2^{1-\alpha}}(t_2+t)^{\alpha}E_a(t;\partial_t u),\end{aligned}$$

we deduce (3.11).

**(iii)** Combining (3.10) with $m=\lambda+\alpha$ and (3.11), we have for $t_3 \ge t_{**}(\varepsilon, R_0, \alpha, \lambda)$ and $t \ge 0$,

$$\begin{aligned}&\frac{d}{dt}\Big(\nu(t_3+t)^{\lambda+\alpha}E_1(t;u)+(t_3+t)^{\lambda}E_2(t;u)\Big)\\ &\le \left(\nu(\lambda+\alpha)(t_3+t)^{\alpha-1}-\frac{1-3\varepsilon}{1-\varepsilon}\right)(t_3+t)^{\lambda}E_{\partial x}(t;u)+\lambda(1+\varepsilon)(t_3+t)^{\lambda-1}E_a(t;u)\\ &\quad+\left(\frac{2}{a_1}+\frac{A_{2\varepsilon}(R_0+1)^2}{\varepsilon a_1^2}+\frac{\lambda}{2\varepsilon a_1^2 t_3^{1-\alpha}}-\frac{\nu}{2}\right)(t_3+t)^{\lambda+\alpha}E_a(t;\partial_t u)\\ &\le -\frac{1-4\varepsilon}{1-\varepsilon}(t_3+t)^{\lambda}E_{\partial x}(t;u)+\lambda(1+\varepsilon)(t_3+t)^{\lambda-1}E_a(t;u).\end{aligned}$$

This proves the assertion. □

*Proof of Proposition 3.5.* Firstly, by (3.7) we observe that

$$\begin{aligned}\nu(t_3+t)^{\alpha}E_1(t;u)+E_2(t;u) &\ge \frac{4}{a_1}(t_3+t)^{\alpha}E_1(t;u)-|E_*(t;u)|+E_a(t;u)\\ &\ge \frac{4}{a_1}(t_3+t)^{\alpha}E_{\partial t}(t;u)-\frac{2}{\sqrt{a_1}}(t_3+t)^{\frac{\alpha}{2}}\sqrt{E_a(t;u)E_{\partial t}(t;u)}+E_a(t;u)\\ &\ge \frac{3}{4}E_a(t;u).\end{aligned}$$

By using the above estimate, we prove the assertion via mathematical induction.

**Step 1** ($k=0$)**.** By (3.12) using Lemma 3.6 implies that

$$\frac{d}{dt}\Big(\nu(t_3+t)^{\lambda+\alpha}E_1(t;u)+(t_3+t)^{\lambda}E_2(t;u)\Big)\le\left(-\frac{1-4\varepsilon}{1-\varepsilon}+\frac{\lambda(1+\varepsilon)(h+2\varepsilon)}{1-\varepsilon}\right)(t_3+t)^{\lambda}E_{\partial x}(t;u).$$

Therefore taking $\lambda_0=\frac{(1-\varepsilon)(1-4\varepsilon)}{(1+\varepsilon)(h+2\varepsilon)}$, ($\lambda_0 \uparrow h^{-1}$ as $\varepsilon \downarrow 0$) we have

$$\frac{d}{dt}\Big(\nu(t_3+t)^{\lambda_0+\alpha}E_1(t;u)+(t_3+t)^{\lambda_0}E_2(t;u)\Big)\le -\frac{\varepsilon(1-4\varepsilon)}{1-\varepsilon}(t_3+t)^{\lambda_0}E_{\partial x}(t;u).$$

Integrating over $(0,t)$ with respect to $t$, we see

$$\frac{3}{4}(t_3+t)^{\lambda_0}E_a(t;u)+\frac{\varepsilon(1-4\varepsilon)}{1-\varepsilon}\int_0^t(t_3+s)^{\lambda_0}E_{\partial x}(s;u)\,ds$$

$$
\begin{aligned}
&\leq \nu(t_3+t)^{\lambda_0+\alpha}E_1(t;u)+(t_3+t)^{\lambda_0}E_2(t;u)+\frac{\varepsilon(1-4\varepsilon)}{1-\varepsilon}\int_0^t (t_3+s)^{\lambda_0}E_{\partial x}(s;u)\,ds\\
&\leq \nu t_3^{\lambda_0+\alpha}E_1(0;u)+t_3^{\lambda_0}E_2(0;u).
\end{aligned}
$$

Using (3.10) with $m=\lambda_0+1$ and integrating over $(0,t)$, we obtain

$$
\begin{aligned}
&(t_3+t)^{\lambda_0+1}E_1(t;u)+\frac{1}{2}\int_0^t (t_3+s)^{\lambda_0+1}E_a(s;\partial_t u)\,ds\\
&\leq t_3^{\lambda_0+1}E_1(0;u)+(\lambda_0+1)\int_0^t (t_3+s)^{\lambda_0}E_{\partial x}(s;u)\,ds\\
&\leq t_3^{\lambda_0+1}E_1(0;u)+\frac{(\lambda_0+1)(1-\varepsilon)}{\varepsilon(1-4\varepsilon)}\Big(\nu t_3^{\lambda_0+\alpha}E_1(0;u)+t_3^{\lambda_0}E_2(0;u)\Big).
\end{aligned}
$$

This proves the desired assertion with $k=0$ and also the integrability of $(t_3+s)^{\lambda_0+1}E_a(s;\partial_t u)$.

**Step 2 ($1<k\leq k_0-1$).** Suppose that for every $t\geq 0$,

$$
(1+t)^{\lambda_0+2k-1}E_1(t;\partial_t^{k-1}u)+(1+t)^{\lambda_0+2k-2}E_a(t;\partial_t^{k-1}u)\leq M_{\varepsilon,k-1}\|(u_0,u_1)\|^2_{H^k\times H^{k-1}(\Omega)}
$$

and additionally,

$$
\int_0^t (1+s)^{\lambda_0+2k-1}E_a(s;\partial_t^k u)\,ds\leq M'_{\varepsilon,k-1}\|(u_0,u_1)\|^2_{H^k\times H^{k-1}(\Omega)}.
$$

Since the initial value $(u_0,u_1)$ satisfies the compatibility condition of order $k$, $\partial_t^k u$ is also a solution of (1.1) with replaced $(u_0,u_1)$ with $(u_{k-1},u_k)$. Applying (3.12) with $\lambda=\lambda_0+2k$, putting $t_{3k}=t_{**}(\varepsilon,R_0,\alpha,\lambda_0+2k)$ (see Lemma 3.9 **(iii)**) and integrating over $(0,t)$, we have

$$
\begin{aligned}
&\frac{3}{4}(t_{3k}+t)^{\lambda_0+2k}E_a(t;\partial_t^k u)+\frac{1-4\varepsilon}{1-\varepsilon}\int_0^t (t_{3k}+s)^{\lambda_0+2k}E_{\partial x}(s;\partial_t^k u)\,ds\\
&\leq \nu(t_{3k}+t)^{\lambda_0+2k+\alpha}E_1(t;\partial_t^k u)+(t_{3k}+t)^{\lambda_0+2k}E_2(t;\partial_t^k u)+\frac{1-4\varepsilon}{1-\varepsilon}\int_0^t (t_{3k}+s)^{\lambda_0+2k}E_{\partial x}(s;\partial_t^k u)\,ds\\
&\leq \nu t_{3k}^{\lambda_0+2k+\alpha}E_1(0;\partial_t^k u)+t_{3k}^{\lambda_0+2k}E_2(0;\partial_t^k u)+(\lambda_0+2k)(1+\varepsilon)\int_0^t (t_{3k}+s)^{\lambda_0+2k-1}E_a(s;\partial_t^k u)\,ds\\
&\leq \nu t_{3k}^{\lambda_0+2k+\alpha}E_1(0;\partial_t^k u)+t_{3k}^{\lambda_0+2k-1}E_2(0;\partial_t^k u)+(\lambda_0+2k)(1+\varepsilon)M'_{\varepsilon,k-1}\|(u_0,u_1)\|^2_{H^k\times H^{k-1}(\Omega)}.
\end{aligned}
$$

Moreover, from (3.10) with $m=\lambda_0+2k+1$ we have

$$
\begin{aligned}
&(t_{3k}+t)^{\lambda_0+2k+1}E_1(t;\partial_t^k u)+\frac{1}{2}\int_0^t (t_{3k}+s)^{\lambda_0+2k+1}E_a(s;\partial_t^{k+1}u)\,ds\\
&\leq t_{3k}^{\lambda_0+2k+1}E_1(0;\partial_t^k u)+(\lambda_0+2k+1)\int_0^t (t_{3k}+s)^{\lambda_0+2k}E_{\partial x}(s;\partial_t^k u)\,ds\\
&\leq M''_{\varepsilon,k}\Big(E_1(0;\partial_t^k u)+E_2(0;\partial_t^k u)+\|(u_0,u_1)\|^2_{H^k\times H^{k-1}(\Omega)}\Big)
\end{aligned}
$$

with some constant $M''_{\varepsilon,k}>0$. By induction we obtain the desired inequalities for all $k\leq k_0-1$. □

## 4. Diffusion phenomena as an application of weighted energy estimates

**Proposition 4.1.** *Assume that* $(u_0, u_1) \in (H^2 \cap H_0^1(\Omega)) \times H_0^1(\Omega)$ *and suppose that* supp $(u_0, u_1) \subset \overline{B}(0, R_0)$. *Let u be the solution of* (1.1). *Then for every* $\varepsilon > 0$, *there exists a constant* $C_{\varepsilon, R_0} > 0$ *such that*

$$\left\| u(\cdot, t) - e^{tL_*}[u_0 + a(\cdot)^{-1} u_1] \right\|_{L^2_{d\mu}} \le C_{\varepsilon, R_0} (1+t)^{-\frac{N-\alpha}{2(2-\alpha)} - \frac{1-\alpha}{2-\alpha} + \varepsilon} \|(u_0, u_1)\|_{H^2 \times H^1}.$$

To prove Proposition 4.1 we use the following lemma stated in [10, Section 4].

**Lemma 4.2.** *Assume that* $(u_0, u_1) \in (H^2 \cap H_0^1(\Omega)) \times H_0^1(\Omega)$ *and suppose that* supp $(u_0, u_1) \subset \{x \in \Omega; |x| \le R_0\}$. *Then for every* $t \ge 0$,

$$\begin{aligned} u(x,t) - e^{tL_*}[u_0 + a(\cdot)^{-1} u_1] = &- \int_{t/2}^{t} e^{(t-s)L_*}[a(\cdot)^{-1} u_{tt}(\cdot, s)] ds \\ &- e^{\frac{t}{2} L_*}[a(\cdot)^{-1} u_t(\cdot, t/2)] \\ &- \int_0^{t/2} L_* e^{(t-s)L_*}[a(\cdot)^{-1} u_t(\cdot, s)] ds, \end{aligned} \tag{4.1}$$

*where* $L_*$ *is the (negative) Friedrichs extension of* $-L = -a(x)^{-1}\Delta$ *in* $L^2_{d\mu}$.

*Proof of Proposition 4.1.* First we show the assertion for $(u_0, u_1)$ satisfying the compatibility condition of order 2. Taking $L^2_{d\mu}$-norm of both side, we have

$$\left\| u(x, \cdot) - e^{tL_*}[u_0 + a(\cdot)^{-1} u_1] \right\|_{L^2_{d\mu}} \le \mathcal{J}_1(t) + \mathcal{J}_2(t) + \mathcal{J}_3(t),$$

where

$$\begin{aligned} \mathcal{J}_1(t) &:= \int_{t/2}^{t} \left\| e^{(t-s)L_*}[a(\cdot)^{-1} u_{tt}(\cdot, s)] \right\|_{L^2_{d\mu}} ds, \\ \mathcal{J}_2(t) &:= \left\| e^{\frac{t}{2} L_*}[a(\cdot)^{-1} u_t(\cdot, t/2)] \right\|_{L^2_{d\mu}}, \\ \mathcal{J}_3(t) &:= \int_0^{t/2} \left\| L_* e^{(t-s)L_*}[a(\cdot)^{-1} u_t(\cdot, s)] \right\|_{L^2_{d\mu}} ds. \end{aligned}$$

Noting that for $x \in \Omega$,

$$a(x)^{-1} \Phi_\varepsilon(x,t)^{-1} \le \frac{1}{a_1} \langle x \rangle^\alpha \exp\left( -\frac{A_{1\varepsilon}}{h + 2\varepsilon} \frac{\langle x \rangle^{2-\alpha}}{1+t} \right) \le \frac{1}{a_1} \left( \frac{\alpha(h+2\varepsilon)}{(2-\alpha) e A_{1\varepsilon}} \right)^{\frac{\alpha}{2-\alpha}} (1+t)^{\frac{\alpha}{2-\alpha}},$$

we see that for $k = 0, 1$,

$$\begin{aligned} \left\| a(\cdot)^{-1} \partial_t^{k+1} u(\cdot, s) \right\|_{L^2_{d\mu}}^2 &= \int_\Omega a(x)^{-1} |\partial_t^{k+1} u(\cdot, s)|^2 \, dx \\ &\le \|a(\cdot)^{-1} \Phi_\varepsilon(\cdot, t)^{-1}\|_{L^\infty(\Omega)} \int_\Omega |\partial_t^{k+1} u(\cdot, s)|^2 \Phi_\varepsilon \, dx \\ &\le \widetilde{C} (1+t)^{\frac{\alpha}{2-\alpha}} E_{\partial t}(t, \partial_t^k u) \end{aligned}$$

$$\leq \widetilde{C}M_{\varepsilon,k}(1+t)^{-\lambda_0-\frac{2-2\alpha}{2-\alpha}-2k}\|(u_0,u_1)\|^2_{H^{k+1}\times H^k}.$$

Therefore from Proposition 3.5 with $k = 1$ and $k = 0$ we have

$$\begin{aligned}
\mathcal{J}_1(t) &\leq \int_{t/2}^{t} \left\|a(\cdot)^{-1}u_{tt}(\cdot,s)\right\|_{L^2_{d\mu}} ds \\
&\leq \sqrt{\widetilde{C}M_1}\|(u_0,u_1)\|_{H^2\times H^1}\int_{t/2}^{t}(1+s)^{-\frac{\lambda_0}{2}-\frac{1-\alpha}{2-\alpha}-1}ds \\
&\leq \frac{2(2-\alpha)}{\lambda_0(2-\alpha)+1-\alpha}\sqrt{\widetilde{C}M_{\varepsilon,1}}(1+t)^{-\frac{\lambda_0}{2}-\frac{1-\alpha}{2-\alpha}}\|(u_0,u_1)\|_{H^2\times H^1}
\end{aligned}$$

and

$$\mathcal{J}_2(t) \leq \left\|a(\cdot)^{-1}u_t(\cdot,t/2)\right\|_{L^2_{d\mu}} \leq \sqrt{\widetilde{C}M_{\varepsilon,0}}(1+t)^{-\frac{\lambda_0}{2}-\frac{1-\alpha}{2-\alpha}}\|(u_0,u_1)\|_{H^1\times L^2}.$$

Moreover, by Lemma 2.2, we see by Cauchy–Schwarz inequality that for $t \geq 1$,

$$\begin{aligned}
\mathcal{J}_3(t) &\leq C\int_0^{t/2}(t-s)^{-\frac{N-\alpha}{2(2-\alpha)}-1}\left\|a(\cdot)^{-1}u_t(\cdot,s)\right\|_{L^1_{d\mu}} ds \\
&\leq C\left(\frac{t}{2}\right)^{-\frac{N-\alpha}{2(2-\alpha)}-1}\int_0^{t/2}\sqrt{\|\Phi_\varepsilon^{-1}(\cdot,s)\|_{L^1(\Omega)}E_{\partial t}(s;u)}\,ds.
\end{aligned}$$

Since

$$\begin{aligned}
\|\Phi^{-1}(\cdot,t)\|_{L^1(\Omega)} &\leq \int_{\mathbb{R}^N}\exp\left(-\frac{A_{1\varepsilon}}{h+2\varepsilon}\frac{|x|^{2-\alpha}}{1+t}\right)dx \\
&= (1+t)^{\frac{N}{2-\alpha}}\int_{\mathbb{R}^N}\exp\left(-\frac{A_{1\varepsilon}}{h+2\varepsilon}|y|^{2-\alpha}\right)dy,
\end{aligned}$$

we deduce

$$\begin{aligned}
\mathcal{J}_3(t) &\leq C'(1+t)^{-\frac{N-\alpha}{2(2-\alpha)}-1}\|(u_0,u_1)\|_{H^1\times L^2}\int_0^{t/2}(1+s)^{\frac{N-\alpha}{2(2-\alpha)}-\frac{\lambda_0}{2}-\frac{1-\alpha}{2-\alpha}}ds \\
&\leq C'\left(\frac{N-\alpha}{2(2-\alpha)}-\frac{\lambda_0}{2}+\frac{1}{2-\alpha}\right)(1+t)^{-\frac{N-\alpha}{2(2-\alpha)}-1}(1+t/2)^{\frac{N-\alpha}{2(2-\alpha)}-\frac{\lambda_0}{2}-\frac{1-\alpha}{2-\alpha}+1}\|(u_0,u_1)\|_{H^1\times L^2} \\
&\leq C''(1+t)^{-\frac{\lambda_0}{2}-\frac{1-\alpha}{2-\alpha}}\|(u_0,u_1)\|_{H^1\times L^2}.
\end{aligned}$$

Consequently, we obtain

$$\left\|u(\cdot,t)-e^{tL_*}[u_0+a(\cdot)^{-1}u_1]\right\|_{L^2_{d\mu}} \leq C'''(1+t)^{-\frac{\lambda_0}{2}-\frac{1-\alpha}{2-\alpha}}\|(u_0,u_1)\|_{H^2\times H^1}.$$

Next we show the assertion for $(u_0,u_1)$ satisfying $(u_0,u_1) \in (H^2\times H^1_0(\Omega))\times H^1_0(\Omega)$ (the compatibility condition of order 1) via an approximation argument. Fix $\phi \in C_c^\infty(\mathbb{R}^N,[0,1])$ such that $\phi \equiv 1$ on $\overline{B}(0,R_0)$ and $\phi \equiv 0$ on $\mathbb{R}^N \setminus B(0,R_0+1)$ and define for $n \in \mathbb{N}$,

$$\begin{pmatrix} u_{0n} \\ u_{1n} \end{pmatrix} = \begin{pmatrix} \phi\tilde{u}_{0n} \\ \phi\tilde{u}_{1n} \end{pmatrix}, \quad \begin{pmatrix} \tilde{u}_{0n} \\ \tilde{u}_{1n} \end{pmatrix} = \left(1+\frac{1}{n}\mathcal{A}\right)^{-1}\begin{pmatrix} u_0 \\ u_1 \end{pmatrix},$$

where $\mathcal{A}$ is an $m$-accretive operator in $\mathcal{H} = H_0^1(\Omega) \times L^2(\Omega)$ associated with (1.1), that is,

$$\mathcal{A} = \begin{pmatrix} 0 & -1 \\ -\Delta & a(x) \end{pmatrix}$$

endowed with domain $D(\mathcal{A}) = (H^2 \cap H_0^1(\Omega)) \times H_0^1(\Omega)$. Then $(u_{0n}, u_{1n})$ satisfies $\mathrm{supp}(u_{0n}, u_{1n}) \subset \overline{B}(0, R_0 + 1)$ and the compatibility condition of order 2. Let $v_n$ be a solution of (1.1) with $(u_{0n}, u_{1n})$. Observe that

$$\begin{aligned} \|(u_{0n}, u_{1n})\|^2_{H^2 \times H^1} &\leq C^2 \|\phi\|^2_{W^{2,\infty}} \|(\tilde{u}_0, \tilde{u}_1)\|^2_{H^2 \times H_1} \\ &\leq C'^2 \|\phi\|^2_{W^{2,\infty}} (\|(\tilde{u}_0, \tilde{u}_1)\|^2_{\mathcal{H}} + \|\mathcal{A}(\tilde{u}_0, \tilde{u}_1)\|^2_{\mathcal{H}}) \\ &\leq C'^2 \|\phi\|^2_{W^{2,\infty}} (\|(u_0, u_1)\|^2_{\mathcal{H}} + \|\mathcal{A}(u_0, u_1)\|^2_{\mathcal{H}}) \\ &\leq C''^2 \|\phi\|^2_{W^{2,\infty}} \|(u_0, u_1)\|^2_{H^2 \times H^1} \end{aligned}$$

with suitable constants $C, C', C'' > 0$, and

$$\begin{pmatrix} u_{0n} \\ u_{1n} \end{pmatrix} \to \begin{pmatrix} \phi u_0 \\ \phi u_1 \end{pmatrix} = \begin{pmatrix} u_0 \\ u_1 \end{pmatrix} \quad \text{in } \mathcal{H}$$

as $n \to \infty$ and also $u_{0n} + a^{-1} u_{1n} \to u_0 + a^{-1} u_1$ in $L^2_{d\mu}$ as $n \to \infty$. Using the result of the previous step, we deduce

$$\left\| v_n(\cdot, t) - e^{tL_*}[u_{0n} + a(\cdot)^{-1} u_{1n}] \right\|_{L^2_{d\mu}} \leq \tilde{C}(1+t)^{-\frac{\lambda_0}{2} - \frac{1-\alpha}{2-\alpha}} \|(u_0, u_1)\|_{H^2 \times H^1}$$

with some constant $\tilde{C} > 0$. Letting $n \to \infty$, by continuity of the $C_0$-semigroup $e^{-t\mathcal{A}}$ in $\mathcal{H}$ we also obtain diffusion phenomena for initial data in $(H^2 \cap H_0^1(\Omega)) \cap H_0^1(\Omega)$. □

## Acknowledgments

This work is supported by Grant-in-Aid for JSPS Fellows 15J01600 of Japan Society for the Promotion of Science and also partially supported by Grant-in-Aid for Young Scientists Research (B), No. 16K17619. The authors would like to thank the referee for giving them valuable comments and suggestions.

## Conflict of Interest

All authors declare no conflicts of interest in this paper.

## References

1. M. Ikawa, *Mixed problems for hyperbolic equations of second order*, J. Math. Soc. Japan, **20** (1968), 580-608.
2. M. Ikawa, M. Ikawa, Hyperbolic partial differential equations and wave phenomena, American Mathematical Society, Providence, RI, 2000.
3. R. Ikehata, *Some remarks on the wave equation with potential type damping coefficients*, Int. J. Pure Appl. Math., **21** (2005), 19-24.

4. A. Matsumura, *On the asymptotic behavior of solutions of semi-linear wave equations*, Publ. Res. Inst. Math. Sci., **12** (1976), 169-189.

5. A. Matsumura, *Energy decay of solutions of dissipative wave equations*, Proc. Japan Acad., Ser. A, **53** (1977), 232-236.

6. K. Mochizuki, *Scattering theory for wave equations with dissipative terms*, Publ. Res. Inst. Math. Sci., **12** (1976), 383-390.

7. K. Nishihara, *$L^p$-$L^q$ estimates of solutions to the damped wave equation in 3-dimensional space and their application*, Math. Z., **244** (2003), 631-649.

8. P. Radu, G. Todorova, and B. Yordanov, *Higher order energy decay rates for damped wave equations with variable coefficients*, Discrete Contin. Dyn. Syst. Ser. S, **2** (2009), 609-629.

9. P. Radu, G. Todorova, and B. Yordanov, *Decay estimates for wave equations with variable coefficients*, Trans. Amer. Math. Soc., **362** (2010), 2279-2299.

10. M. Sobajima and Y. Wakasugi, *Diffusion phenomena for the wave equation with space-dependent damping in an exterior domain*, J. Differential Equations, **261** (2016), 5690-5718.

11. G. Todorova, and B. Yordanov, *Critical exponent for a nonlinear wave equation with damping*, J. Differential Equations, **174** (2001), 464-489.

12. G. Todorova, and B. Yordanov, *Weighted $L^2$-estimates for dissipative wave equations with variable coefficients*, J. Differential Equations, **246** (2009), 4497-4518.

13. Y. Wakasugi, *On diffusion phenomena for the linear wave equation with space-dependent damping*, J. Hyp. Diff. Eq., **11** (2014), 795-819.

# 2

# A regularity criterion of weak solutions to the 3D Boussinesq equations

**Ahmad Mohammed Alghamdi**[1], **Sadek Gala**[2,3,*] **and Maria Alessandra Ragusa**[3,4]

[1] Department of Mathematical Science , Faculty of Applied Science, Umm Alqura University, P. O. Box 14035, Makkah 21955, Saudi Arabia

[2] Department of Mathematics, University of Mostaganem, Algeria

[3] Dipartimento di Mathematica e Informatica, Università di Catania, Viale Andrea Doria, 6, 95125 Catania, Italy

[4] RUDN University, 6 Miklukho - Maklay St, Moscow, 117198, Russia

[*] **Correspondence:** sadek.gala@gmail.com

**Abstract:** In this note, a regularity criterion of weak solutions to the 3D-Boussinesq equations with respect to Serrin type condition under the framework of Besov space $\dot{B}^{r}_{\infty,\infty}$. It is shown that the weak solution $(u, \theta)$ is regular on $(0, T]$ if $u$ satisfies

$$\int_0^T \|u(\cdot,t)\|^{\frac{2}{1+r}}_{\dot{B}^{r}_{\infty,\infty}}\, dt < \infty,$$

for $0 < r < 1$. This result improves some previous works.

**Keywords:** Regularity criterion; Boussinesq equations; a priori estimates
**Mathematics Subject Classification:** 35Q35, 76D03

---

## 1. Introduction and main result

In this work, we consider the Cauchy problem of 3D viscous incompressible Boussinesq equations [17]:

$$\begin{cases} \partial_t u + u\cdot\nabla u - \Delta u + \nabla\pi = \theta e_3, \\ \partial_t\theta + u\cdot\nabla\theta - \Delta\theta = 0, \\ \nabla\cdot u = 0, \\ (u,\theta)(x,0) = (u_0,\theta_0)(x), \quad x\in\mathbb{R}^3, \end{cases} \tag{1.1}$$

where $u = u(x,t)$ and $\theta = \theta(x,t)$ denote the unknown velocity vector field and the scalar function temperature, while $u_0, \theta_0$ with $\nabla \cdot u_0 = 0$ in the sense of distribution are given initial data. $e_3 = (0,0,1)^T$. $\pi = \pi(x,t)$ the pressure of fluid at the point $(x,t) \in \mathbb{R}^3 \times (0,\infty)$. There are a huge literatures on the incompressible Boussinesq equations such as [1,3–5,7,9–11,22,24–27] and the references therein.

When $\theta = 0$, (1.1) reduces to the well-known incompressible Navier-Stokes equations and many results are available. Since Leray [16] and Hopf [12] constructed the so-called well-known Leray-Hopf weak solution $u(x,t)$ of the incompressible Navier-Stokes equation for arbitrary $u_0 \in L^2(\mathbb{R}^3)$ with $\nabla \cdot u_0(x) = 0$ in last century, the problem on the uniqueness and regularity of the Leray-Hopf weak solutions is one of the most challenging problem of the mathematical community. There are two approaches to tackle this problem : The first is to study the partial regularity of suitable weak solutions to Navier-Stokes equation which was initiated by L. Caffarelli, R. Kohn and L. Nirenberg [2]. The other way is to propose different criteria to guarantee the regularity of the weak solutions which was studied by G. Prodi [19], J. Serrin [20], Struwe [21], etc. However, similar to the Navier-Stokes equations, the question of global regularity of the weak solutions of the 3D Boussinesq equations still remains a big open problem. This paper is concerned with the second approach and is devoted to presenting an improved regularity criterion of weak solutions for the 3D Boussinesq equations in the Besov space.

There has been a lot of work on the regularity theory of Boussinesq equations [6,22,24,25,27,28]. In particular, Fan and Ozawa [6] showed that the weak solution becomes regular if the velocity satisfies

$$\int_0^T \|u(\cdot,t)\|^2_{\dot{B}^0_{\infty,\infty}} \, dt < \infty.$$

Before stating our main result, let us first recall the definition of the homogeneous Besov space (see e.g. [23]).

**Definition 1.1.** *Let* $\{\varphi_j\}_{j\in\mathbb{Z}}$ *be the Littlewood-Paley dyadic decomposition of unity that satisfies* $\widehat{\varphi} \in C_0^\infty\left(B_2 \backslash B_{\frac{1}{2}}\right)$, $\widehat{\varphi}_j(\xi) = \widehat{\varphi}\left(2^{-j}\xi\right)$ *and*

$$\sum_{j\in\mathbb{Z}} \widehat{\varphi}_j(\xi) = 1 \quad for\ any\ \ \xi \neq 0,$$

*where* $B_R$ *is the ball in* $\mathbb{R}^3$ *centered at the origin with radius* $R > 0$. *The homogeneous Besov spaces* $\dot{B}^s_{p,q}(\mathbb{R}^3)$ *are defined to be*

$$\dot{B}^s_{p,q}(\mathbb{R}^3) = \left\{ f \in \mathcal{S}'(\mathbb{R}^3)/\mathcal{P}(\mathbb{R}^3) : \|f\|_{\dot{B}^s_{p,q}} < \infty \right\}$$

*where*

$$\|f\|_{\dot{B}^s_{p,q}} = \begin{cases} \left( \sum\limits_{j\in\mathbb{Z}} \left\|2^{js}\varphi_j * f\right\|^q_{L^p} \right)^{\frac{1}{q}} & if\ \ 1 < q < \infty, \\ \sup\limits_{j\in\mathbb{Z}} 2^{js} \left\|\varphi_j * f\right\|_{L^p} & if\ \ q = \infty, \end{cases}$$

*for* $s \in \mathbb{R}$, $1 \le p,q \le \infty$, *where* $\mathcal{S}'$ *is the space of tempered distributions and* $\mathcal{P}$ *is the space of polynomials.*

To aid the introduction of our main result, we recall the definition of weak solutions.

**Definition 1.2.** *Let* $(u_0, \theta_0) \in L^2(\mathbb{R}^3)$ *with* $\operatorname{div} u_0 = 0$ *in the sense of distributions. A measurable pair* $(u, \theta)$ *is said to be a weak solution of (1.1) on* $(0, T)$*, provided that*

**a)** $(u, \theta) \in L^\infty(0, T; L^2(\mathbb{R}^3)) \cap L^2(0, T; H^1(\mathbb{R}^3))$*;*

**b)** *(1.1)*$_{1,2,3}$ *are satisfied in the sense of distributions;*

**c)** *the strong energy inequality*

$$\|u(\cdot,t)\|_{L^2}^2 + \|\theta(\cdot,t)\|_{L^2}^2 + 2\int_\epsilon^t (\|\nabla u(\cdot,\tau)\|_{L^2}^2 + \|\nabla\theta(\cdot,\tau)\|_{L^2}^2)d\tau \leq \|u(\cdot,\epsilon)\|_{L^2}^2 + \|\theta(\cdot,\epsilon)\|_{L^2}^2,$$

*for all* $0 \leq \epsilon \leq t \leq T$.

By a strong solution we mean that a weak solution $(u, \theta)$ of the Boussinesq equations (1.1) satisfies

$$(u, \theta) \in L^\infty(0, T; H^1(\mathbb{R}^3)) \cap L^2(0, T; H^2(\mathbb{R}^3)).$$

It is well known that the strong solution is regular and unique.

The main result on the regularity criterion of the weak solutions now reads :

**Theorem 1.3.** *Suppose* $(u_0, \theta_0) \in L^2(\mathbb{R}^3)$ *with* $\operatorname{div} u_0 = 0$ *in the sense of distributions. Assume that* $(u(x,t), \theta(x,t))$ *is a weak solution of (1.1) on* $\mathbb{R}^3 \times (0, T)$ *and satisfies the strong energy inequality. If* $u$ *satisfies*

$$\int_0^T \|u(\cdot,t)\|_{\dot{B}^r_{\infty,\infty}}^{\frac{2}{1+r}} dt < \infty \quad \text{with } 0 < r < 1, \tag{1.2}$$

*then the weak solution* $(u, \theta)$ *becomes a regular solution on* $(0, T]$.

**Remark 1.1.** *If* $r > 0$*, we have*

$$B^r_{\infty,\infty} = L^\infty \cap \dot{B}^r_{\infty,\infty} \text{ and } \|f\|_{B^r_{\infty,\infty}} \approx \|f\|_{\dot{B}^r_{\infty,\infty}} + \|f\|_{L^\infty}.$$

*Here* $B^r_{\infty,\infty}$ *is the inhomogeneous Besov space. Definitions and basic properties of the inhomogeneous Besov spaces can be find in [23]. For concision, we omit them here. So this result is an improvement of the earlier regularity criterion.*

In order to prove our main result, we need the following lemma.

**Lemma 1.4.** *Let* $f : \mathbb{R}_+ \to \mathbb{R}_+$ *be a function such that* $f(x) = ax^{1-r} + bx^{-r}$*, for all* $0 < r < 1$ *and* $a, b \in \mathbb{R}_+$*. Then there holds*

$$f(x) \leq \left[\left(\frac{r}{1-r}\right)^{1-r} + \left(\frac{1-r}{r}\right)^r\right] a^r b^{1-r}.$$

The proof of this lemma is straight forward and can be obtained by simple calculations.

## 2. Proof of Theorem 1.3

**Proof:** Apply $\nabla$ operator to the equation of $(1.1)_1$ and $(1.1)_2$, then taking the inner product with $\nabla u$ and $\nabla\theta$, respectively and using integration by parts, we get

$$\begin{aligned}
&\frac{1}{2}\frac{d}{dt}(\|\nabla u\|_{L^2}^2+\|\nabla\theta\|_{L^2}^2)+\|\Delta u\|_{L^2}^2+\|\Delta\theta\|_{L^2}^2\\
=&-\int_{\mathbb{R}^3}\nabla u\cdot\nabla u\cdot\nabla u dx+\int_{\mathbb{R}^3}\nabla(\theta e_3)\cdot\nabla u dx-\int_{\mathbb{R}^3}\nabla u\cdot\nabla\theta\cdot\nabla\theta dx\\
=&\ \mathcal{I}_1+\mathcal{I}_2+\mathcal{I}_3. \qquad (2.1)
\end{aligned}$$

According to the homogeneous Littlewood-Paley decomposition, $\nabla u$ can be written as

$$\nabla u=\sum_{j=-\infty}^{+\infty}\Delta_j(\nabla u)=\sum_{j=-\infty}^{N}\Delta_j(\nabla u)+\sum_{j=N+1}^{+\infty}\Delta_j(\nabla u),$$

where $N$ is a positive integer to be chosen later. We decompose $\mathcal{I}_1$ as follows

$$\begin{aligned}
\mathcal{I}_1 =&-\int_{\mathbb{R}^3}\sum_{j=-\infty}^{N}\Delta_j(\nabla u)\cdot\nabla u\cdot\nabla u dx-\int_{\mathbb{R}^3}\sum_{j=N+1}^{+\infty}\Delta_j(\nabla u)\cdot\nabla u\cdot\nabla u dx\\
\le&\left|\int_{\mathbb{R}^3}\sum_{j=-\infty}^{N}\Delta_j(\nabla u)\cdot\nabla u\cdot\nabla u dx\right|+\left|\int_{\mathbb{R}^3}\sum_{j=N+1}^{+\infty}\Delta_j(\nabla u)\cdot\nabla u\cdot\nabla u dx\right|\\
\le&\sum_{j=-\infty}^{N}\int_{\mathbb{R}^3}|\Delta_j(\nabla u)|\,|\nabla u|^2dx+2\sum_{j=-\infty}^{N}\int_{\mathbb{R}^3}|\Delta_j u|\,|\nabla u|\,|\Delta u|\,dx\\
=&\ \mathcal{I}_{11}+\mathcal{I}_{12}.
\end{aligned}$$

For $\mathcal{I}_{11}$, by Hölder inequality and the definition of Besov space, for $0<r<1$, we derive that

$$\begin{aligned}
\mathcal{I}_{11}\le&\sum_{j=-\infty}^{N}\left\|\Delta_j(\nabla u)\right\|_{L^\infty}\|\nabla u\|_{L^2}^2\\
=&\ \|\nabla u\|_{L^2}^2\sum_{j=-\infty}^{N}2^{(-1+r)j}\left\|\Delta_j(\nabla u)\right\|_{L^\infty}2^{(1-r)j}\\
\le&\ C\left(\sum_{j=-\infty}^{N}2^{(1-r)j}\right)\sup_{j\in\mathbb{Z}}\left(2^{(-1+r)j}\left\|\Delta_j(\nabla u)\right\|_{L^\infty}\right)\|\nabla u\|_{L^2}^2\\
\le&\ C2^{(1-r)N}\|\nabla u\|_{\dot{B}^{-1+r}_{\infty,\infty}}\|\nabla u\|_{L^2}^2\\
\le&\ C2^{(1-r)N}\|u\|_{\dot{B}^{r}_{\infty,\infty}}\|\nabla u\|_{L^2}^2. \qquad (2.2)
\end{aligned}$$

For $\mathcal{I}_{12}$, in view of the definition of Besov space, it follows that

$$\mathcal{I}_{12}\le\sum_{j=N+1}^{+\infty}\left\|\Delta_j u\right\|_{L^\infty}\|\nabla u\|_{L^2}\|\Delta u\|_{L^2}$$

$$
\begin{aligned}
&= \|\nabla u\|_{L^2}\|\Delta u\|_{L^2}\sum_{j=N+1}^{+\infty} 2^{-rj}\left(2^{rj}\left\|\Delta_j u\right\|_{L^\infty}\right)\\
&\le C\left(\sum_{j=N+1}^{+\infty} 2^{-rj}\right)\left(\sup_{j\in\mathbb{Z}} 2^{rj}\left\|\Delta_j u\right\|_{L^\infty}\right)\|\nabla u\|_{L^2}\|\Delta u\|_{L^2}\\
&\le C2^{-Nr}\|u\|_{\dot B^r_{\infty,\infty}}\|\nabla u\|_{L^2}\|\Delta u\|_{L^2}. \qquad (2.3)
\end{aligned}
$$

It follows from (2.2)-(2.3) and Lemma 1.4 with $x = 2^N$, $a = \|\nabla u\|_{L^2}$ and $b = \|\Delta u\|_{L^2}$ that

$$
\begin{aligned}
\mathcal{I}_1 &\le C2^{(1-r)N}\|u\|_{\dot B^r_{\infty,\infty}}\|\nabla u\|_{L^2}^2 + C2^{-Nr}\|u\|_{\dot B^r_{\infty,\infty}}\|\nabla u\|_{L^2}\|\Delta u\|_{L^2}\\
&= C\|u\|_{\dot B^r_{\infty,\infty}}\|\nabla u\|_{L^2}\left(2^{(1-r)N}\|\nabla u\|_{L^2} + 2^{-rN}\|\Delta u\|_{L^2}\right)\\
&\le C\|u\|_{\dot B^r_{\infty,\infty}}\|\nabla u\|_{L^2}\left[\left(\frac{r}{1-r}\right)^{1-r} + \left(\frac{1-r}{r}\right)^{r}\right]\|\nabla u\|_{L^2}^r\|\Delta u\|_{L^2}^{1-r}\\
&\le C\|u\|_{\dot B^r_{\infty,\infty}}\|\nabla u\|_{L^2}^{1+r}\|\Delta u\|_{L^2}^{1-r},
\end{aligned}
$$

by choosing

$$
N = \left[\frac{1}{\ln 2}\ln\left(\frac{r}{1-r}\frac{\|\Delta u\|_{L^2}}{\|\nabla u\|_{L^2}}\right)\right].
$$

By Young's inequality, we get

$$
\mathcal{I}_1 \le \frac{1}{2}\|\Delta u\|_{L^2}^2 + C\|u\|_{\dot B^r_{\infty,\infty}}^{\frac{2}{1+r}}\|\nabla u\|_{L^2}^2.
$$

We estimate $\mathcal{I}_3$ in the same way as $\mathcal{I}_1$. We decompose $\mathcal{I}_3$ as follows

$$
\begin{aligned}
\mathcal{I}_3 &= -\int_{\mathbb{R}^3}\sum_{j=-\infty}^{N}\Delta_j(\nabla u)\cdot\nabla\theta\cdot\nabla\theta dx - \int_{\mathbb{R}^3}\sum_{j=N+1}^{+\infty}\Delta_j(\nabla u)\cdot\nabla\theta\cdot\nabla\theta dx\\
&\le \sum_{j=-\infty}^{N}\int_{\mathbb{R}^3}\left|\Delta_j(\nabla u)\right||\nabla\theta|^2 dx + 2\sum_{j=-\infty}^{N}\int_{\mathbb{R}^3}\left|\Delta_j u\right||\nabla\theta||\Delta\theta| dx\\
&= \mathcal{I}_{31} + \mathcal{I}_{32}.
\end{aligned}
$$

Then, by using Lemma 1.4, $\mathcal{I}_3$ can be estimated as

$$
\begin{aligned}
\mathcal{I}_3 &\le C2^{(1-r)N}\|u\|_{\dot B^r_{\infty,\infty}}\|\nabla\theta\|_{L^2}^2 + C2^{-Nr}\|u\|_{\dot B^r_{\infty,\infty}}\|\nabla\theta\|_{L^2}\|\Delta\theta\|_{L^2}\\
&= C\|u\|_{\dot B^r_{\infty,\infty}}\|\nabla\theta\|_{L^2}\left(2^{(1-r)N}\|\nabla\theta\|_{L^2} + 2^{-rN}\|\Delta\theta\|_{L^2}\right)\\
&\le C\|u\|_{\dot B^r_{\infty,\infty}}\|\nabla\theta\|_{L^2}^{1+r}\|\Delta\theta\|_{L^2}^{1-r}\\
&\le \frac{1}{2}\|\Delta\theta\|_{L^2}^2 + C\|u\|_{\dot B^r_{\infty,\infty}}^{\frac{2}{1+r}}\|\nabla\theta\|_{L^2}^2
\end{aligned}
$$

The term $\mathcal{I}_2$ can be estimated by Cauchy's inequality as

$$
\mathcal{I}_2 \le \|\nabla u\|_{L^2}\|\nabla\theta\|_{L^2} \le \frac{1}{2}(\|\nabla u\|_{L^2}^2 + \|\nabla\theta\|_{L^2}^2).
$$

Plugging all the estimates into (2.1) yields that

$$\frac{d}{dt}(\|\nabla u\|_{L^2}^2 + \|\nabla\theta\|_{L^2}^2) + \|\Delta u\|_{L^2}^2 + \|\Delta\theta\|_{L^2}^2 \le C(1 + \|u\|_{\dot{B}^{r}_{\infty,\infty}}^{\frac{2}{1+r}})(\|\nabla u\|_{L^2}^2 + \|\nabla\theta\|_{L^2}^2).$$

Applying Gronwall's inequality, we get

$$(u, \theta) \in L^\infty(0, T; H^1(\mathbb{R}^3)) \cap L^2(0, T; H^2(\mathbb{R}^3)).$$

Therefore, by the standard regularity arguments of weak solutions to drive high-order derivative bounds, which would imply

$$(u, \theta) \in C^\infty(\mathbb{R}^3 \times (0, T))$$

by Sobolev imbedding theorems, as desired. The proof of Theorem 1.3 is completed. □

## Acknowledgments

Part of the work was carried out while the second author was long term visitor at University of Catania. The hospitality and support of Catania University are graciously acknowledged.

Maria Alessandra Ragusa is supported by the second author Ministry of Education and Science of the Russian Federation (Agreement number N. 02. 03.21.0008)

## Conflict of Interest

All authors declare no conflicts of interest in this paper.

## References

1. J. R. Cannon and E. Dibenedetto, *The initial problem for the Boussinesq equation with data in $L^p$*, Lecture Notes in Mathematics, Springer, Berlin, **771** (1980), 129-144.
2. L. Caffarelli, R. Kohn, and L. Nirenberg, *Partial regularity of suitable weak solutions of the navier-stokes equations,* Comm. Pure Appl. Math., **35** (1982), 771-831.
3. D. Chae and H.-S. Nam, *Local existence and blow-up criterion for the Boussinesq equations,* Proc. Roy. Soc. Edinburgh, Sect. A, **127** (1997), 935-946.
4. B. Q. Dong, J. Song, and W. Zhang, *Blow-up criterion via pressure of three-dimensional Boussinesq equations with partial viscosity (in Chinese),* Sci. Sin. Math., **40** (2010), 1225-1236.
5. J. Fan and Y. Zhou, *A note on regularity criterion for the 3D Boussinesq system with partial viscosity,* Appl. Math. Lett., **22** (2009), 802-805.
6. J. Fan and T. Ozawa, *Regularity criteria for the 3D density-dependent Boussinesq equations,* Nonlinearity, **22** (2009), 553-568.
7. S. Gala, *On the regularity criterion of strong solutions to the 3D Boussinesq equations,* Applicable Analysis, **90** (2011), 1829-1835.

8. S. Gala and M.A. Ragusa, *Logarithmically improved regularity criterion for the Boussinesq equations in Besov spaces with negative indices,* Applicable Analysis, **95** (2016), 1271-1279.
9. S. Gala, Z. Guo, and M. A. Ragusa, *A remark on the regularity criterion of Boussinesq equations with zero heat conductivity,* Appl. Math. Lett., **27** (2014), 70-73.
10. Z. Guo and S. Gala, *Regularity criterion of the Newton-Boussinesq equations in* $\mathbb{R}^3$, Commun. Pure Appl. Anal., **11** (2012), 443-451.
11. J. Geng and J. Fan, *A note on regularity criterion for the 3D Boussinesq system with zero thermal conductivity,* Appl. Math. Lett., **25** (2012), 63-66.
12. E. Hopf, *Über die Anfangswertaufgabe für die hydrodynamichen Grundgleichungen,* Math. Nach., **4** (1950/1951), 213-231.
13. Y. Jia, X. Zhang, and B. Dong, *Remarks on the blow-up criterion for smooth solutions of the Boussinesq equations with zero diffusion,* C.P.A.A., **12** (2013), 923-937.
14. T. Kato and G. Ponce, *Commutator estimates and the Euler and Navier-Stokes equations,* Commun. Pure Appl. Math., **41** (1988), 891-907.
15. C. Kenig, G. Ponce, and L. Vega, *Well-posedness of the initial value problem for the Korteweg-de-Vries equation,* J. Amer. Math. Soc., **4** (1991), 323-347.
16. J. Leray, *Sur le mouvement d'un liquide visqueux emplissant l'espace,* Acta. Math., **63** (1934), 183-248.
17. A. Majda, *Introduction to PDEs and Waves for the Atmosphere and Ocean,* Courant Lecture Notes in Mathematics, AMS/CIMS, **9** (2003).
18. M. Mechdene, S. Gala, Z. Guo, and M.A. Ragusa, *Logarithmical regularity criterion of the three-dimensional Boussinesq equations in terms of the pressure,* Z. Angew. Math. Phys., **67** (2016), 67-120.
19. G. Prodi, *Un teorema di unicità per le equazioni di Navier-Stokes*, Ann. Mat. Pura Appl., **48** (1959), 173-182.
20. J. Serrin, *The initial value problem for the Navier-Stokes equations,* In Nonlinear Problems (Proc. Sympos., Madison, Wis.), Univ. of Wisconsin Press, Madison, Wis., 1963, 69-98.
21. M. Struwe, *On partial regularity results for the Navier-Stokes equations,* Comm. Pure Appl. Math., **41** (1988), 437-458.
22. N. Ishimura and H. Morimoto, *Remarks on the blow-up criterion for the 3D Boussinesq equations,* Math. Meth. Appl. Sci., **9** (1999), 1323-1332.
23. H. Triebel, *Theory of Function Spaces,* Birkhäuser Verlag, Basel, 1983.
24. H. Qiu, Y. Du, and Z. Yao, *Blow-up criteria for 3D Boussinesq equations in the multiplier space,* Comm. Nonlinear Sci. Num. Simulation, **16** (2011), 1820-1824.
25. H. Qiu, Y. Du, and Z. Yao, *A blow-up criterion for 3D Boussinesq equations in Besov spaces,* Nonlinear Analysis TMA, **73** (2010), 806-815.
26. Z. Xiang, *The regularity criterion of the weak solution to the 3D viscous Boussinesq equations in Besov spaces*, Math. Methods Appl. Sci., **34** (2011), 360-372.

27. F. Xu, Q. Zhang, and X. Zheng, *Regularity Criteria of the 3D Boussinesq Equations in the Morrey-Campanato Space,* Acta Appl. Math., **121** (2012), 231-240.

28. Z. Ye, *A Logarithmically improved regularity criterion of smooth solutions for the 3D Boussinesq equations,* Osaka J. Math., **53** (2016), 417-423.

# 3

# Surface tension, higher order phase field equations, dimensional analysis and Clairaut's equation

**Gunduz Caginalp***

Mathematics Department University of Pittsburgh, Pittsburgh, PA 15260, USA

* **Correspondence:** Email: caginalp@pitt.edu

**Abstract:** A higher order phase field free energy leads to higher order differential equations. The surface tension involves $L^2$ norms of higher order derivatives. An analysis of dimensionless variables shows that the surface tension satisfies a Clairaut's equation in terms of the coefficients of the higher order phase field equations. The Clairaut's equation can be solved by characteristics on a suitable surface in the $\mathbb{R}^N$ space of coefficients. This perspective may also be regarded as interpreting dimensional analysis through Clairaut's equation. The surface tension is shown to be a homogeneous function of monomials of the coefficients.

**Keywords:** surface tension; phase field equations; Clairaut's equation; homogeneous functions; dimensional analysis

## 1. Introduction

A major challenge in many applied mathematical problems, such as those arising from materials science, is the need to compute large systems of equations. When the problems are cast in terms of ordinary differential equations (usually by imposing some symmetry), this is equivalent to studying equations with higher order derivatives. If one could approximate these higher order differential equations by second order equations, it would be a considerable simplification. Many problems in applied mathematics are associated with interfaces or moving boundaries, and surface tension often plays an important role in those problems [11] as a stabilizing agent for interfaces that would otherwise be highly unstable (e.g., dendritic behavior [2, 9, 4]). In a recent paper, quantum field theoretic renormalization methods were used to approximate higher order ODEs by second order counterparts [6]. A related issue is whether one can approximate solutions of one higher order equation with those of another with different, e.g., smaller coefficients, that can be used in conjunction with the renormalization methods cited above.

In this paper, we consider a prototype free energy of the phase field type that has been studied in several papers including [5, 10], and the resulting differential equations that follow from minimization of this free energy. In particular we show that the suface tension $\sigma$ as a function of rescaled coefficients satisfies Clairaut's equation that can be solved by the method of charactertics. This means that if one knows the surface tension for a suitable surface $\Gamma$ in the coefficient space, then one can solve for $\sigma$ for all values of the parameters.

This methodology can be viewed as complementary to those of [6] involving reduction in the order of the differential equation.

## 2. The Free Energy and Higher Order Phase Field Equations.

We consider the simplest free energy functional that has the necessary features (see [10, 5]). Let $\phi \in C^N(\mathbb{R})$ and let the $j^{th}$ derivative of $\phi$ in $x$ by denoted by either $D^j\phi$ or $\phi^{(j)}$. Let $\phi$ satisfy the boundary conditions

$$\phi^{(j)}(\pm\infty) = 0 \tag{2.1}$$

and define $W$ as the standard double well potential with minima at $\phi := \pm\phi_0$, e.g.,

$$W(\phi) := \left(\phi^2 - 1\right)^2.$$

A free energy, $\mathcal{F}[\phi]$, is defined by

$$\mathcal{F}[\phi] := \int_{\mathbb{R}} F\left(x, \phi(x), ..., D^N\phi(x)\right) dx \tag{2.2}$$

$$F\left[x, \phi(x), ..., \phi^{(N)}(x)\right] = \frac{1}{2}\sum_{j=1}^{N} (-1)^{j+1} c_{2j} \left\{D^j\phi(x)\right\}^2 + W(\phi(x)).$$

This free energy arises from an averaging process in the statistical mechanics of a two state system, where $\phi$ represents an order parameter that makes a transition from the lower energy phase (e.g., solid) at $-1$ (more generally $\phi_-$) to $+1$ (more generally $\phi_+$).

The higher order phase field equations are obtained from this free energy. For a smooth test function that also satisfies the same boundary conditions as $\phi$, the functional derivative is evaluated as

$$\begin{aligned}\frac{\partial}{\partial\varepsilon}\mathcal{F}[\phi + \varepsilon\eta]\,|_{\varepsilon=0} &= \frac{\partial}{\partial\varepsilon}\int_{\mathbb{R}} F\left[x, \phi(x) + \varepsilon\eta(x), ..., \phi^{(N)}(x) + \varepsilon\eta^{(N)}(x)\right] dx|_{\varepsilon=0} \\ &= \int_{\mathbb{R}} \sum_{j=1}^{N} c_{2j}(-1)^{j+1} D^j\phi(x) D^j\eta(x) + W'(\phi(x))\eta(x)\, dx.\end{aligned}$$

Upon performing integration by parts and setting this expression to zero, one has

$$0 = \frac{\partial}{\partial\varepsilon}\mathcal{F}[\phi + \varepsilon\eta]\,|_{\varepsilon=0} = \int_{\mathbb{R}} \left\{-\sum_{j=1}^{N} c_{2j}D^{2j}\phi(x) + W'(\phi(x))\right\}\eta(x)\, dx.$$

Since this is true for all test functions $\eta$, one has the ODE,

$$-\sum_{j=1}^{N} c_{2j}D^{2j}\phi(x) + W'(\phi(x)) = 0. \tag{2.3}$$

The surface tension in physical terms is usually described in (in arbitrary physical dimension) as the difference in a small cylindrical volume along the surface, normalized by the cross-sectional area of the cylinder. The difference between the free energy obtained by integrating over the cylinder minus the average of the free energy of the two phases is normalized by the area. With $\mathcal{F}_{cyl}[\phi]$ denoting this free energy one can write (see e.g., [3] or [5])

$$\sigma \sim \frac{\mathcal{F}_{cyl}[\phi] - \frac{1}{2}\left\{\mathcal{F}_{cyl}[\phi_+] + \mathcal{F}_{cyl}[\phi_-]\right\}}{Cross\ Sectional\ Area}. \tag{2.4}$$

In the context of our one-dimensional analysis using a symmetric $W$, we note that the two terms involving the pure phases vanish, and the integral over this cylinder can be regarded as one-dimensional after the division.

Recalling the comment above that the $W(\phi_\pm)$ terms vanish, it is evident that in one-dimension (or a physical setting with this symmetry) the definition can be interpreted as the free energy of the transition layer solution, $\phi$. Thus we can write the mathematical definition as follows.

**Definition.** Given a set of non-negative coefficients, let $\phi$ be a solution to (2.3) subject to boundary conditions (2.1), the *surface tension* $\sigma(c_2, ..., c_{2N})$ is defined by

$$\sigma(c_2, ..., c_{2N}) = \mathcal{F}[\phi]. \tag{2.5}$$

**Proposition.** The surface tension defined by (2.5) can be expressed as

$$\sigma(c_2, ..., c_{2N}) = \sum_{j=1}^{N} (-1)^{j+1} jc_{2j} \left\|D^j\phi\right\|^2_{L^2(\mathbb{R})} \tag{2.6}$$

where $\|f\|_{L^2(\mathbb{R})}$ is the usual $L^2$ norm, i.e., $\|f\|_{L^2(\mathbb{R})} := \int_{\mathbb{R}} |f|^2 \, dx$.

**Remarks.** Mathematically, the surface tension is well-defined so long as one has a solution to the equation (2.3). The connection between the two definitions is easily understood (see [7, 5]) in light of the calculations in the proof below, which was presented in [7].

Proof. First we derive an identity by muliplying (2.3) by $D\phi$ and integrating over $(-\infty, x)$

$$\int_{-\infty}^{x} -\sum_{j=1}^{N} c_{2j}\left\{D^{2j}\phi(z)\right\}\{D\phi(z)\} + D\phi(z)\,W'(\phi(z))\,dz = 0. \tag{2.7}$$

The term involving $W'$ is an exact differential, $\frac{d}{dz}W(\phi(z))$. Noting the identity [7]

$$D^{2j}\phi D\phi = D\left\{\frac{(-1)^{j-1}}{2}\left(D^j\phi\right)^2 + (-1)\sum_{k=1}^{j-1}(-1)^{k-1}D^{2j-k}\phi D^k\phi\right\}$$

one observes that the left hand side is also an exact identity, yielding,

$$\sum_{j=1}^{N} c_{2j}\left\{\frac{(-1)^{j-1}}{2}\left(D^j\phi\right)^2 + (-1)\sum_{k=1}^{j-1}(-1)^{k-1}D^{2j-k}\phi D^k\phi\right\} = W(\phi).$$

Note that all terms vanish at $-\infty$ due to the boundary conditions.

Now we integrate this expression over $(-\infty, \infty)$ and obtain after integrating by parts $j-k$ times in the second term on the left hand side

$$\sum_{j=1}^{N}\left\{c_{2j}\frac{(-1)^{j-1}}{2}\left\|D^j\phi\right\|^2 + c_{2j}\sum_{k=1}^{j-1}(-1)^{k-1}(-1)^{j-k}\left\|D^j\phi\right\|^2\right\}$$
$$= \int_{-\infty}^{\infty} W(\phi(x))\,dx.$$

Simplifying this expression yields

$$\sum_{j=1}^{N}\left(j-\frac{1}{2}\right)(-1)^{j-1}c_{2j}\left\|D^j\phi\right\|^2 = \int_{-\infty}^{\infty} W(\phi(x))\,dx. \tag{2.8}$$

Using this identity, the free energy (2.2) and the expression (2.5), $\sigma$ can be written as

$$\sigma(c_2, ..., c_{2N}) = \int_{-\infty}^{\infty}\frac{1}{2}\sum_{j=1}^{N}(-1)^{j+1}c_{2j}\left\{D^j\phi(x)\right\}^2 + W(\phi(x))\,dx$$
$$= \sum_{j=1}^{N}(-1)^{j+1}jc_{2j}\left\|D^j\phi\right\|^2, \tag{2.9}$$

where the last expression is obtained by substituting (2.8) for the $W$ term. ///

## 3. Surface tension and dimensional analysis.

Within this mathematical setting we have been using reduced dimensional parameters. The order parameter, $\phi$, is assumed to be dimensionless, as usual. With $x$ having units of length (we write $x \sim L$), the coefficients $c_{2j}$ have units of $L^{2j}$ as is clear from (2.3). The surface tension in the form (2.6) is defined as having units of length (as is typical in the reduced units used in physics) since

$$c_{2j}\left\|D^j\phi\right\|^2_{L^2(\mathbb{R})} = c_{2j}\int_{\mathbb{R}}\left|D^j\phi\right|^2 dx \sim L^{2j}L^{-2j+1} = L.$$

The variables with units of length are $c_{2n}^{1/(2n)}$ and $\sigma$. We define

$$z_1 := c_2^{1/2}, ..., \quad z_N := c_{2N}^{1/(2N)}$$

A set of dimensionless variables are

$$\Pi_0 = \frac{\sigma}{z_1}, \quad \Pi_1 = \frac{z_2}{z_1}, ..., \Pi_{N-1} = \frac{z_N}{z_1}. \tag{3.1}$$

The basic principle of dimensional analysis [1] is that a dimensionless quantity such as $\Pi_0$ can only depend on other dimensionless quantities, namely, $\Pi_1, ..., \Pi_{N-1}$ through some function $G$ :

$$\Pi_0 = G(\Pi_1, ..., \Pi_{N-1}) \tag{3.2}$$

Assuming a smooth solution $\phi$ to the phase field equation, (2.3), one can see from (2.5) or (2.6) that $G$ is differentiable in $z_1$ so we can write

$$\frac{\partial \Pi_0}{\partial z_1} = \sum_{j=1}^{N-1} \frac{\partial G}{\partial \Pi_j} \frac{\partial \Pi_j}{\partial z_1} = \sum_{j=1}^{N-1} \frac{\partial G}{\partial \Pi_j} \left( -\frac{z_{j+1}}{z_1^2} \right). \tag{3.3}$$

For $k \geq 2$, we have similarly

$$\frac{\partial \Pi_0}{\partial z_k} = \frac{\partial G}{\partial \Pi_{k-1}} \frac{\partial \Pi_{k-1}}{\partial z_k} = \frac{\partial G}{\partial \Pi_{k-1}} \left( \frac{1}{z_1} \right) \tag{3.4}$$

Using (3.3) and (3.4) together, one obtains

$$\frac{\partial \Pi_0}{\partial z_1} = \sum_{j=1}^{N-1} z_1 \frac{\partial \Pi_0}{\partial z_{j+1}} \left( -\frac{z_{j+1}}{z_1^2} \right). \tag{3.5}$$

At this point we regard $\sigma$ as a function of the $z_j$ rather than the $c_{2j}$ . In other words, we define $\tilde{\sigma}(z_1, ..., z_N) = \sigma(c_2, ..., c_{2N})$ and subsequently drop the tilda, as we will only use surface tension as a function of the $z_j$ below. Also, by using the definition of $\Pi_0$ we have for $k \geq 2$, we use the identity expressed by the definition of $\Pi_0$ so $\Pi_0 = \sigma(z_1, ..., z_n) / z_1$ with derivatives

$$\frac{\partial \Pi_0}{\partial z_1} = \frac{\partial \left( \frac{\sigma}{z_1} \right)}{\partial z_1} = -\frac{1}{z_1^2} \sigma + \frac{1}{z_1} \frac{\partial \sigma}{\partial z_1}, \tag{3.6}$$

$$\frac{\partial \Pi_0}{\partial z_k} = \frac{\partial \left( \frac{\sigma}{z_1} \right)}{\partial z_k} = \frac{1}{z_1} \frac{\partial \sigma}{\partial z_k}.$$

Thus one obtains

$$-\frac{1}{z_1^2} \sigma + \frac{1}{z_1} \frac{\partial \sigma}{\partial z_1} = -\frac{1}{z_1} \sum_{j=1}^{N-1} z_{j+1} \frac{\partial \Pi_0}{\partial z_{j+1}} = -\frac{1}{z_1} \sum_{j=1}^{N-1} z_{j+1} \frac{1}{z_1} \frac{\partial \sigma}{\partial z_{j+1}}$$

$$= -\frac{1}{z_1^2} \sum_{j=2}^{N} z_j \frac{\partial \sigma}{\partial z_j}.$$

Rewriting this by multiplying by $z_1^2$ we have

$$\sum_{j=1}^{N} z_j \frac{\partial \sigma}{\partial z_j} = \sigma, \quad or, \; z \cdot \nabla \sigma = \sigma \tag{3.7}$$

Hence this is in the form of a Clairaut's equation [8].

## 4. Solution to Clairaut's equation for surface tension.

Clairaut's equation, (3.7) is a first order nonlinear partial differential equation that can be solved by the well-known method of characteristics (see e.g., [8]). In order to obtain a solution in $\mathbb{R}^N$ (actually the portion of $\mathbb{R}^N$ where each $z_i$ is non-negative) we will need to have a surface of "initial conditions" $\Gamma$ described below.

**Definition**. An acceptable manifold $\Gamma$ is a smooth surface in $\mathbb{R}^N$ such that the tangent plane to $\Gamma$ at any point $\vec{s} \in \Gamma$ intersects at a nonzero angle with any ray $\vec{v}$ emanating from the origin and going through an arbitrary point $\left(z_1^0, z_2^0, ..., z_N^0\right)$ such that $z_i^0 \geq 0$.

We define the coordinates $(r, s)$ with $r \in \mathbb{R}$, $s \in \mathbb{R}^{N-1}$ that form a new coordinate system

$$(r(z_1, .., z_N), s(z_1, .., z_N)), \quad \hat{\sigma}(r, s) = \sigma(z_1, ..., z_N) \tag{4.1}$$

and use the chain rule to obtain,

$$\frac{\partial \hat{\sigma}}{\partial r} = \frac{\partial \sigma}{\partial z_1}\frac{\partial z_1}{\partial r} + ... + \frac{\partial \sigma}{\partial z_N}\frac{\partial z_N}{\partial r}.$$

One has then, from the standard methods of characteristics,

$$\begin{aligned} \frac{dz_1}{dr} &= z_1, \ so, \ z_1(r; s) = C_1(s)e^r, \\ &... \\ \frac{dz_N}{dr} &= z_N, \ so, \ z_N(r; s) = C_N(s)e^r, \\ \frac{d\sigma}{dr} &= \sigma, \ so, \ \sigma(r; s) = C_{N+1}(s)e^r. \end{aligned} \tag{4.3}$$

Next, we need to satisfy the "initial conditions" i.e., the values on the surface $(z_1, ..., z_n) \in \Gamma \subset \mathbb{R}^N$ where $\sigma$ is specified. For the $z_i$ variables, we make the choice that when $s \in \Gamma$, we have $r = 0$. Hence, we have $\sigma(0, s) =: \sigma_0(s)$ as the conditions on the surface $\Gamma$, yielding the solutions

$$\begin{aligned} &z_1(r; s) = C_1(s)e^r, ..., z_N(r, s) = C_N(s)e^r, \\ &\sigma(r; s) = \sigma_0(s)e^r. \end{aligned}$$

with characteristics defined by

$$\frac{z_1(r; s)}{z_2(r; s)} = \frac{C_1(s)}{C_2(s)}, ....$$

This means that there is a fixed ratio of $z_i$ to $z_j$ for all $i, j \in \{1, ..., N\}$.

In summary, one can solve for all $\sigma(z_1, ..., z_N)$ in the subset $\mathbb{R}_+^N \subset \mathbb{R}^N$ for which all of the $z_j$ are nonnegative provided we specify the values of $\sigma(z_1, ..., z_N)$ on a surface $\Gamma \subset \mathbb{R}_+^N$ that is convex, and each point of $\Gamma$ intersects at a non-zero angle with each ray that emanates from the origin. If the surface $\Gamma$ is not convex, then the characteristics may intersect so that we obtain only local solutions.

## 5. Surface tension and homogeneous functions

We let $\Omega \subset \mathbb{R}_+^N$ be an open cone, i.e., if $z \in \Omega$ is in the set then so is $tz$, where $\mathbb{R}_+^N := \left\{z \in \mathbb{R}^N : z_j > 0\right\}$.

**Definition.** For any scalar $k$ a real-valued function $f(z_1, ..., z_n)$ with $(z_1, ..., z_n) \in \Omega$ is *homogeneous of degree k* if

$$f(tz_1, ..., tz_n) = t^k f(z_1, ..., z_n) \quad for\ all\ t > 0. \tag{5.1}$$

We recall two classical results.

**Theorem.** Let $f$ be a $C^1$ function on an open cone in $\mathbb{R}^n$. If $f$ is homogeneous of degree $k$ then its first order partial derivatives are homogeneous of degree $k - 1$.

**Theorem.** Let $f : \Omega \to \mathbb{R}$ be a continuously differentiable function. Then the following are equivalent:

(1) $f$ is a homogeneous function of degee $k$;

(2) $f$ satisfies for $z \in \Omega$, the equation

$$\sum_{j=1}^{n} z_j \frac{\partial f(z)}{\partial z_j} = kf(z). \tag{5.2}$$

We now apply these concepts to the surface tension. Since $\sigma(z)$ satisfies (3.7), the theorem implies that $\sigma$ is homogeneous of degree 1. Hence, all of its partial derivatives are of degree 0. Hence we have the relation, for any $j \in \{1, ..., n\}$

$$z_1 \frac{\partial}{\partial z_1}\left(\frac{\partial \sigma}{\partial z_j}\right) + ... + z_n \frac{\partial}{\partial z_n}\left(\frac{\partial \sigma}{\partial z_j}\right) = 0. \tag{5.3}$$

Hence, each of the partial derivatives $\partial\sigma/\partial z_j$ satisfies the homogeneous Clairot's equation, $\vec{z} \cdot \vec{\nabla}\sigma = 0$ . Stated differently, the fact that the derivatives are homogeneous of order 0 means that for any $t > 0$ one has

$$\frac{\partial \sigma}{\partial z_j}(tz_1, ..., tz_n) = \frac{\partial \sigma}{\partial z_j}(z_1, ..., z_n),$$

i.e., the partial derivative is constant along the entire ray $\{tz : t > 0, z \in \mathbb{R}^n_+\}$.

One can also obtain similar results for a particular subspace. For example, if $n = 3$, and we set $z_2 = 0$ then one has similar results on the $z_1 z_3$ plane.

## Conflict of Interest

The author declares no conflicts of interest in this paper.

## References

1. G. Barenblatt and P. Makinen, *Dimensional Analysis*, Routledge, Taylor and Francis, Abingdon, UK, 1987.
2. M. Ben Amar, *Dendritic growth rate at arbitrary undercooling*, Phys. Rev. A., **41** (1990), 2080-2092.
3. G. Caginalp, *A microscopic derivation of macroscopic phase boundary problems*, Journal of Statistical Physics, **59** (1990), 869-884.

4. M. Conti, *Thermal and chemical diffusion in the rapid solidification of binary alloys*, Phys. Rev. E., **61** (2000), 642-650.
5. G. Caginalp and E. Esenturk, *Anisotropic Phase Field Equations of Arbitrary Order*, Discrete and Continuous Dynamical Systems, Series S **4** (2011), 311-350.
6. G. Caginalp and E. Esenturk, *Renormalization Methods for Higher Order Differential Equations*, J. Physics A., **47** (2014), 315004.
7. G. Caginalp and P. Fife, *Higher order phase field models and detailed anisotropy*, Physical Review B., **34** (1986), 4940-4943.
8. L.C. Evans, *Partial Differential Equations*, American Math Soc., Providence, RI, 2010.
9. N. Goldenfeld, Lectures on phase transitions and renormalization group, Perseus Books, 1992.
10. A. Miranville, *Higher-order anisotropic Caginalp phase-field systems* Mediterr. J. Math., **13** (2016), 4519-4535.
11. M. Niezgodka and P. Strzelecki, *Free Boundary Problems*, Theory and Applications Proceedings of the Zakopane Conference, 1995.

# 4

# Coefficient bounds for a subclass of multivalent functions of reciprocal order

**Khalida Inayat Noor**[1] **, Nazar Khan**[2,*]**and Qazi Zahoor Ahmad**[2]

[1] Department of Mathematics Comsats Institute of Information Technology Park Road, Islamabad, Pakistan

[2] Department of Mathematics Abbottabad University of Science and Technology, Abbottabad, Pakistan

[*] **Correspondence:** nazarmaths@hotmail.com

**Abstract:** The aim of this paper is to introduce a new subclass of multivalent functions of complex order and to study some interesting properties such as coefficient estimates, sufficiency criteria, Fekete-Szego inequality, inclusion result and integral preserving property for this newly defined class.

**Keywords:** Multivalent functions; Convolution; Carlson Shaffer operator
**2010 Mathematics Subject Classification:** Primary 30C45, 30C10; Secondary 47B38.

## 1. Introduction

Let $\mathcal{A}_p$ denote the class of $p$-valent functions $f$ which are analytic in the regions $\mathbb{U} = \{z \in \mathbb{C} : |z| < 1\}$ and normalized by

$$f(z) = z^p +_{k=1}^{\infty} a_{p+k}z^{p+k}. \tag{1.1}$$

We note that $\mathcal{A}_1 = \mathcal{A}$. Also let $\mathcal{N}(\alpha)$ and $\mathcal{M}(\alpha)$ denote the usual classes of starlike and convex functions of reciprocal order $\alpha$, $\alpha > 1$, and are defined by

$$\mathcal{N}(\alpha) = \left\{ f(z) \in \mathcal{A} : Re\frac{zf'(z)}{f(z)} < \alpha, \ (z \in \mathbb{U}) \right\}, \tag{1.2}$$

$$\mathcal{M}(\alpha) = \left\{ f(z) \in \mathcal{A} : 1 + Re\frac{zf''(z)}{f'(z)} < \alpha, \ (z \in \mathbb{U}) \right\}. \tag{1.3}$$

These classes were introduced by Uralegaddi et.al [21] in 1994 and then studied by the authors in [12]. After that Nunokawa and his coauthors [11] proved that for $f \in \mathcal{N}(\alpha)$, $0 < \alpha < \frac{1}{2}$, if and only if the following inequality holds

$$\left| \frac{zf'(z)}{f(z)} - \frac{1}{2\alpha} \right| < \frac{1}{2\alpha}, \ (z \in \mathbb{U}).$$

In 2002, Owa and Srivastava [13] generalized this idea for the classes of $p$-valent starlike and $p$-valent convex functions of reciprocal order $\alpha$ with $\alpha > p$, and further investigated by Polatoglu et.al [14]. Recently in 2011, Uyanik et.al [22] extended this idea to the classes of $p$-valently spirallike and $p$-valently Robertson functions and discussed coefficient inequalities and sufficient conditions for the functions of these classes.

The convolution (Hadamard product) of functions $f, g \in \mathcal{A}_p$ is defined by

$$(f * g)(z) = z^p + \sum_{n=1}^{\infty} a_{n+p} b_{k+p} z^{n+p}, \quad z \in \mathbb{U},$$

where $f$ is given by (1.1) and

$$g(z) = z^p + \sum_{n=1}^{\infty} b_{n+p} z^{n+p}, \quad z \in \mathbb{U}.$$

The incomplete beta function $\phi_p$ defined by

$$\phi_p(a, c; z) = z^p + \sum_{n=1}^{\infty} \frac{(a)_n}{(c)_n} z^{n+p}, \quad (a \in \mathbb{R},\ c \in \mathbb{R} \setminus (0, -1, ...),\ z \in \mathbb{U}),$$

where $(\alpha)_n$ is the pochhamer symbol defined in terms of Gamma function by

$$(\alpha)_n = \frac{\Gamma(\alpha + n)}{\Gamma(\alpha)} = \begin{cases} \alpha(\alpha + 1)...(\alpha + n - 1), & \text{if } n \in \mathbb{N} \\ 1, & \text{if } n = 0. \end{cases}$$

With the help of incomplete beta function $\phi_p$ and concepts of convolution, Saitoh in [18, 19] introduced the operator $\mathcal{L}_p : \mathcal{A}_p \to \mathcal{A}_p$ and is defined by

$$\begin{aligned} \mathcal{L}_p(a, c) f(z) &= \phi_p(a, c; z) * f(z), \\ &= z^p + \sum_{n=1}^{\infty} \varphi_n(a)\, a_{n+p}\, z^{n+p} 1.4 \end{aligned} \tag{1.4}$$

with $a > -p$ and

$$\varphi_n(a) = \frac{\Gamma(a + n)\Gamma(c)}{\Gamma(a)\Gamma(c + n)}. \tag{1.5}$$

This operator is an extension of the familiar Carlson-Shaffer operator, which has been used widely on the space of analytic and univalent functions in $\mathbb{U}$, see [3, 20]. The following identity can be easily derived

$$z\left(\mathcal{L}_p(a, c) f(z)\right)' = a\mathcal{L}_p(a + 1, c) f(z) - (a - p)\mathcal{L}_p(a, c) f(z). \tag{1.6}$$

Motivated from the above mentioned work, we now introduce a new subclass of multivalent functions of reciprocal order using the operator defined in (1.4).

An analytic multivalent function $f$ of the form (1.1) belongs to the class $\mathcal{SC}_p^{\lambda}(a,b,c,\beta)$, if and only if

$$Re\left\{\frac{2e^{i\lambda}}{b}\left(\frac{\mathcal{L}_p(a+1,c)f(z)}{\mathcal{L}_p(a,c)f(z)}-1\right)\right\} < (\beta-1)\cos\lambda,$$

where $b\in\mathbb{C}\setminus\{0\}$, $\lambda$ is real with $|\lambda|<\frac{\pi}{2}, \beta>1$.

It is noticed that, by giving specific values to $a,b,c,\beta$ and $\lambda$ in $\mathcal{SC}_p^{\lambda}(a,b,c,\beta)$, we obtain many well-known as well as new subclasses of analytic, univalent and multivalent functions, for example;

(i). For $\lambda=0$, $a=1=c$ and $b=2$, we obtain $\mathcal{SC}_p^{0}(1,2,1,\beta)=\mathcal{M}_p(\beta)$ studied in [13] and further for $p=1$, we have the class $\mathcal{M}(\beta)$ introduced and studied in [7, 21].

(ii). For $p=c=b=1$, $\lambda=0$ and $a=2$, we have $\mathcal{SC}_1^{0}(1,2,1,\beta)=\mathcal{N}(\beta)$ studied in [7, 21].

(iii). For $a=1=c$ and $b=2$, we get the class $\mathcal{SC}_p^{\lambda}(1,2,1,\beta)=\mathcal{S}_p(\lambda,\beta)$ and for $a=2$, $b=c=1$, we obtain $\mathcal{SC}_p^{\lambda}(1,2,1,\beta)=\mathcal{C}_p(\lambda,\beta)$, introduced and studied in [22].

(iv). For $\lambda=0$, $a=2$, $b=p=1$, and $c=2-\alpha$, we obtain the class $\mathcal{SC}_1^{0}(2,1,2-\alpha,\beta)=\mathcal{P}_\alpha(\beta)$ [4].

The $q^{th}$ Hankel determinant $H_q(n)$, $q\geq 1, n\geq 1$, stated by Pommerenke [15] and furter investigated by Noonan and Thomas [10] as

$$H_q(n)=\begin{vmatrix} a_n & a_{n+1} & \cdots & a_{n+q+1} \\ a_{n+1} & a_{n+2} & \cdots & a_{n+q} \\ \vdots & \vdots & \ddots & \vdots \\ a_{n+q-1} & a_{n+q} & \cdots & a_{n+2q-2} \end{vmatrix}$$

This Hankel determinant is useful and has also been studied by several authors for details see [1, 2]. The growth rate of Hankel determinant $H_q(n)$ as $n\to\infty$ was investigated, respectively,when $f$ is a member of certain subclass of analytic functions, such as the class of $p$-valent functions [15, 10], the class of starlike functions [15], the class of univalent functions [16], the class of close-to-convex functions [8] , a new class $V_k$ [9]. It is well known that the Fekete-Szegö functional is $H_2(1)=\left|a_3-a_2^2\right|$. Fekete and Szegö further generalized the estimate $\left|a_3-\mu a_2^2\right|$ where $\mu$ is real and $f\in S$, the class of univalent functions. For our discussion in this paper we investigate the coeficient bound, the upper bounds of the Hankel determinant $H_3(1)$ for a subclass of multivalent functions.

We will need the following lemma's for our work.

**Lemma 1.2.** [17]. *If $q$ is an analytic function with $Req(z)>0$ and*

$$q(z)=1+_{n=2}^{\infty}d_nz^n,\ z\in\mathbb{U}, \tag{1.7}$$

*then for $n\geq 1$,*

$$|d_n|\leq 2.$$

**Lemma 1.3.** [6]. *If $q$ is of the form* (1.7) *with positive real part, then the following sharp estimate holds*

$$\left|d_2\_vd_1^2\right|\leq 2\max\{1,|2v-1|\},\ \text{for all } v\in\mathbb{C}.$$

**Lemma 1.4.** [5]. *If $q$ is of the form* (1.7) *with positive real part, then*

$$\begin{aligned} 2d_2 &= d_1^2+x(4-d_1^2). \\ 4d_3 &= d_1^3+2(4-d_1^2)d_1x-d_1(4-d_1^2)x^2+2(4-d_1^2)\left(1-|x|^2\right)z. \end{aligned}$$

*for some $x, z$ with $|x|\leq 1$ and $|z|\leq 1$.*

## 2. Main Results

**Theorem 2.1.** *If* $f \in \mathcal{SC}_p^{\lambda}(a,b,c,\beta)$, *then*

$$\left|a_{p+1}\right| \leq \frac{|b|\,|\eta|\,c}{p},$$

*and*

$$\left|a_{n+p-1}\right| \leq \frac{a\,|b|\,|\eta|}{(n+p-2)\varphi_{n-1}(a)} \prod_{j=1}^{n-2}\left(1+\frac{a\,|b|\,|\eta|}{(p+j)}\right), \quad n \geq 3, \tag{2.1}$$

*where* $\varphi_{n-1}(\delta)$ *is given by* (1.5) *and*

$$\eta = (1-\beta)\cos\lambda + i\sin\lambda. \tag{2.2}$$

*Proof.* Let $f \in \mathcal{SC}_p^{\lambda}(a,b,c,\beta)$. Then we have

$$Re\left\{\frac{2e^{i\lambda}}{b}\left(\frac{\mathcal{L}_p(a+1,c)f(z)}{\mathcal{L}_p(a,c)f(z)}-1\right)\right\} < (\beta-1)\cos\lambda, \quad z \in \mathbb{U}.$$

Now let us define a function $q$ by

$$e^{i\lambda}\left(1-\frac{2}{b}+\frac{2}{b}\frac{\mathcal{L}_p(a+1,c)f(z)}{\mathcal{L}_p(a,c)f(z)}\right) = ((1-\beta)q(z)+\beta)\cos\lambda + i\sin\lambda. \tag{2.3}$$

where $q$ is analytic in $\mathbb{U}$ with $q(0)=1$ and $Re q(z) > 0$, $z \in \mathbb{U}$. Then (2.3) can be written as

$$1-\frac{2}{b}+\frac{2}{b}\frac{\mathcal{L}_p(a+1,c)f(z)}{\mathcal{L}_p(a,c)f(z)} = 1+\frac{(1-\beta)\cos\lambda + i\sin\lambda}{e^{i\lambda}}\sum_{n=1}^{\infty} c_n z^n,$$

or equivalently

$$2e^{i\lambda}\left(\mathcal{L}_p(a+1,c)f(z) - \mathcal{L}_p(a,c)f(z)\right) = b\eta\mathcal{L}_p(a,c)f(z)\sum_{n=1}^{\infty} c_n z^n, \tag{2.4}$$

where $\eta$ is given by (2.2). From (2.4) and (1.6) we have

$$2e^{i\lambda}\left[z\left(\mathcal{L}_p(a,c)f(z)\right)' - p\mathcal{L}_p(a,c)f(z)\right] = ab\eta\mathcal{L}_p(a,c)f(z)\sum_{n=1}^{\infty} c_n z^n,$$

that is,

$$2e^{i\lambda}\left[\sum_{k=1}^{\infty}(k+p-1)\varphi_k(a)a_{k+p}z^{k+p}\right] = ab\eta\left[z^p+\sum_{k=1}^{\infty}\varphi_k(a)a_{k+p}z^{k+p}\right]\left(\sum_{k=1}^{\infty} d_n z^n\right).$$

Comparing the coefficients of $z^{n+p-1}$ on both sides, we obtain

$$2e^{i\lambda}(n+p-2)\varphi_{n-1}(a)a_{n+p-1} = ab\eta\left\{d_1 a_{n-2}\varphi_{n+p-2}(a) + \ldots + d_{n-1}\right\}. \tag{2.5}$$

Taking absolute on both sides and then applying Lemma 1.1, we have

$$\left|a_{n+p-1}\right| \leq \frac{a\,|b|\,|\eta|}{(n+p-2)\varphi_{n-1}(a)}\left\{1+\varphi_1(a)\left|a_{p+1}\right|+\ldots+\varphi_{n-2}(a)\left|a_{n+p-2}\right|\right\} \tag{2.6}$$

We now apply mathematical induction on (2.6). So for $n = 2$

$$\left|a_{p+1}\right| \leq \frac{|b|\,|\eta|\,c}{p}.$$

which shows that the result is true for $n = 2$. For $n = 3$

$$\left|a_{p+2}\right| \leq \frac{a\,|b|\,|\eta|}{(p+1)\varphi_2(a)}\left\{1+\varphi_1(a)\left|a_{p+1}\right|\right\} \tag{2.7}$$

and using the bound of $\left|a_{p+1}\right|$ in (2.7), we have

$$\left|a_{p+2}\right| \leq \frac{a\,|b|\,|\eta|}{(p+1)\varphi_2(a)}\left\{1+\frac{a\,|b|\,|\eta|}{p}\right\}.$$

Therefore (2.1) holds for $n = 3$.

Assume that (2.1) is true for $n = k$, that is,

$$\left|a_{k+p-1}\right| \leq \frac{a\,|b|\,|\eta|}{(k+p-2)\varphi_{k-1}(a)}\prod_{j=1}^{k-2}\left(1+\frac{a\,|b|\,|\eta|}{(p+j)}\right).$$

Consider

$$\begin{aligned}\left|a_{k+p}\right| &\leq \frac{a\,|b|\,|\eta|}{(k+p-1)\varphi_k(a)}\left\{\left(1+\frac{a\,|b|\,|\eta|}{p}\right)+\frac{a\,|b|\,|\eta|}{(p+1)}\left(1+\frac{a\,|b|\,|\eta|}{p}\right)\right.\\ &\qquad\left.+\ldots+\frac{a\,|b|\,|\eta|}{(p+k-1)}\prod_{j=1}^{k-2}\left(1+\frac{a\,|b|\,|\eta|}{(p+j)}\right)\right\}\\ &= \frac{a\,|b|\,|\eta|}{(k+p-1)\varphi_k(a)}\prod_{j=1}^{k-1}\left(1+\frac{a\,|b|\,|\eta|}{(p+j)}\right).\end{aligned}$$

Therefore, the result is true for $n = k + 1$ and hence by using mathematical induction, (2.1) holds true for all $n \geq 3$. □

The following corollaries which were proved by owa and Nishawski [12] comes as a special case from Theorem 2.1 by varying the parameter $a, b, c, p$ and $\lambda$.

**Corollary 2.2.** *If $f \in \mathcal{M}(\beta)$, then*

$$|a_n| \leq_{l=2}^{n} \frac{(l+2\beta-4)}{(n-1)!}, \quad \text{for all } n \geq 2.$$

**Corollary 2.3.** *If $f \in \mathcal{N}(\beta)$, then*

$$|a_n| \leq_{l=2}^{n} \frac{(l+2\beta-4)}{n!}, \quad \text{for all } n \geq 2.$$

**Theorem 2.4.** *Let $f \in \mathcal{A}_p$ and satisfies*

$$\sum_{n=1}^{\infty}\left|\left(1+\frac{2n}{ab}\right)e^{i\lambda}-\beta\cos\lambda\right|\varphi_n(a)\left|a_{n+p}\right| < \left|\beta\cos\lambda - e^{i\lambda}\right|. \tag{2.8}$$

*Then $f \in \mathcal{SC}_p^{\lambda}(a,b,c,\beta)$.*

*Proof.* To prove that $f$ belongs to $\mathcal{SC}_p^{\lambda}(a,b,c,\beta)$ we need to prove that

$$\left|\frac{e^{i\lambda}\left(1-\frac{2}{b}+\frac{2}{b}\frac{\mathcal{L}_p(a+1,c)f(z)}{\mathcal{L}_p(a,c)f(z)}\right)-1}{e^{i\lambda}\left(1-\frac{2}{b}+\frac{2}{b}\frac{\mathcal{L}_p(a+1,c)f(z)}{\mathcal{L}_p(a,c)f(z)}\right)-(2\beta\cos\lambda-1)}\right| < 1. \tag{2.9}$$

For this consider the left hand side of (2.9), we have

$$\begin{aligned}
LHS &= \left|\frac{e^{i\lambda}\left(1-\frac{2}{b}+\frac{2}{b}\frac{\mathcal{L}_p(a+1,c)f(z)}{\mathcal{L}_p(a,c)f(z)}\right)-1}{e^{i\lambda}\left(1-\frac{2}{b}+\frac{2}{b}\frac{\mathcal{L}_p(a+1,c)f(z)}{\mathcal{L}_p(a,c)f(z)}\right)-(2\beta\cos\lambda-1)}\right| \\
&= \left|\frac{\left(e^{i\lambda}-1\right)bz^p+\sum_{n=1}^{\infty}\left(\left(b+\frac{2n}{a}\right)e^{i\lambda}-1\right)\varphi_n(a)a_{n+p}z^{n+p}}{(be^{i\lambda}-2b\beta\cos\lambda+b)+\sum_{n=1}^{\infty}\left(\left(b+\frac{2n}{a}\right)e^{i\lambda}-2b\beta\cos\lambda+b\right)\varphi_n(a)a_{n+p}z^{n+p}}\right| \\
&\le \frac{\left|\left(e^{i\lambda}-1\right)\right|+\sum_{n=1}^{\infty}\left|\left(1+\frac{2n}{ab}\right)e^{i\lambda}-1\right|\varphi_n(a)\left|a_{n+p}\right|}{|2\beta\cos\lambda-e^{i\lambda}-1|-\sum_{n=1}^{\infty}\left|\left(1+\frac{2n}{ab}\right)e^{i\lambda}-2\beta\cos\lambda+1\right|\varphi_n(a)\left|a_{n+p}\right|}.
\end{aligned}$$

The last expression is bounded above by 1 if

$$\begin{aligned}
\left|e^{i\lambda}-1\right|+\sum_{n=1}^{\infty}\left|\left(1+\frac{2n}{ab}\right)e^{i\lambda}-1\right|\varphi_n(a)\left|a_{n+p}\right| &\le \left|2\beta\cos\lambda-e^{i\lambda}-1\right| \\
&\quad -\sum_{n=1}^{\infty}\left|\left(1+\frac{2n}{ab}\right)e^{i\lambda}-2\beta\cos\lambda+1\right|\varphi_n(a)\left|a_{n+p}\right|
\end{aligned}$$

which is equivalent to the condition (2.8) and so $f \in \mathcal{SC}_p^{\lambda}(a,b,c,\beta)$. □

If we set $p = 1, \lambda = 0$, $a = 1$, $b = 2$ and $c = 1$, in above Theorem we have the following result by [12].

**Corollary 2.5.** *If $f \in A$ satisfies*

$$\sum_{n=2}^{\infty}\{(n-1)+|n-2\beta+1|\}\left|a_n\right| \le 2(\beta-1)$$

*for some $\beta(\beta>1)$, then $f \in \mathcal{M}(\beta)$.*

**Corollary 2.6.** [4]. *A function $f \in \mathcal{P}_{\alpha}(\beta)$ if and only if*

$$\sum_{n=2}^{\infty}\frac{\Gamma(n+1)\Gamma(2-\alpha)}{\Gamma(n+1-\alpha)}(n-\beta)\left|a_n\right| \le (\beta-1).$$

*The result is sharp.*

**Theorem 2.7.** *Let $f \in \mathcal{SC}_p^0(a, b, c, \beta)$ and of the form* (1.1). *Then*

$$\left|a_{p+2} - \mu a_{p+1}^2\right| \leq \frac{|b|\, c(c+1)(1-\beta)}{(a+1)(p+1)} \max\{1, |2\upsilon - 1|\}, \tag{2.10}$$

*where*

$$\upsilon = \frac{\mu bc(a+1)(p+1)(1-\beta)}{2p^2(c+1)} - \frac{ab(1-\beta)}{2p}. \tag{2.11}$$

*Proof.* Let $f \in \mathcal{SC}_p^0(a, b, c, \beta)$. Then from (2.5) with $\lambda = 0$, we have

$$\begin{aligned} a_{p+1} &= \frac{bc(1-\beta)}{2p} d_1 \\ a_{p+2} &= \frac{bc(c+1)(1-\beta)}{2(a+1)(p+1)} \left\{ d_2 + \frac{ab(1-\beta)}{2p} d_1^2 \right\}. \end{aligned}$$

For any complex number $\mu$, we have

$$\begin{aligned} a_{p+2} - \mu a_{p+1}^2 &= \frac{bc(c+1)(1-\beta)}{2(a+1)(p+1)} \left[ d_2 - \frac{b(1-\beta)}{2p} \left\{ \frac{\mu c(a+1)(p+1)}{p(c+1)} - a \right\} d_1^2 \right] \\ &= \frac{bc(c+1)(1-\beta)}{2(a+1)(p+1)} \left[ d_2 - \upsilon d_1^2 \right], \end{aligned}$$

where $\upsilon$ is given by (2.11)

Taking modulus on both sides and applying Lemma 1.2, we have

$$\begin{aligned} \left|a_{p+2} - \mu a_{p+1}^2\right| &= \left| \frac{bc(c+1)(1-\beta)}{2(a+1)(p+1)} \right| \left| d_2 - \upsilon d_1^2 \right| \\ &\leq \frac{|b|\, c(c+1)(1-\beta)}{(a+1)(p+1)} \max\{1, |2\upsilon - 1|\}. \end{aligned}$$

This proves the required result. □

Taking $\mu = 1$, we obtain the following result.

**Corollary 2.8.** *Let $f \in \mathcal{SC}_p^0(a, b, c, \beta)$ and of the form* (1.1). *Then*

$$\left|a_{p+2} - a_{p+1}^2\right| \leq \frac{|b|\, c(c+1)(1-\beta)}{(a+1)(p+1)} \max\{1, |2\upsilon - 1|\},$$

*where*

$$\upsilon = \frac{bc(a+1)(p+1)(1-\beta)}{2p^2(c+1)} - \frac{ab(1-\beta)}{2p}.$$

**Theorem 2.9.** *Let $f \in \mathcal{SC}_p^\lambda(a, b, c, \beta)$. Then*

$$\left|a_{p+1}a_{p+3} - a_{p+2}^2\right| \leq \left[ \frac{4\,|b|\, c(c+1)(1-\beta)}{(p+1)(a+1)} \right]^2. \tag{2.12}$$

*Proof.* Let $f \in \mathcal{SC}_p^{\lambda}(a,b,c,\beta)$. Then from (2.5) we have

$$a_{p+1} = \frac{ab(1-\beta)}{2p\varphi_1(a)}d_1. \tag{2.13}$$

$$a_{p+2} = \frac{ab(1-\beta)}{2(p+1)\varphi_2(a)}\left\{d_2 + \frac{ab(1-\beta)}{2p}d_1^2\right\}. \tag{2.14}$$

$$a_{p+3} = \frac{ab(1-\beta)}{2(p+2)\varphi_3(a)}\left\{d_3 + \frac{ab(1-\beta)}{2(p+1)}d_2^2 + \frac{\{ab(1-\beta)\}^2}{4p(p+1)}d_1^2 d_2 + \frac{ab(1-\beta)}{2p}d_1^2\right\}. \tag{2.15}$$

From (2.13), (2.14) and (2.15) we obtain

$$\begin{aligned}\left|a_{p+1}a_{p+3} - a_{p+2}^2\right| &= \frac{\{ab(1-\beta)\}^2}{4p(p+2)\varphi_1(a)\varphi_3(a)} \times \\ &\left\{d_3 + \frac{ab(1-\beta)}{2(p+1)}d_2^2 + \frac{\{ab(1-\beta)\}^2}{4p(p+1)}d_1^2 d_2 + \frac{ab(1-\beta)}{2p}d_1^2\right\} \\ &- \frac{\{ab(1-\beta)\}^2}{4(p+1)^2\varphi_2^2(a)}\left\{d_2^2 + \frac{\{ab(1-\beta)\}^2}{4p^2}d_1^4 + \frac{ab(1-\beta)}{p}d_1^2 d_2\right\}.\end{aligned}$$

After some simplification we have

$$\begin{aligned}\left|a_{p+1}a_{p+3} - a_{p+2}^2\right| = \frac{|A|^2}{4}\Big[&Bd_1d_3 + Cd_1d_2^2 + \\ &Ed_1^3d_2 + Fd_1^3 - Gd_2^2 - Hd_1^4 - Kd_1^2d_2,\Big]\end{aligned} \tag{2.16}$$

where

$$A = ab(1-\beta), \quad B = \frac{1}{p(p+1)\varphi_1(a)\varphi_3(a)}, \quad C = \frac{A}{2p(p+1)(p+2)\varphi_1(a)\varphi_3(a)},$$
$$E = \frac{A^2}{4p^2(p+1)(p+2)\varphi_1(a)\varphi_3(a)}, \quad F = \frac{A}{2p^2(p+1)\varphi_1(a)\varphi_3(a)}, \quad G = \frac{1}{(p+1)^2\varphi_2^2(a)},$$
$$H = \frac{A^2}{4p^2(p+1)^2\varphi_2^2(a)}, \quad K = \frac{A}{p(p+1)^2\varphi_2^2(a)}.$$

Substituting the values of $d_2$ and $d_3$ frome Lemma 1.3 in (2.16) we have

$$\begin{aligned}&\left|Bd_1d_3 + Cd_1d_2^2 + Ed_1^3d_2 + Fd_1^3 - Gd_2^2 - Hd_1^4 - Kd_1^2d_2\right| \\ = &\Big|\frac{1}{4}Bd_1\left\{d_1^3 + 2d_1(4-d_1^2)x - d_1(4-d_1^2)x^2 + 2(4-d_1^2)\left(1-|x|^2\right)z\right\} \\ &+\frac{1}{4}Cd_1\left\{d_1^4 + 2d_1^2(4-d_1^2)x + (4-d_1^2)^2x^2\right\} + \frac{1}{2}Ed_1^3\left\{d_1^2 + (4-d_1^2)x\right\} \\ &+Fd_1^3 - \frac{1}{4}G\left\{d_1^4 + 2d_1^2(4-d_1^2)x + (4-d_1^2)^2x^2\right\} - Hd_1^4 \\ &-\frac{1}{2}Kd_1^2\left\{d_1^2 + (4-d_1^2)x\right\}\Big|.\end{aligned}$$

Simple computation gives

$$4\left|Bd_1d_3 + Cd_1d_2^2 + Ed_1^3d_2 + Fd_1^3 - Gd_2^2 - Hd_1^4 - Kd_1^2d_2\right| = \left|(C+2E)\,d_1^5 + (B-G-H-2K)\,d_1^4 + Fd_1^3 + (2B+2Cd_1+2Ed_1-2G-2K)\,d_1^2\left(4-d_1^2\right)|x| + 2Bd_1\left(4-d_1^2\right)\left(1-|x|^2\right)|z| + \left\{Cd_1\left(4-d_1^2\right) - Bd_1^2 - G\left(4-d_1^2\right)\right\}\left(4-d_1^2\right)|x|^2\right| \tag{2.17}$$

Applying triangle inequality and replacing $|x|$ by $\rho$ in (2.17) we have

$$\begin{aligned} 4\left|Bd_1d_3 + Cd_1d_2^2 + Ed_1^3d_2 + Fd_1^3 - Gd_2^2 - Hd_1^4 - Kd_1^2d_2\right| &\le (|C|+2|E|)\,d_1^5 + |F|\,d_1^3 + \\ & (B-G+|H|+2|K|)\,d_1^4 + \{2B+2|C|\,d_1+2|E|\,d_1-2G+2|K|\}\,d_1^2\left(4-d_1^2\right)\rho \\ & +2Bd_1\left(4-d_1^2\right)\left(1-\rho^2\right) + \left\{|C|\,d_1\left(4-d_1^2\right) - Bd_1^2 - G\left(4-d_1^2\right)\right\}\left(4-d_1^2\right)\rho^2 \\ &= F(d_1,\rho). \end{aligned} \tag{2.18}$$

Taking partial derivative of $F(d_1,\rho)$ with respect to $\rho$, we have

$$\begin{aligned} \frac{\partial F(d_1,\rho)}{\partial\rho} &= \{2B+2|C|\,d_1+2|E|\,d_1-2G+2|F|\}\,d_1^2\left(4-d_1^2\right) - 4Bd_1\left(4-d_1^2\right)\rho \\ & +2\left\{|C|\,d_1\left(4-d_1^2\right) - Bd_1^2 - G\left(4-d_1^2\right)\right\}\left(4-d_1^2\right)\rho \end{aligned}$$

Clearly $\frac{\partial F(d_1,\rho)}{\partial\rho} > 0$, for $0 < \rho < 1$ and $0 < d_1 < 2$. Therefore, $F(d_1,\rho)$ is an increasing function of $\rho$. Also for a fixed $d_1 \in [0,2]$, we have

$$\max_{0\le\rho\le1} F(d_1,\rho) = F(d_1,\rho) = J(d_1).$$

Therefore by putting $\rho = 1$ in (2.18) we have

$$\begin{aligned} J(d_1) &= \{|C|+2|E|\}\,d_1^5 + \{B-G+|H|+2|K|\}\,d_1^4 + |F|\,d_1^3 \\ & + \{2B+2|C|\,d_1+2|E|\,d_1-2G+2|K|\}\,d_1^2\left(4-d_1^2\right) \\ & + \left\{|C|\,d_1\left(4-d_1^2\right) - Bd_1^2 - G\left(4-d_1^2\right)\right\}\left(4-d_1^2\right) \end{aligned}$$

Differentiating with respect to $d_1$, we have

$$\begin{aligned} J'(d_1) &= 5\{|C|+2|E|\}\,d_1^4 + 4\{B-G+|H|+2|K|\}\,d_1^3 + 3|F|\,d_1^2 \\ & +4\{B-G+2|K|\}\,d_1\left(4-d_1^2\right) - 4\{B-G+2|K|\}\,d_1^3 \\ & +6\{|C|+|E|\}\,d_1^2\left(4-d_1^2\right) - 4\{|C|+|E|\}\,d_1^4 + |C|\,d_1^2\left(4-d_1^2\right)^2 \\ & -4|C|\,d_1^2\left(4-d_1^2\right) - 2Bd_1\left(4-d_1^2\right) + 2Bd_1^3 - 4Gd_1\left(4-d_1^2\right) \end{aligned}$$

Again differentiating with respect to $d_1$ we have

$$J''(d_1) = 20\{|C|+2|E|\}\,d_1^3 + 12\{B-G+|H|+2|K|\}\,d_1^2$$

$$+6|F|d_1+4\{B-G+2|K|\}\left(4-d_1^2\right)-8\{B-G+2|K|\}d_1^2$$
$$-12\{B-G+2|K|\}d_1^2+126\{|C|+|E|\}\left(4-d_1^2\right)-126\{|C|+|E|\}d_1^3$$
$$-16\{|C|+|E|\}d_1^3+2|C|d_1\left(4-d_1^2\right)^2-4|C|d_1^3\left(4-d_1^2\right)^2-8|C|d_1\left(4-d_1^2\right)$$
$$+8|C|d_1^3-2B\left(4-d_1^2\right)+4Bd_1^2+6Bd_1^2-4G\left(4-d_1^2\right)+8Gd_1^2.$$

For maximum value of $J(d_1)$, clearly $J'(d_1)=0$ for $d_1=0$ and $J''(0)<0$, so $J(d_1)$ has maximum value at $d_1=0$ hence

$$\left|a_{p+1}a_{p+3}-a_{p+2}^2\right|\leq\left[\frac{4|b|c(c+1)(1-\beta)}{(p+1)(a+1)}\right]^2.$$

□

## 3. Subordination Results for the Function Class $\mathcal{SC}_p^\lambda(a,b,c,\beta)$

Given functions $f,g\in A$, $f$ is said to subordinate to $g$ denoted by $f\prec g$, $z\in\mathbb{U}$, if there exist a function $w\in V$, where

$$V=\{w\in A: w(0)=0,\ |w(z)|<1,\ z\in\mathbb{U}\}$$

such that $f(z)=g(w(z))$.

**Lemma 3.1.** [18]. *Let $q(z)$ be convex in $\mathbb{U}$ and $Re\,(\mu_1 q(z)+\mu_2)>0$, where $\mu_1,\mu_2\in\mathbb{C}\setminus\{0\}$, $z\in\mathbb{U}$. If $h(z)$ is analytic in $\mathbb{U}$ with $q(0)=h(0)$ and*

$$h(z)+\frac{zh'(z)}{\mu_1 h(z)+\mu_2}\prec q(z),\quad z\in\mathbb{U},$$

*then $h(z)\prec q(z)$.*

**Lemma 3.2.** *A function $f\in\mathcal{SC}_p^\lambda(a,b,c,\beta)$, if and only if*

$$e^{i\lambda}\left(1-\frac{2}{b}+\frac{2}{b}\frac{\mathcal{L}_p(a+1,c)f(z)}{\mathcal{L}_p(a,c)f(z)}\right)\prec q(z),\quad z\in\mathbb{U},$$

*where*

$$q(z)=\frac{\cos\lambda-\{2\beta\cos\lambda+i\sin\lambda-\cos\lambda\}z}{1-z}. \tag{3.1}$$

*for some real $\lambda(|\lambda|<\frac{\pi}{2})$ and $\beta>p$.*

The proof of above lemma is similier to that of Theorem 1 in [22] so we omit the proof.

**Theorem 3.3.** *Let $\beta>p$, $b\in\mathbb{C}\setminus\{0,-1\}$. Then*

$$\mathcal{SC}_p^0(a+1,b,c,\beta)\subset\mathcal{SC}_p^0(a,b+1,c,\beta_1),$$

*where*

$$\beta_1=\frac{b(a+1)}{a(b+1)}\beta-\frac{b-a}{a(b+1)}.$$

*Proof.* Suppose $f \in \mathcal{SC}_p^0(a+1,b,c,\beta)$ and set

$$1-\frac{2}{b+1}+\frac{2}{b+1}\frac{\mathcal{L}_p(a+1,c)f(z)}{\mathcal{L}_p(a,c)f(z)}=h(z), \tag{3.2}$$

where $h$ is analytic in $\mathbb{U}$ and $h(0)=1$.

Logarithmic differentiation of (3.2), gives

$$\frac{z\left(\mathcal{L}_p(a+1,c)f(z)\right)'}{\mathcal{L}_p(a+1,c)f(z)}-\frac{z\left(\mathcal{L}_p(a,c)f(z)\right)'}{\mathcal{L}_p(a,c)f(z)}=\frac{(b+1)\,zh'(z)}{(b+1)\{h(z)-1\}+2}.$$

Using the identity(1.6) we have

$$1-\frac{2}{b}+\frac{2}{b}\frac{\mathcal{L}_p(a+2,c)f(z)}{\mathcal{L}_p(a+1,c)f(z)}=1-\frac{a}{a+1}\frac{b+1}{b}+\frac{a}{a+1}\frac{b+1}{b}h(z)+\frac{2}{b(a+1)}\frac{zh'(z)}{h(z)-1+\frac{2}{b+1}}. \tag{3.3}$$

Let

$$1-\frac{a}{a+1}\frac{b+1}{b}+\frac{a}{a+1}\frac{b+1}{b}h(z)=H(z),$$

where $H$ is analytic in $\mathbb{U}$ and $H(0)=1$. From (3.3) we have

$$1-\frac{2}{b}+\frac{2}{b}\frac{\mathcal{L}_p(a+2,c)f(z)}{\mathcal{L}_p(a+1,c)f(z)}=H(z)+\frac{zH'(z)}{\mu_1 H(z)+\mu_2},$$

where $\mu_1=\frac{b(a+1)}{2}$ and $\mu_2=\frac{2a-ab-b}{2}$. Since $f(z)\in VD_p^0(a+1,b,c,\beta)$, so from Lemma 3.2 we have

$$H(z)+\frac{zH'(z)}{\mu_1 H(z)+\mu_2}=1-\frac{2}{b}+\frac{2}{b}\frac{\mathcal{L}_p(a+2,c)f(z)}{\mathcal{L}_p(a+1,c)f(z)}\prec q(z),$$

where $q$ is given by

$$q(z)=\frac{1-(2\beta-1)z}{1-z}. \tag{3.4}$$

Applying Lemma 3.1 we have

$$H(z)\prec q(z)$$

or equivalently

$$h(z)\prec\frac{1-(2\beta_1-1)z}{1-z},$$

where

$$\beta_1=\frac{b(a+1)}{a(b+1)}\beta-\frac{b-a}{a(b+1)}.$$

This complete the proof. □

**Theorem 3.4.** *Let $f\in\mathcal{SC}_p^0(a,b,c,\beta)$. Then $F\in\mathcal{SC}_p^0(a,b,c,\beta)$, where $F$ is Bernardi integral operator defined by*

$$F(z)=\frac{p+1}{z^c}\int_0^z t^{c-1}f(t)dt,\quad c>-1. \tag{3.5}$$

*Proof.* Suppose

$$1-\frac{2}{b}+\frac{2}{b}\frac{\mathcal{L}_p(a+2,c)F(z)}{\mathcal{L}_p(a+1,c)F(z)}=h(z), \tag{3.6}$$

where $h$ is analytic in $\mathbb{U}$ and $h(0)=1$.

Now differentiating (3.5) we have

$$(c+p)f(z)=cF(z)+zF'(z)$$

Applying the operator $\mathcal{L}_p(a,c)$ we have

$$(c+p)\mathcal{L}_p(a,c)f(z)=c\mathcal{L}_p(a,c)F(z)+\mathcal{L}_p(a+1,c)F(z) \tag{3.7}$$

and

$$(c+p)\mathcal{L}_p(a+1,c)f(z)=c\mathcal{L}_p(a+1,c)F(z)+\mathcal{L}_p(a+2,c)F(z) \tag{3.8}$$

From (3.7) and (3.8) we have

$$\frac{\mathcal{L}_p(a+1,c)f(z)}{\mathcal{L}_p(a,c)f(z)}=\frac{c\frac{\mathcal{L}_p(a+1,c)f(z)}{\mathcal{L}_p(a,c)f(z)}+\frac{\mathcal{L}_p(a+2,c)f(z)}{\mathcal{L}_p(a+1,c)f(z)}\frac{\mathcal{L}_p(a+1,c)f(z)}{\mathcal{L}_p(a,c)f(z)}}{c+\frac{\mathcal{L}_p(a+1,c)f(z)}{\mathcal{L}_p(a,c)f(z)}}. \tag{3.9}$$

Logarithmic differentiation of (3.6) , totgether with (1.6) and (3.9) we have

$$1-\frac{2}{b}+\frac{2}{b}\frac{\mathcal{L}_p(a+1,c)f(z)}{\mathcal{L}_p(a,c)f(z)}=h(z)+\frac{zh'(z)}{\mu_3h(z)+\mu_4},$$

where $\mu_3=\frac{ab}{2}$ and $\mu_4=c+p-\frac{ab}{2}$

Since $f\in\mathcal{SC}_p^0(a,b,c,\beta)$, so from Lemma 3.2 we have

$$h(z)+\frac{zh'(z)}{\mu_3h(z)+\mu_4}=1-\frac{2}{b}+\frac{2}{b}\frac{\mathcal{L}_p(a+1,c)f(z)}{\mathcal{L}_p(a,c)f(z)}\prec q(z),$$

where $q$ is given by (3.4)

Applying Lemma 3.1 we have

$$h(z)\prec q(z),$$

which implies that $F\in\mathcal{SC}_p^0(a,b,c,\beta)$. □

**Conflict of Interest**

All authors declare no conflicts of interest in this paper.

### References


1. M. Arif, K. I. Noor, M. Raza, *Hankel determinant problem of a subclass of analytic fucntions*, J. Ineq. Appl, **22** (2012) , 1-7.
2. M. Arif, K. I. Noor, M. Raza, W. Haq, *Some properties of a generalized class of analytic functions related with Janowski functions,* Abst. Appl. Anal, (2012), 1-11.
3. B. C. Carlson, D. B. Shaffer, *Starlike and prestarlike hypergeometric functions,* SIAM J. Math. Anal., **15** (1984), 737-745.
4. K. K. Dixit, A. L. Pathak, *A new class of analytic functions with positive coefficients*, Ind. J. Pure. Appl. Math., **34** (2003), 209-218.
5. U. Grenander, G. Szego, *Toeplitz Forms and Their Applications,* University of California Press, Berkeley, 1958.
6. F. R. Keogh, E. P. Merkes, *A coefficient inequality for certain class of analytic functions*, Proc. Amer. Math. Soc., **20** (1969), 8-12.
7. J. Nishiwaki, S. Owa, *Coefficient inequalities for certain analytic functios*, Int. J. Math. Math Sci, **29** (2002), 285-290.
8. K. I. Noor, *On the Hankel determinants of close-to-convex univalent functions*, Int. J. Math. Math. Sci, **3** (1980), 447-481.
9. K. I. Noor, *Hankel determinant problem for the class of functions with bounded boundary rotation*, Revue Roumaine de Mathématiques Pures et Appliquées, **28** (1983), 731-739.
10. J. W. Noonan, D. K. Thomas, *On second Hankel determinant of a really mean p-valent functions*, Trans. Amer. Math. Soc, (1976), 337-346.
11. M. Nunokawa, S. Owa, J. Nishiwaki, K. Kuroki, T. Hayatni, *Differential subordination and argumental property,* Comput. Math. Appl., **56** (2008), 2733-2736.
12. S. Owa, J. Nishiwaki, *Coefficient estimates for certain classes of analytic functions*, J. Ineq. Pure Appl. Math, **3** (2002), 1-5.
13. S. Owa, H. M. Srivastava, *Some generalized convolution properties associated with certain subclasses of analytic functions*, J. Ineq. Pure Appl. Math, **3** (2002), 1-13.
14. Y. Polatoglu, M. Blocal, A. Sen, E. Yavuz, *An investigation on a subclass of p-valently starlike functions in the unit disc*, Turk. J. Math., **31** (2007), 221-228.
15. C. Pommerenke, *On the coefficients and Hankel determinants of univalent functions*, J. Lond. Math. Soc., **1** (1966), 111-122.
16. C. Pommerenke, *On the Hankel determinants of univalent functions*, Mathematika, **14** (1967), 108-112.
17. C. Pommerenke, *Univalent Functions,* Vandenhoeck and Ruprechet, Gottingen, 1975.
18. H. Saitoh, *A linear operator and its application of first order differential subordinations*, Math. Japon., **44** (1996), 31-38.
19. H. Saitoh, *On certain subclasses of analytic functions involving a linear operator*, in: Proceedings of the Second International Workshop, Japan, ( 1996) , 401-411.

20. H. M. Srivastava, S. Owa, *Some characterizations and distortions theorems involving fractional calculus, generalized hypergeometric functions, Hadamard products, linear operators, and certain subclasses of analytic functions*, Nagoya Math. J., **106** (1987), 1-28.

21. B. A. Uralegaddi, M. D. Ganigi, S. M. Sarangi, *Univalent functions with positive coefficients,* Tamkang J. Math., **25** (1994), 225-230.

22. N. Uyanik, H. Shiraishi, S. Owa, Y. Polatoglu, *Reciprocal classes of p-valently spirallike and p-valently Robertson functions*, J. Ineq. Appl., (2011) , 1-10.

# 5

# Large Deviations for Stochastic Fractional Integrodifferential Equations

**Murugan Suvinthra***, **Krishnan Balachandran and Rajendran Mabel Lizzy**

Department of Mathematics, Bharathiar University, Coimbatore 641046, India

* **Correspondence:** suvinthra@gmail.com

**Abstract:** In this work we establish a Freidlin-Wentzell type large deviation principle for stochastic fractional integrodifferential equations by using the weak convergence approach. The compactness argument is proved on the solution space of corresponding skeleton equation and the weak convergence is done for Borel measurable functions whose existence is asserted from Yamada-Watanabe theorem. Examples are included which illustrate the theory and also depict the link between large deviations and optimal controllability.

**Keywords:** Fractional differential equations; Large deviation principle; Stochastic integrodifferential equations
**Mathematics subject classification:** 34A08, 45J05, 60F10, 60H10

---

## 1. Introduction

The subject of fractional calculus deals with the investigations of derivatives and integrals, of any arbitrary real or complex order, which unify and extend the notions of integer-order derivative and $n$-fold integral. It can be considered as a branch of mathematical analysis which deals with integrodifferential operators and equations where the integrals are of convolution type and exhibit (weakly singular) kernels of power-law type. It is strictly related to the theory of pseudo-differential operators. Fractional order models have the tendency to capture non-local relations in space and time, thus forming an improvised model for analyzing complex phenomena. It is a successful tool for describing complex quantum field dynamical systems, dissipation and long-range phenomena that cannot be well illustrated using ordinary differential and integral operators. For an introductory study on fractional calculus and fractional derivatives, see the literatures [19, 21, 25].

Inducing randomness into the model helps us to analyze better by taking into consideration the effect of uncertainty, thus leading to stochastic fractional differential equations (refer [24] and references therein). The theory of existence, controllability and stability of fractional differential equations has been studied by many authors (for instance, see [1, 2, 15, 16]). However there seems to be possibly limited literature to the study of large deviations for stochastic fractional differential equations.

Large deviation theory is a branch of probability theory that deals with the study of rare events. Though the probability of occurrence of rare events is too small, their impact may be large and so it is significant to study such rare events. Large deviation theory finds its application in many areas such as mathematical finance, statistical mechanics and various fields ranging from physics to biology. The origin of large deviations dates back to the 1930s where there was a necessity to solve the problem of total claim exceeding the reserve fund set aside in an insurance company. The solution was discovered by the Swedish mathematician Cramer via refinement of the central limit theorem. Subsequent developments has been made since then and there was major breakthrough into the subject after Varadhan [31] established a general framework for large deviation principle and formulated the Varadhan's lemma in 1966. In 1970, Wentzell and Freidlin [13] developed a theory to enhance the large deviation principle for differential equations with small stochastic perturbations, which involves time discretization of the original problem and then analyzing the large deviation principle in the limit. Fleming [12] developed a stochastic control approach to establish large deviation principle and then Dupuis and Ellis [11] combined the weak convergence approach with the theory of Fleming. These developments indeed explore the close association of large deviation theory with optimal controllability problems.

Using the weak convergence approach, the large deviations for homeomorphism flows of non-Lipschitz Stochastic Differential Equations (SDEs) was studied by Ren and Zhang [27]; the large deviations for two-dimensional stochastic Navier-Stokes equations by Sritharan and Sundar [28], and for stochastic evolution equations with small multiplicative noise by Liu [18]. For more references on this approach, one may refer [5, 6, 11, 14, 26]. By using the approximating method, Mohammed and Zhang [23] established a Freidlin-Wentzell type large deviation principle for the stochastic delay differential equations. Mo and Luo [22] also studied the large deviations for the stochastic delay differential equations by employing the weak convergence approach. Bo and Jiang [4] analyzed the large deviation for Kuramoto-Sivashinsky stochastic partial differential equation. A large deviation principle for stochastic differential equations with deviating arguments is dealt with in [30].

A Freidlin-Wentzell type large deviation principle is discussed in Dembo and Zeitouni [8] for the following stochastic differential equation:

$$\left.\begin{aligned} dX(t) &= b(t, X(t))\mathrm{d}t + \sqrt{\epsilon}\sigma(t, X(t))\mathrm{d}W(t), \quad t \in (0, T], \\ X(0) &= X_0. \end{aligned}\right\} \tag{1}$$

In the case that the system is affected by hereditary influences, the drift and diffusion coefficients ($b$ and $\sigma$) also depend on an integral component, thus giving rise to stochastic integrodifferential equations. The large deviations for stochastic integrodifferential equations has been carried out in [29]. In this paper, we consider the stochastic fractional integrodifferential equations with Gaussian noise perturbation of multiplicative type and establish the large deviation principle by using the results developed by Budhiraja and Dupuis [7]. The compactness argument is done with the associated control equation and weak convergence result is obtained by observing the nature of the solution of the stochastic control equation as the perturbation of the noise term tends to zero.

## 2. Preliminaries

Let $\mathbb{X}$ and $\mathbb{H}$ be separable Hilbert spaces. Denote by $L(\mathbb{X})$ the space of all bounded linear operators from $\mathbb{X}$ to $\mathbb{X}$. Denote by $J$ the time interval $[0, T]$. Let $\{\Omega, \mathcal{F}, \mathbb{P}\}$ be a complete filtered probability space equipped with a complete family of right continuous increasing sub $\sigma$-algebras $\{\mathcal{F}_t, t \in J\}$ satisfying $\{\mathcal{F}_t \subset \mathcal{F}\}$. Let $Q$ be a symmetric, positive, trace class operator on $\mathbb{H}$ and $W(\cdot)$ be a $\mathbb{H}$-valued Wiener process with covariance operator $Q$. Denote the space $\mathcal{H}_0 := Q^{1/2}\mathbb{H}$. Then $\mathcal{H}_0$ is a Hilbert space with the inner product $(X, Y)_0 := (Q^{-1/2}X, Q^{-1/2}Y)$ for all $X, Y \in \mathcal{H}_0$ and the corresponding norm is denoted by $\|\cdot\|_0$. Let $L_Q$ denote the space of all Hilbert-Schmidt operators from $\mathcal{H}_0$ to $\mathbb{X}$. Consider the nonlinear stochastic fractional integrodifferential equation in $\mathbb{X}$ of the form

$$\left.\begin{aligned} {}^C D^\alpha X(t) &= AX(t) + b\left(t, X(t), \int_0^t f(t,s,X(s))\mathrm{d}s\right) + \sigma\left(t, X(t), \int_0^t g(t,s,X(s))\mathrm{d}s\right)\frac{\mathrm{d}W(t)}{\mathrm{d}t},\ t \in J, \\ X(0) &= X_0, \end{aligned}\right\} \tag{2}$$

where $1/2 < \alpha \leq 1, X_0 \in \mathbb{X}$ and $A : \mathbb{X} \to \mathbb{X}$ is a bounded linear operator. Also the drift coefficient $b : J \times \mathbb{X} \times \mathbb{X} \to \mathbb{X}$, the noise coefficient $\sigma : J \times \mathbb{X} \times \mathbb{X} \to L_Q(\mathcal{H}_0; \mathbb{X})$ and $f, g : J \times J \times \mathbb{X} \to \mathbb{X}$. Assume the following Lipschitz conditions on the drift and noise coefficients: For all $x_1, x_2, y_1, y_2 \in \mathbb{X}$ and $0 \leq s \leq t \leq T$, there exist constants $L_b, L_\sigma, L_f, L_g > 0$ such that

$$\left.\begin{aligned} \|b(t,x_1,y_1) - b(t,x_2,y_2)\|_{\mathbb{X}} &\leq L_b[\|x_1 - x_2\|_{\mathbb{X}} + \|y_1 - y_2\|_{\mathbb{X}}], \\ \|\sigma(t,x_1,y_1) - \sigma(t,x_2,y_2)\|_{L_Q} &\leq L_\sigma[\|x_1 - x_2\|_{\mathbb{X}} + \|y_1 - y_2\|_{\mathbb{X}}], \\ \|f(t,s,x_1) - f(t,s,x_2)\|_{\mathbb{X}} &\leq L_f\|x_1 - x_2\|_{\mathbb{X}}, \\ \|g(t,s,x_1) - g(t,s,x_2)\|_{\mathbb{X}} &\leq L_g\|x_1 - x_2\|_{\mathbb{X}}. \end{aligned}\right\} \tag{3}$$

Also assume the following linear growth assumptions on the coefficients: For all $x, y \in \mathbb{X}$ and $0 \leq s \leq t \leq T$, there exist positive constants $K_b, K_\sigma, K_f, K_g > 0$ such that

$$\left.\begin{aligned} \|b(t,x,y)\|_{\mathbb{X}}^2 &\leq K_b[1 + \|x\|_{\mathbb{X}}^2 + \|y\|_{\mathbb{X}}^2], \\ \|\sigma(t,x,y)\|_{L_Q}^2 &\leq K_\sigma[1 + \|x\|_{\mathbb{X}}^2 + \|y\|_{\mathbb{X}}^2], \\ \|f(t,s,x)\|_{\mathbb{X}}^2 &\leq K_f[1 + \|x\|_{\mathbb{X}}^2], \\ \|g(t,s,x)\|_{\mathbb{X}}^2 &\leq K_g[1 + \|x\|_{\mathbb{X}}^2]. \end{aligned}\right\} \tag{4}$$

Let us first quote some basic definitions from fractional calculus. For $\alpha, \beta > 0$, with $n - 1 < \alpha < n$, $n - 1 < \beta < n$ and $n \in \mathbb{N}$, $D$ is the usual differential operator and suppose $f \in L_1(\mathbb{R}_+)$, $\mathbb{R}_+ = [0, \infty)$.

(i) Caputo Fractional Derivative:
The Riemann Liouville fractional integral of a function $f$ is defined as

$$I^\alpha f(t) = \frac{1}{\Gamma(\alpha)}\int_0^t (t-s)^{\alpha-1} f(s)\mathrm{d}s,$$

and the Caputo derivative of $f$ is ${}^C D^\alpha f(t) = I^{n-\alpha} f^{(n)}(t)$, that is,

$${}^C D^\alpha f(t) = \frac{1}{\Gamma(n-\alpha)}\int_0^t (t-s)^{n-\alpha-1} f^{(n)}(s)\mathrm{d}s,$$

where the function $f(t)$ has absolutely continuous derivatives up to order $n - 1$.

(ii) Mittag-Leffler Operator Function: Two parameter family of Mittag-Leffler operator functions is defined as

$$E_{\alpha,\beta}(A) = \sum_{k=0}^{\infty} \frac{A^k}{\Gamma(k\alpha + \beta)}, \alpha, \beta > 0.$$

Here $A$ is the bounded linear operator. In particular, for $\beta = 1$, the one parameter Mittag-Leffler operator function is

$$E_\alpha(A) = \sum_{k=0}^{\infty} \frac{A^k}{\Gamma(k\alpha + 1)}.$$

The Mittag-Leffler functions are in fact generalizations of the exponential function and are applicable in varied situations involving fractional derivatives, see for example [9]. Assume the following boundedness on the Mittag-Leffler operator functions with one and two parameters:

$$M_1 = \sup_{t\in J} \left\|E_\alpha(At^\alpha)\right\|_{L(\mathbb{X})}, \quad M_2 = \sup_{t\in J} \left\|E_{\alpha,\alpha}(At^\alpha)\right\|_{L(\mathbb{X})}. \tag{5}$$

In order to find the solution representation, we need the following hypothesis and make use of the Lemma that follows.

(H1) The operator $A \in L(\mathbb{X})$ commutes with the fractional integral operator $I^\alpha$ on $\mathbb{X}$ and $\|A\|^2_{L(\mathbb{X})} < \frac{(2\alpha-1)(\Gamma(\alpha))^2}{T^{2\alpha}}$.

**Lemma 2.1.** *[17] Suppose that A is a linear bounded operator defined on $\mathbb{X}$ (more generally, $\mathbb{X}$ may be a Banach space) and assume that $\|A\|_{L(\mathbb{X})} < 1$. Then $(I - A)^{-1}$ is linear and bounded. Also*

$$(I - A)^{-1} = \sum_{k=0}^{\infty} A^k.$$

*The convergence of the above series is in the operator norm and $\|(I - A)^{-1}\|_{L(\mathbb{X})} \leq (1 - \|A\|_{L(\mathbb{X})})^{-1}$.*

We next show that $\|I^\alpha A\|_{L(\mathbb{X})} < 1$ and, by the Lemma, we obtain $(I - I^\alpha A)^{-1}$ is bounded and linear. Let $X \in \mathbb{X}$; then by (H1), we have

$$\begin{aligned}
\mathbb{E}\Big[\|(I^\alpha A)X\|^2_{\mathbb{C}(J;\mathbb{X})}\Big] &\leq \frac{T}{(\Gamma(\alpha))^2}\mathbb{E}\Big[\sup_{t\in J}\int_0^t (t-s)^{2\alpha-2}\|AX(s)\|^2_{\mathbb{X}}\mathrm{d}s\Big] \\
&\leq \frac{T^{2\alpha}}{(2\alpha-1)(\Gamma(\alpha))^2}\mathbb{E}\Big[\sup_{t\in J}\|AX(t)\|^2_{\mathbb{X}}\Big] < \mathbb{E}\|X\|^2_{\mathbb{C}(J;\mathbb{X})},
\end{aligned}$$

hence yielding the desired inequality. On the other hand, defining the random differential operator

$$\mathrm{d}F(t, X(t)) := b\left(t, X(t), \int_0^t f(t, s, X(s))\mathrm{d}s\right)\mathrm{d}t + \sigma\left(t, X(t), \int_0^t g(t, s, X(s))\mathrm{d}s\right)\mathrm{d}W(t)$$

and operating by $I^\alpha$ on both sides of (2), we have

$$\begin{aligned}
X(t) &= X_0 + I^\alpha AX(t) + I^\alpha\frac{\mathrm{d}F(t, X(t))}{\mathrm{d}t}, \\
X(t) &= (I - I^\alpha A)^{-1}\left(X_0 + I^\alpha\frac{\mathrm{d}F(t, X(t))}{\mathrm{d}t}\right).
\end{aligned}$$

Therefore, using Lemma 2.1 and the fact that $I^\alpha$ commutes with A, we obtain (see [3, 20])

$$X(t) = \sum_{k=0}^{\infty}(I^\alpha A)^k\left(X_0 + I^\alpha\frac{\mathrm{d}F(t, X(t))}{\mathrm{d}t}\right)$$

$$
\begin{aligned}
&= \sum_{k=0}^{\infty} I^{k\alpha} A^k X_0 + I^{k\alpha} A^k I^{\alpha} \frac{\mathrm{d}F(t, X(t))}{\mathrm{d}t} \\
&= \sum_{k=0}^{\infty} I^{k\alpha} A^k X_0 + I^{k\alpha+\alpha} A^k \frac{\mathrm{d}F(t, X(t))}{\mathrm{d}t} \\
&= \sum_{k=0}^{\infty} \frac{A^k t^{\alpha k}}{\Gamma(k\alpha+1)} X_0 + \int_0^t (t-s)^{\alpha-1} \left( \sum_{k=0}^{\infty} \frac{A^k (t-s)^{\alpha k}}{\Gamma(k\alpha+\alpha)} \right) \mathrm{d}F(s, X(s)), \\
&= E_{\alpha}(At^{\alpha}) X_0 + \int_0^t (t-s)^{\alpha-1} E_{\alpha,\alpha}(A(t-s)^{\alpha}) \mathrm{d}F(s, X(s)).
\end{aligned}
$$

Thus we obtain the solution representation of (2) as

$$
\begin{aligned}
X(t) &= E_{\alpha}(At^{\alpha}) X_0 + \int_0^t (t-s)^{\alpha-1} E_{\alpha,\alpha}(A(t-s)^{\alpha}) b\left(s, X(s), \int_0^s f(s,\tau,X(\tau))\mathrm{d}\tau\right) \mathrm{d}s \\
&\quad + \int_0^t (t-s)^{\alpha-1} E_{\alpha,\alpha}(A(t-s)^{\alpha}) \sigma\left(s, X(s), \int_0^s g(s,\tau,X(\tau))\mathrm{d}\tau\right) \mathrm{d}W(s). \qquad (6)
\end{aligned}
$$

We now present some basic definitions and results from large deviation theory. For this, let $\{X^{\epsilon}\}$ be a family of random variables defined on the space $\mathbb{X}$ and taking values in a Polish space $\mathcal{Z}$ (i.e., a complete separable metric space $\mathcal{Z}$).

**Definition 2.1.** *(Rate Function). A function $\mathcal{I} : \mathcal{Z} \to [0, \infty]$ is called a rate function if $\mathcal{I}$ is lower semicontinuous. A rate function $\mathcal{I}$ is called a good rate function if for each $N < \infty$, the level set $K_N = \{f \in \mathcal{Z} : \mathcal{I}(f) \leq N\}$ is compact in $\mathcal{Z}$.*

**Definition 2.2.** *(Large Deviation Principle). Let $\mathcal{I}$ be a rate function on $\mathcal{Z}$. We say the family $\{X^{\epsilon}\}$ satisfies the large deviation principle with rate function $\mathcal{I}$ if the following two conditions hold:*
*(i) Large deviation upper bound. For each closed subset $F$ of $\mathcal{Z}$,*

$$\limsup_{\epsilon \to 0} \epsilon \log \mathbb{P}(X^{\epsilon} \in F) \leq -\mathcal{I}(F).$$

*(ii) Large deviation lower bound.For each open subset $G$ of $\mathcal{Z}$,*

$$\liminf_{\epsilon \to 0} \epsilon \log \mathbb{P}(X^{\epsilon} \in G) \geq -\mathcal{I}(G).$$

**Definition 2.3.** *(Laplace Principle). Let $\mathcal{I}$ be a rate function on $\mathcal{Z}$. We say the family $\{X^{\epsilon}\}$ satisfies the Laplace principle with rate function $\mathcal{I}$ if for all real-valued bounded continuous functions $h$ defined on $\mathcal{Z}$,*

$$\lim_{\epsilon \to 0} \epsilon \log \mathbf{E} \left\{ \exp\left[ -\frac{1}{\epsilon} h(X^{\epsilon}) \right] \right\} = - \inf_{f \in \mathcal{Z}} \{ h(f) + \mathcal{I}(f) \}.$$

One of the main results of the theory of large deviations is the equivalence between the Laplace principle and the large deviation principle when the underlying space is Polish. For a proof we refer the reader to Theorem 1.2.1 and Theorem 1.2.3 in [11].

**Theorem 2.1.** *The family $\{X^{\epsilon}\}$ satisfies the Laplace principle with good rate function $\mathcal{I}$ on a Polish space $\mathcal{Z}$ if and only if $\{X^{\epsilon}\}$ satisfies the large deviation principle with the same rate function $\mathcal{I}$.*

## 3. Large Deviation Principle

In this section, we consider the stochastic fractional integrodifferential equation (2) with the random noise term being perturbed by a small parameter $\epsilon > 0$ in the form

$$\left.\begin{aligned} {}^{C}D^{\alpha}X^{\epsilon}(t) &= AX^{\epsilon}(t) + b\left(t, X^{\epsilon}(t), \int_0^t f(t,s,X^{\epsilon}(s))\mathrm{d}s\right) \\ &\quad + \sqrt{\epsilon}\,\sigma\left(t, X^{\epsilon}(t), \int_0^t g(t,s,X^{\epsilon}(s))\mathrm{d}s\right)\frac{\mathrm{d}W(t)}{\mathrm{d}t},\ t \in (0,T], \\ X^{\epsilon}(0) &= X_0. \end{aligned}\right\} \tag{7}$$

Let $\mathcal{G}^{\epsilon} : \mathbb{C}(J : \mathbb{H}) \to \mathcal{Z}$ be a measurable map defined by $\mathcal{G}^{\epsilon}(W(\cdot)) := X^{\epsilon}(\cdot)$, where $X^{\epsilon}$ is the solution of the above equation (7). We implement the variational representation developed by Budhiraja and Dupuis to study the large deviation principle for the solution processes $\{X^{\epsilon}\}$. Let

$$\mathcal{A} = \left\{ v : v \text{ is } \mathcal{H}_0 \text{ - valued } \mathcal{F}_t \text{ - predictable process and } \int_0^T \|v(s,\omega)\|_0^2\,\mathrm{d}s < \infty \text{ a.s.} \right\},$$

$$S_N = \left\{ v \in L^2(J;\mathcal{H}_0) : \int_0^T \|v(s)\|_0^2\,\mathrm{d}s \le N \right\},$$

where $L^2(J;\mathcal{H}_0)$ is the space of all $\mathcal{H}_0$ -valued square integrable functions on $J$. Then $S_N$ endowed with the weak topology in $L^2(J;\mathcal{H}_0)$ is a compact Polish space (see [10]). Let us also define

$$\mathcal{A}_N = \{ v \in \mathcal{A} : v(\omega) \in S_N\ \mathbb{P} - a.s \}.$$

We now state the variational representation developed by Budhiraja and Dupuis [7, Theorem 4.4] that provides sufficient conditions under which Laplace principle (equivalently, large deviation principle) holds for the family $\{X^{\epsilon}\}$:

**Proposition 3.1.** *Suppose that there exists a measurable map $\mathcal{G}^0 : \mathbb{C}(J : \mathbb{H}) \to \mathcal{Z}$ such that the following hold:*

*(i) Let $\{v^{\epsilon} : \epsilon > 0\} \subset \mathcal{A}_N$ for some $N < \infty$. Let $v^{\epsilon}$ converge in distribution as $S_N$-valued random elements to $v$. Then $\mathcal{G}^{\epsilon}\left(W(\cdot) + \frac{1}{\sqrt{\epsilon}}\int_0^{\cdot} v^{\epsilon}(s)\,\mathrm{d}s\right)$ converges in distribution to $\mathcal{G}^0\left(\int_0^{\cdot} v(s)\,\mathrm{d}s\right)$.*

*(ii) For every $N < \infty$, the set*

$$K_N := \left\{ \mathcal{G}^0\left(\int_0^{\cdot} v(s)\,\mathrm{d}s\right) : v \in S_N \right\}$$

*is a compact subset of $\mathcal{Z}$.*

*For each $h \in \mathcal{Z}$, define*

$$I(h) := \inf_{\left\{ v \in \mathbb{L}^2(J:\mathcal{H}_0) : h = \mathcal{G}^0\left(\int_0^{\cdot} v(s)\,\mathrm{d}s\right)\right\}} \left\{ \frac{1}{2}\int_0^T \|v(s)\|_0^2\,\mathrm{d}s \right\}, \tag{8}$$

*where the infimum over an empty set is taken as $\infty$. Then the family $\{X^{\epsilon} : \epsilon > 0\} = \mathcal{G}^{\epsilon}(W(\cdot))$ satisfies the Laplace principle in $\mathcal{Z}$ with the rate function $I$ given by (8).*

In Proposition 3.1, (ii) is a compactness criterion and it is to be noticed that it has a coincidence with the fact that the level set for a good rate function is compact. Thanks to the variational representation prescribed by Budhiraja and Dupuis, the study of large deviation principle for any stochastic differential equation can now be simplified to the problem of identifying Borel measurable function $\mathcal{G}^0$ so that the hypothesis in the above proposition is satisfied.

Consider the controlled equation associated to (7) with control $v \in S_N$.

$$\left.\begin{aligned} {}^C D^\alpha X_v(t) &= AX_v(t) + b\left(t, X_v(t), \int_0^t f(t,s,X_v(s))ds\right) \\ &\quad + \sigma\left(t, X_v(t), \int_0^t g(t,s,X_v(s))ds\right)v(t), \ t \in (0,T], \\ X_v(0) &= X_0, \end{aligned}\right\} \tag{9}$$

and let $X_v(t)$ denote the solution of the equation (9). The main result in this chapter is the following Freidlin-Wentzell type theorem:

**Theorem 3.1.** *With the assumption (H1) on the bounded linear operator A, the family $\{X^\epsilon(t)\}$ of solutions of (7) satisfies the large deviation principle (equivalently, Laplace principle) in $C(J;\mathbb{X})$ with the good rate function*

$$\mathcal{I}(h) := \inf\left\{\frac{1}{2}\int_0^T \|v(t)\|_0^2\, dt; X_v = h\right\}, \tag{10}$$

*where $v \in L^2(J;\mathcal{H}_0)$ and $X_v$ denotes the solution of the control equation (9) with the convention that the infimum of an empty set is infinity.*

In order to prove the theorem, the main work is to verify the sufficient conditions in Proposition 3.1. Initially we formulate the following perturbed controlled stochastic equation corresponding to (7):

$$\left.\begin{aligned} {}^C D^\alpha X_v^\epsilon(t) &= AX_v^\epsilon(t) + b\left(t, X_v^\epsilon(t), \int_0^t f(t,s,X_v^\epsilon(s))ds\right) + \sigma\left(t, X_v^\epsilon(t), \int_0^t g(t,s,X_v^\epsilon(s))ds\right)v(t) \\ &\quad + \sqrt{\epsilon}\sigma\left(t, X_v^\epsilon(t), \int_0^t g(t,s,X_v^\epsilon(s))ds\right)\frac{dW(t)}{dt}, \ t \in (0,T], \\ X_v^\epsilon(0) &= X_0. \end{aligned}\right\} \tag{11}$$

The solution representation is given by

$$\begin{aligned} X_v^\epsilon(t) &= E_\alpha(At^\alpha)X_0 + \int_0^t (t-s)^{\alpha-1}E_{\alpha,\alpha}(A(t-s)^\alpha)b\left(s, X_v^\epsilon(s), \int_0^s f(s,\tau,X_v^\epsilon(\tau))d\tau\right)ds \\ &\quad + \int_0^t (t-s)^{\alpha-1}E_{\alpha,\alpha}(A(t-s)^\alpha)\sigma\left(s, X_v^\epsilon(s), \int_0^s g(s,\tau,X_v^\epsilon(\tau))d\tau\right)v(s)ds \\ &\quad + \sqrt{\epsilon}\int_0^t (t-s)^{\alpha-1}E_{\alpha,\alpha}(A(t-s)^\alpha)\sigma\left(s, X_v^\epsilon(s), \int_0^s g(s,\tau,X_v^\epsilon(\tau))d\tau\right)dW(s). \end{aligned} \tag{12}$$

Before proceeding further analysis, we show that the solution $X_v^\epsilon(t)$ obeys the following energy estimate:

**Theorem 3.2.** *The solution $X_v^\epsilon(t)$ of (11) is bounded in the space $\mathbb{L}^2(\Omega;\mathbb{C}(J;\mathbb{X}))$, that is, there exists a positive constant $K > 0$ such that*

$$\mathbb{E}\left[\sup_{t\in J}\|X_v^\epsilon(t)\|_{\mathbb{X}}^2\right] \le K. \tag{13}$$

*Proof.* First we define the stopping time $\tau_N := \inf\{t : \|X_v^\epsilon(t)\|^2 \geq N\}$. And, for any $t \in [0, T \wedge \tau_N]$, consider the solution representation of (11) given by (12), take $\|\cdot\|_{\mathbb{X}}^2$ on both sides and use the algebraic identity $(a+b+c+d)^2 \leq 4(a^2+b^2+c^2+d^2)$ to get

$$\|X_v^\epsilon(t)\|_{\mathbb{X}}^2 \leq 4\|E_\alpha(At^\alpha)\|_{L(\mathbb{X})}^2 \|X_0\|_{\mathbb{X}}^2 + 4\left\|\int_0^t (t-s)^{\alpha-1} E_{\alpha,\alpha}(A(t-s)^\alpha) b\left(s, X_v^\epsilon(s), \int_0^s f(s,\tau,X_v^\epsilon(\tau))\mathrm{d}\tau\right)\mathrm{d}s\right\|_{\mathbb{X}}^2$$
$$+ 4\left\|\int_0^t (t-s)^{\alpha-1} E_{\alpha,\alpha}(A(t-s)^\alpha)\sigma\left(s, X_v^\epsilon(s), \int_0^s g(s,\tau,X_v^\epsilon(\tau))\mathrm{d}\tau\right)v(s)\mathrm{d}s\right\|_{\mathbb{X}}^2$$
$$+ 4\epsilon\left\|\int_0^t (t-s)^{\alpha-1} E_{\alpha,\alpha}(A(t-s)^\alpha)\sigma\left(s, X_v^\epsilon(s), \int_0^s g(s,\tau,X_v^\epsilon(\tau))\mathrm{d}\tau\right)\mathrm{d}W(s)\right\|_{\mathbb{X}}^2.$$

Using the Holder inequality and the bounds on $\|E_\alpha(\cdot)\|_{L(\mathbb{X})}$ and $\|E_{\alpha,\alpha}(\cdot)\|_{L(\mathbb{X})}$ given by (5), we obtain the estimate

$$\begin{aligned}\|X_v^\epsilon(t)\|_{\mathbb{X}}^2 \leq\ & 4M_1^2\|X_0\|_{\mathbb{X}}^2 + 4M_2^2 \int_0^t (t-s)^{2\alpha-2}\mathrm{d}s \int_0^t \left\|b\left(s, X_v^\epsilon(s), \int_0^s f(s,\tau,X_v^\epsilon(\tau))\mathrm{d}\tau\right)\right\|_{\mathbb{X}}^2 \mathrm{d}s \\ & + 4M_2^2 \int_0^t (t-s)^{2\alpha-2} \left\|\sigma\left(s, X_v^\epsilon(s), \int_0^s g(s,\tau,X_v^\epsilon(\tau))\mathrm{d}\tau\right)\right\|_{L_Q}^2 \mathrm{d}s \int_0^t \|v(s)\|_0^2 \mathrm{d}s \\ & + 4\epsilon\left\|\int_0^t (t-s)^{2\alpha-2} E_{\alpha,\alpha}(A(t-s)^\alpha)\sigma\left(s, X_v^\epsilon(s), \int_0^s g(s,\tau,X_v^\epsilon(\tau))\mathrm{d}\tau\right)\mathrm{d}W(s)\right\|_{\mathbb{X}}^2.\end{aligned}$$

Now using the linear growth property of '$b$' and '$\sigma$' given by (3) results in

$$\begin{aligned}\|X_v^\epsilon(t)\|_{\mathbb{X}}^2 \leq\ & 4M_1^2\|X_0\|_{\mathbb{X}}^2 + 4K_b M_2^2 \frac{T^{2\alpha-1}}{2\alpha-1}\int_0^t \left[1 + \|X_v^\epsilon(s)\|_{\mathbb{X}}^2 + \left\|\int_0^s f(s,\tau,X_v^\epsilon(\tau))\mathrm{d}\tau\right\|_{\mathbb{X}}^2\right]\mathrm{d}s \\ & + 4K_\sigma M_2^2 N \int_0^t (t-s)^{2\alpha-2}\left[1 + \|X_v^\epsilon(s)\|_{\mathbb{X}}^2 + \left\|\int_0^s g(s,\tau,X_v^\epsilon(\tau))\mathrm{d}\tau\right\|_{\mathbb{X}}^2\right]\mathrm{d}s \\ & + 4\epsilon\left\|\int_0^t (t-s)^{\alpha-1} E_{\alpha,\alpha}(A(t-s)^\alpha)\sigma\left(s, X_v^\epsilon(s), \int_0^s g(s,\tau,X_v^\epsilon(\tau))\mathrm{d}\tau\right)\mathrm{d}W(s)\right\|_{\mathbb{X}}^2.\end{aligned}$$

Using Holder's inequality for the integrands $\|\int_0^s f(s,\tau,X_v^\epsilon(\tau))\mathrm{d}\tau\|_{\mathbb{X}}^2$ and $\|\int_0^s g(s,\tau,X_v^\epsilon(\tau))\|_{\mathbb{X}}^2$ and also making use of the linear growth property of '$f$' and '$g$' given by (4), we get, on simplifying,

$$\begin{aligned}\|X_v^\epsilon(t)\|_{\mathbb{X}}^2 \leq\ & 4M_1^2\|X_0\|_{\mathbb{X}}^2 + 4K_b M_2^2 \frac{T^{2\alpha-1}}{2\alpha-1}\int_0^t \left[1 + \|X_v^\epsilon(s)\|_{\mathbb{X}}^2 + K_f T\int_0^t \left[1 + \|X_v^\epsilon(\tau)\|_{\mathbb{X}}^2\right]\mathrm{d}\tau\right]\mathrm{d}s \\ & + 4K_\sigma M_2^2 N \int_0^t (t-s)^{2\alpha-2}\left[1 + \|X_v^\epsilon(s)\|_{\mathbb{X}}^2 + K_g T\int_0^t \left[1 + \|X_v^\epsilon(\tau)\|_{\mathbb{X}}^2\right]\mathrm{d}\tau\right]\mathrm{d}s \\ & + 4\epsilon\left\|\int_0^t (t-s)^{\alpha-1} E_{\alpha,\alpha}(A(t-s)^\alpha)\sigma\left(s, X_v^\epsilon(s), \int_0^s g(s,\tau,X_v^\epsilon(\tau))\mathrm{d}\tau\right)\mathrm{d}W(s)\right\|_{\mathbb{X}}^2. \qquad (14)\end{aligned}$$

The stochastic integral term can be estimated by means of the Burkholder-Davis-Gundy inequality as

$$\mathbb{E}\left\{\sup_{0\leq t\leq T\wedge\tau_N}\left[\left\|\int_0^t (t-s)^{\alpha-1} E_{\alpha,\alpha}(A(t-s)^\alpha)\,\sigma\left(s, X_v^\epsilon(s), \int_0^s g(s,\tau,X_v^\epsilon(\tau))\mathrm{d}\tau\right)\mathrm{d}W(s)\right\|_{\mathbb{X}}^2\right]\right\}$$

$$\le M_2^2 \int_0^T (T-s)^{2\alpha-2} \left\| \sigma\left(s, X_v^\epsilon(s), \int_0^s g(s,\tau,X_v^\epsilon(\tau))\mathrm{d}\tau\right)\right\|_{L_Q}^2 \mathrm{d}s$$
$$\le K_\sigma M_2^2 \int_0^T (T-s)^{2\alpha-2}\left[1+\left\|X_v^\epsilon(s)\right\|_{\mathbb{X}}^2 + K_g T\int_0^T\left[1+\left\|X_v^\epsilon(\tau)\right\|_{\mathbb{X}}^2\right]\mathrm{d}\tau\right]\mathrm{d}s$$
$$\le K_\sigma M_2^2 \int_0^T (T-s)^{2\alpha-2}\left[1+\left\|X_v^\epsilon(s)\right\|_{\mathbb{X}}^2\right]\mathrm{d}s + K_\sigma M_2^2 K_g T\frac{T^{2\alpha-1}}{2\alpha-1}\int_0^T\left[1+\left\|X_v^\epsilon(s)\right\|_{\mathbb{X}}^2\right]\mathrm{d}s.$$

Hence (14) becomes, after taking supremum and expectation on both sides and simplifying,

$$\begin{aligned}\mathbb{E}\Big[\sup_{0\le t\le T\wedge\tau_N}\left\|X_v^\epsilon(t)\right\|_{\mathbb{X}}^2\Big] \le\; & 4M_1^2\mathbb{E}\|X_0\|_{\mathbb{X}}^2 + 4K_b M_2^2(1+K_f T^2)\frac{T^{2\alpha-1}}{2\alpha-1}\mathbb{E}\int_0^T\left[1+\left\|X_v^\epsilon(s)\right\|_{\mathbb{X}}^2\right]\mathrm{d}s\\ &+4K_\sigma M_2^2(N+\epsilon)\,\mathbb{E}\int_0^T(T-s)^{2\alpha-2}\left[1+\left\|X_v^\epsilon(s)\right\|_{\mathbb{X}}^2\right]\mathrm{d}s\\ &+4K_\sigma M_2^2 K_g T\frac{T^{2\alpha-1}}{2\alpha-1}(N+\epsilon)\,\mathbb{E}\int_0^T\left[1+\left\|X_v^\epsilon(s)\right\|_{\mathbb{X}}^2\right]\mathrm{d}s.\end{aligned}$$

Further simplifying and applying the well known Gronwall inequality, we end up with

$$\mathbb{E}\left[\sup_{0\le t\le T\wedge\tau_N}\left\|X_v^\epsilon(t)\right\|_{\mathbb{X}}^2\right] \le \left(4M_1^2\mathbb{E}\|X_0\|_{\mathbb{X}}^2 + C_T\right)e^{C_T} = K, \tag{15}$$

where $C_T = 4M_2^2\frac{T^{2\alpha-1}}{2\alpha-1}\left[K_b\left(1+K_fT^2\right)T + K_\sigma\left(1+K_gT\right)(N+\epsilon)\right]$. Observe that $T\wedge\tau_N \to T$ as $N\to\infty$, hence resulting in (13). □

**Lemma 3.1** (Compactness)**.** *Define* $\mathcal{G}^0 : C(J;\mathbb{H}) \to C(J;\mathbb{X})$ *by*

$$\mathcal{G}^0(h) := \begin{cases} X_v, & \text{if } h = \int_0^{\cdot} v(s)\mathrm{d}s \text{ for some } v\in S_N,\\ 0\;, & \text{otherwise.}\end{cases}$$

*Then, for each* $N<\infty$*, the set*

$$K_N = \left\{\mathcal{G}^0\left(\int_0^{\cdot} v(s)\mathrm{d}s\right) : v\in S_N\right\}$$

*is a compact subset of* $C(J;\mathbb{X})$*.*

*Proof.* Let $\{v_n\}$ be a sequence of controls from $S_N$ that converge weakly to $v$ in $\mathbb{L}^2(J;\mathcal{H}_0)$ and let $X_{v_n}(t)$ denote the solution of (9) with control $v$ replaced by $v_n$. Take $Y_n(t) = X_{v_n}(t) - X_v(t)$. Then the equation corresponding to $Y_n(t)$ would be

$$\left.\begin{aligned} {}^C D^\alpha Y_n(t) &= AY_n(t) + b\left(t, X_{v_n}(t), \int_0^t f(t,s,X_{v_n}(s))\mathrm{d}s\right) - b\left(t, X_v(t), \int_0^t f(t,s,X_v(s))\mathrm{d}s\right)\\ &\quad+\sigma\left(t, X_{v_n}(t), \int_0^t g(t,s,X_{v_n}(s))\mathrm{d}s\right)v_n(t) - \sigma\left(t, X_v(t), \int_0^t g(t,s,X_v(s))\mathrm{d}s\right)v(t),\\ Y_n(0) &= 0.\end{aligned}\right\} \tag{16}$$

The solution representation is

$$
\begin{aligned}
Y_n(t) = & \int_0^t (t-s)^{\alpha-1} E_{\alpha,\alpha}(A(t-s)^{\alpha}) \left[ b\left(s, X_{v_n}(s), \int_0^s f(s,\tau,X_{v_n}(\tau))\mathrm{d}\tau\right) \right. \\
& \left. -b\left(s, X_v(s), \int_0^s f(s,\tau,X_v(\tau))\mathrm{d}\tau\right)\right] \mathrm{d}s \\
& + \int_0^t (t-s)^{\alpha-1} E_{\alpha,\alpha}(A(t-s)^{\alpha}) \left[ \sigma\left(s, X_{v_n}(s), \int_0^s g(s,\tau,X_{v_n}(\tau))\mathrm{d}\tau\right) v_n(s) \right. \\
& \left. -\sigma\left(s, X_v(s), \int_0^s g(s,\tau,X_v(\tau))\mathrm{d}\tau\right) v(s)\right] \mathrm{d}s \\
=: & \, I_1(t) + I_2(t) + I_3(t), \qquad (17)
\end{aligned}
$$

where

$$
\begin{aligned}
I_1(t) := & \int_0^t (t-s)^{\alpha-1} E_{\alpha,\alpha}(A(t-s)^{\alpha}) \left[ b\left(s, X_{v_n}(s), \int_0^s f(s,\tau,X_{v_n}(\tau))\mathrm{d}\tau\right) \right. \\
& \left. -b\left(s, X_v(s), \int_0^s f(s,\tau,X_v(\tau))\mathrm{d}\tau\right)\right] \mathrm{d}s, \qquad (18) \\
I_2(t) := & \int_0^t (t-s)^{\alpha-1} E_{\alpha,\alpha}(A(t-s)^{\alpha}) \left[ \sigma\left(s, X_{v_n}(s), \int_0^s g(s,\tau,X_{v_n}(\tau))\mathrm{d}\tau\right) \right. \\
& \left. -\sigma\left(s, X_v(s), \int_0^s g(s,\tau,X_v(\tau))\mathrm{d}\tau\right)\right] v_n(s)\mathrm{d}s, \qquad (19) \\
I_3(t) := & \int_0^t (t-s)^{\alpha-1} E_{\alpha,\alpha}(A(t-s)^{\alpha})\, \sigma\left(s, X_v(s), \int_0^s g(s,\tau,X_v(\tau))\mathrm{d}\tau\right)(v_n(s)-v(s))\mathrm{d}s. \qquad (20)
\end{aligned}
$$

First consider the integral $I_1(t)$ and taking $\|\cdot\|_{\mathbb{X}}$ on both sides, we get

$$
\begin{aligned}
\left\|I_1(t)\right\|_{\mathbb{X}} \leq & \int_0^t (t-s)^{\alpha-1} \|E_{\alpha,\alpha}(A(t-s)^{\alpha})\|_{L(\mathbb{X})} \left\| b\left(s, X_{v_n}(s), \int_0^s f(s,\tau,X_{v_n}(\tau))\mathrm{d}\tau\right) \right. \\
& \left. - b\left(s, X_v(s), \int_0^s f(s,\tau,X_v(\tau))\mathrm{d}\tau\right)\right\|_{\mathbb{X}} \mathrm{d}s.
\end{aligned}
$$

Using the boundedness of $\|E_{\alpha,\alpha}(\cdot)\|$ given by (5) and Lipschitz continuity of '$b$' from (3), we obtain

$$
\left\|I_1(t)\right\|_{\mathbb{X}} \leq L_b M_2 \int_0^t (t-s)^{\alpha-1}\left[\left\|Y_n(s)\right\|_{\mathbb{X}} + \int_0^s \left\|f(s,\tau,X_{v_n}(\tau)) - f(s,\tau,X_v(\tau))\right\|_{\mathbb{X}} \mathrm{d}\tau\right]\mathrm{d}s.
$$

Using the Lipschitz continuity of '$f$', we get subsequently

$$
\begin{aligned}
\left\|I_1(t)\right\|_{\mathbb{X}} & \leq L_b\, M_2 \int_0^t (t-s)^{\alpha-1}\left[\left\|Y_n(s)\right\|_{\mathbb{X}} + L_f \int_0^s \left\|Y_n(\tau)\right\|_{\mathbb{X}} \mathrm{d}\tau\right]\mathrm{d}s \\
& \leq L_b\, M_2 \int_0^t (t-s)^{\alpha-1}\left\|Y_n(s)\right\|_{\mathbb{X}} \mathrm{d}s + L_b L_f M_2 \int_0^t (t-s)^{\alpha-1} \int_0^t \left\|Y_n(\tau)\right\|_{\mathbb{X}} \mathrm{d}\tau\, \mathrm{d}s \\
& = L_b\, M_2 \int_0^t (t-s)^{\alpha-1}\left\|Y_n(s)\right\|_{\mathbb{X}} \mathrm{d}s + L_b L_f M_2 \frac{T^{\alpha}}{\alpha} \int_0^t \left\|Y_n(s)\right\|_{\mathbb{X}} \mathrm{d}s. \qquad (21)
\end{aligned}
$$

In a similar way, consider the integral $I_2(t)$ and estimating using the boundedness of $\|E_{\alpha,\alpha}(\cdot)\|$ and Lipschitz continuity of '$\sigma$', we get

$$\left\|I_2(t)\right\|_{\mathbb{X}} \leq L_\sigma M_2 \int_0^t (t-s)^{\alpha-1}\Big[\left\|Y_n(s)\right\|_{\mathbb{X}} + \int_0^s \left\|g(s,\tau,X_{v_n}(\tau)) - g(s,\tau,X_v(\tau))\right\|_{\mathbb{X}} \mathrm{d}\tau\Big]\|v_n(s)\|_0 \,\mathrm{d}s.$$

Now, using the Lipschitz continuity of '$g$', we obtain

$$\begin{aligned}\left\|I_2(t)\right\|_{\mathbb{X}} &\leq L_\sigma M_2 \int_0^t (t-s)^{\alpha-1}\left\|Y_n(s)\right\|_{\mathbb{X}}\|v_n(s)\|_0 \,\mathrm{d}s \\ &\quad + L_\sigma L_g M_2 \int_0^t \left\|Y_n(\tau)\right\|_{\mathbb{X}} \mathrm{d}\tau \int_0^t (t-s)^{\alpha-1}\|v_n(s)\|_0 \,\mathrm{d}s. \qquad (22)\end{aligned}$$

For the third integral, applying Holder's inequality, one gets

$$\begin{aligned}\left\|I_3(t)\right\|_{\mathbb{X}} &\leq \int_0^t (t-s)^{\alpha-1} \left\|E_{\alpha,\alpha}(A(t-s)^\alpha)\right\|_{L(\mathbb{X})}\left\|\sigma\Big(s, X_v(s), \int_0^s g(s,\tau,X_v(\tau))\,\mathrm{d}\tau\Big)(v_n(s)-v(s))\right\|_{\mathbb{X}} \mathrm{d}s \\ &\leq M_2\Big(\int_0^t (t-s)^{2\alpha-2}\,\mathrm{d}s\Big)^{1/2}\Big(\int_0^t \left\|\sigma\Big(s, X_v(s), \int_0^s g(s,\tau,X_v(\tau))\,\mathrm{d}\tau\Big)(v_n(s)-v(s))\right\|_{\mathbb{X}}^2 \mathrm{d}s\Big)^{1/2} \\ &\leq M_2 T_\alpha\Big(\int_0^t \left\|\sigma\Big(s, X_v(s), \int_0^s g(s,\tau,X_v(\tau))\,\mathrm{d}\tau\Big)(v_n(s)-v(s))\right\|_{\mathbb{X}}^2 \mathrm{d}s\Big)^{1/2}, \qquad (23)\end{aligned}$$

where $T_\alpha = \frac{T^{\alpha-1/2}}{\sqrt{2\alpha-1}}$. Now (17) becomes, after substituting (21) - (23) and applying Gronwall's inequality,

$$\begin{aligned}\left\|Y_n(t)\right\|_{\mathbb{X}} &\leq M_2 T_\alpha\Big(\int_0^t \left\|\sigma\Big(s, X_v(s), \int_0^s g(s,\tau,X_v(\tau))\,\mathrm{d}\tau\Big)(v_n(s)-v(s))\right\|_{\mathbb{X}}^2 \mathrm{d}s\Big)^{1/2} \\ &\quad\times \exp\Big\{L_b M_2 \int_0^t (t-s)^{\alpha-1}\mathrm{d}s + L_b L_f M_2 \frac{T^{\alpha+1}}{\alpha} + L_\sigma M_2 \int_0^t (t-s)^{\alpha-1}\|v_n(s)\|_0 \mathrm{d}s \\ &\quad + L_\sigma L_g M_2 T \int_0^t (t-s)^{\alpha-1}\|v_n(s)\|_0 \,\mathrm{d}s\Big\}.\end{aligned}$$

Applying Holder's inequality to the last two integral terms on the exponential index, one gets

$$\begin{aligned}\left\|Y_n(t)\right\|_{\mathbb{X}} &\leq M_2 T_\alpha\Big[\int_0^t \left\|\sigma\Big(s, X_v(s), \int_0^s g(s,\tau,X_v(\tau))\,\mathrm{d}\tau\Big)(v_n(s)-v(s))\right\|_{\mathbb{X}}^2 \mathrm{d}s\Big]^{1/2} \\ &\quad\times \exp\Big\{L_b M_2 \frac{T^\alpha}{\alpha} + L_b L_f M_2 \frac{T^{\alpha+1}}{\alpha} \\ &\quad + (L_\sigma M_2 + L_\sigma L_g M_2 T)\Big(\int_0^t (t-s)^{2\alpha-2}\mathrm{d}s\Big)^{1/2}\Big(\int_0^t \|v_n(s)\|_0^2 \mathrm{d}s\Big)^{1/2}\Big\}.\end{aligned}$$

On simplifying and taking supremum over $t \in J$, we get

$$\sup_{t\in J}\left\|Y_n(t)\right\|_{\mathbb{X}} \leq M_2 T_\alpha\Big[\int_0^T \left\|\sigma\Big(s, X_v(s), \int_0^s g(s,\tau,X_v(\tau))\,\mathrm{d}\tau\Big)(v_n(s)-v(s))\right\|_{\mathbb{X}}^2 \mathrm{d}s\Big]^{1/2}$$

$$\times \exp\left\{L_b M_2 \frac{T^{\alpha}}{\alpha}(1 + L_f T) + L_{\sigma} M_2 \frac{T^{2\alpha-1}}{2\alpha - 1}(1 + L_g T)\sqrt{N}\right\}. \tag{24}$$

Since $v_n \rightharpoonup v$ weakly in $\mathbb{L}^2(J; \mathcal{H}_0)$ and $\sigma$ is a Hilbert-Schmidt operator and hence compact, we have that $\sigma v_n \to \sigma v$ strongly in $\mathbb{L}^2(J; \mathbb{X})$ and so $Y_n = X_{v_n} - X_v \to 0$ in $\mathbb{C}(J; \mathbb{X})$, thereby proving the compactness. □

**Lemma 3.2** (Weak Convergence). *Let $\{v^{\epsilon} : \epsilon > 0\} \subset \mathcal{A}_N$ for some $N < \infty$. Assume that $v^{\epsilon}$ converge to $v$ in distribution as $S_N$-valued random elements; then*

$$\mathcal{G}^{\epsilon}\left(W(\cdot) + \frac{1}{\sqrt{\epsilon}} \int_0^{\cdot} v^{\epsilon}(s)\mathrm{d}s\right) \to \mathcal{G}^0\left(\int_0^{\cdot} v(s)\mathrm{d}s\right)$$

*in distribution as $\epsilon \to 0$.*

*Proof.* Consider the nonlinear stochastic fractional integrodifferential equation (11) with control $v^{\epsilon} \in \mathbb{L}^2(J; \mathcal{H}_0)$ and let the solution be denoted by $X^{\epsilon}_{v^{\epsilon}}(t)$. Take $Y^{\epsilon}(t) = X^{\epsilon}_{v^{\epsilon}}(t) - X_v(t)$. Then

$$\begin{aligned}
Y^{\epsilon}(t) = &\int_0^t (t-s)^{\alpha-1} E_{\alpha,\alpha}(A(t-s)^{\alpha})\left[b\left(s, X^{\epsilon}_{v^{\epsilon}}(s), \int_0^s f(s,\tau,X^{\epsilon}_{v^{\epsilon}}(\tau))\mathrm{d}\tau\right)\right.\\
&\left. - b\left(s, X_v(s), \int_0^s f(s,\tau,X_v(\tau))\mathrm{d}\tau\right)\right]\mathrm{d}s\\
&+ \int_0^t (t-s)^{\alpha-1} E_{\alpha,\alpha}(A(t-s)^{\alpha})\left[\sigma\left(s, X^{\epsilon}_{v^{\epsilon}}(s), \int_0^s g(s,\tau,X^{\epsilon}_{v^{\epsilon}}(\tau))\mathrm{d}\tau\right)\right.\\
&\left. - \sigma\left(s, X_v(s), \int_0^s g(s,\tau,X_v(\tau))\mathrm{d}\tau\right)\right]v^{\epsilon}(s)\mathrm{d}s\\
&+ \int_0^t (t-s)^{\alpha-1} E_{\alpha,\alpha}(A(t-s)^{\alpha})\sigma\left(s, X_v(s), \int_0^s g(s,\tau,X_v(\tau))\mathrm{d}\tau\right)(v^{\epsilon}(s) - v(s))\mathrm{d}s\\
&+ \sqrt{\epsilon}\int_0^t (t-s)^{\alpha-1} E_{\alpha,\alpha}(A(t-s)^{\alpha})\sigma\left(s, X^{\epsilon}_{v^{\epsilon}}(s), \int_0^s g(s,\tau,X_v(\tau))\mathrm{d}\tau\right)\mathrm{d}W(s).
\end{aligned}$$

Taking $\|\cdot\|^2$ on both sides and using the algebraic inequality $(a+b+c+d)^2 \leq 4(a^2+b^2+c^2+d^2)$, we obtain

$$\left\|Y^{\epsilon}(t)\right\|^2_{\mathbb{X}} \leq \mathcal{I}_1(t) + \mathcal{I}_2(t) + \mathcal{I}_3(t) + \mathcal{I}_4(t), \tag{25}$$

where

$$\begin{aligned}
\mathcal{I}_1(t) := 4&\left\|\int_0^t (t-s)^{\alpha-1} E_{\alpha,\alpha}(A(t-s)^{\alpha})\left[b\left(s, X^{\epsilon}_{v^{\epsilon}}(s), \int_0^s f(s,\tau,X^{\epsilon}_{v^{\epsilon}}(\tau))\mathrm{d}\tau\right)\right.\right.\\
&\left.\left. - b\left(s, X_v(s), \int_0^s f(s,\tau,X_v(\tau))\mathrm{d}\tau\right)\right]\mathrm{d}s\right\|^2_{\mathbb{X}},
\end{aligned} \tag{26}$$

$$\mathcal{I}_2(t) := 4\left\|\int_0^t (t-s)^{\alpha-1} E_{\alpha,\alpha}(A(t-s)^{\alpha})\left[\sigma\left(s, X^{\epsilon}_{v^{\epsilon}}(s), \int_0^s g(s,\tau,X^{\epsilon}_{v^{\epsilon}}(\tau))\mathrm{d}\tau\right)\right.\right.$$

$$- \sigma\left(s, X_v(s), \int_0^s g(s,\tau,X_v(\tau))\mathrm{d}\tau\right)\bigg] v^\epsilon(s)\mathrm{d}s\bigg\|_{\mathbb{X}}^2, \tag{27}$$

$$\mathcal{I}_3(t) := 4\left\| \int_0^t (t-s)^{\alpha-1}E_{\alpha,\alpha}(A(t-s)^\alpha)\sigma\left(s, X_v(s), \int_0^s g(s,\tau,X_v(\tau))\mathrm{d}\tau\right)(v^\epsilon(s)-v(s))\mathrm{d}s\right\|_{\mathbb{X}}^2, \tag{28}$$

$$\mathcal{I}_4(t) := 4\,\epsilon\left\| \int_0^t (t-s)^{\alpha-1}E_{\alpha,\alpha}(A(t-s)^\alpha)\sigma\left(s, X_{v^\epsilon}^\epsilon(s), \int_0^s g(s,\tau,X_v(\tau))\mathrm{d}\tau\right)\mathrm{d}W(s)\right\|_{\mathbb{X}}^2. \tag{29}$$

First consider the integral $\mathcal{I}_1(t)$ and applying Holder's inequality along with the bound for $\|E_{\alpha,\alpha}(\cdot)\|_{L(\mathbb{X})}$ given by (5) and the Lipschitz continuity of '$b$' given by (3), one gets

$$\begin{aligned}
\mathcal{I}_1(t) &\leq 4\int_0^t (t-s)^{2\alpha-2}\|E_{\alpha,\alpha}(A(t-s)^\alpha)\|_{L(\mathbb{X})}^2 \mathrm{d}s \\
&\quad\times \int_0^t \left\| b\left(s, X_{v^\epsilon}^\epsilon(s), \int_0^s f(s,\tau,X_{v^\epsilon}^\epsilon(\tau))\mathrm{d}\tau\right) - b\left(s, X_v(s), \int_0^s f(s,\tau,X_v(\tau))\mathrm{d}\tau\right)\right\|_{\mathbb{X}}^2 \mathrm{d}s \\
&\leq 4L_b^2M_2^2 \int_0^t (t-s)^{2\alpha-2}\mathrm{d}s \int_0^t \left[\|Y^\epsilon(s)\|_{\mathbb{X}} + \int_0^s \|f(s,\tau,X_{v^\epsilon}^\epsilon(\tau)) - f(s,\tau,X_v(\tau))\|\mathrm{d}\tau\right]^2 \mathrm{d}s.
\end{aligned}$$

Using the algebraic identity $(a+b)^2 \leq 2(a^2+b^2)$ and Holder's inequality to the last integral term on the right hand side and then using the Lipschitz continuity of '$f$', we obtain simultaneously

$$\begin{aligned}
\mathcal{I}_1(t) &\leq 8L_b^2M_2^2\frac{T^{2\alpha-1}}{2\alpha-1}\int_0^t \left[\|Y^\epsilon(s)\|_{\mathbb{X}}^2 + T\int_0^s \|f(s,\tau,X_{v^\epsilon}^\epsilon(\tau)) - f(s,\tau,X_v(\tau))\|_{\mathbb{X}}^2\right]\mathrm{d}s \\
&\leq 8L_b^2M_2^2\frac{T^{2\alpha-1}}{2\alpha-1}\int_0^t \left[\|Y^\epsilon(s)\|_{\mathbb{X}}^2 + L_f^2T\int_0^s \|Y^\epsilon(\tau)\|_{\mathbb{X}}^2\mathrm{d}\tau\right]\mathrm{d}s.
\end{aligned}$$

On simplifying, the integral $\mathcal{I}_1(t)$ can be estimated as

$$\mathcal{I}_1(t) \leq 8L_b^2M_2^2(1+L_f^2T^2)\frac{T^{2\alpha-1}}{2\alpha-1}\int_0^t \|Y^\epsilon(s)\|_{\mathbb{X}}^2\mathrm{d}s. \tag{30}$$

Similarly consider the integral $\mathcal{I}_2(t)$, apply Holder's inequality followed by the bound for $\|E_{\alpha,\alpha}(\cdot)\|_{\mathbb{X}}$ and the Lipschitz continuity of '$\sigma$' to get

$$\begin{aligned}
\mathcal{I}_2(t) &\leq 4\int_0^t (t-s)^{2\alpha-2}\|E_{\alpha,\alpha}(A(t-s)^\alpha)\|_{L(\mathbb{X})}^2 \mathrm{d}s \\
&\quad\times \int_0^t \left\| \sigma\left(s, X_{v^\epsilon}^\epsilon(s), \int_0^s g(s,\tau,X_{v^\epsilon}^\epsilon(\tau))\mathrm{d}\tau\right) - \sigma\left(s, X_v(s), \int_0^s g(s,\tau,X_v(\tau))\mathrm{d}\tau\right)\right\|_{L_Q}^2 \|v^\epsilon(s)\|_0^2\mathrm{d}s \\
&\leq 8L_\sigma^2M_2^2\frac{T^{2\alpha-1}}{2\alpha-1}\int_0^t \left[\|Y^\epsilon(s)\|_{\mathbb{X}}^2 + L_g^2T\int_0^s \|Y^\epsilon(\tau)\|_{\mathbb{X}}^2\mathrm{d}\tau\right]\|v^\epsilon(s)\|_0^2\mathrm{d}s.
\end{aligned}$$

On further simplifying and making use of the fact that the control variable $v \in S_N$, we obtain

$$\mathcal{I}_2(t) \leq 8L_\sigma^2M_2^2\frac{T^{2\alpha-1}}{2\alpha-1}\left[\int_0^t \|Y^\epsilon(s)\|_{\mathbb{X}}^2\|v^\epsilon(s)\|_0^2\mathrm{d}s + L_g^2NT\int_0^t \|Y^\epsilon(\tau)\|_{\mathbb{X}}^2\mathrm{d}\tau\right]. \tag{31}$$

Now consider the integral $\mathcal{I}_3(t)$, apply Holder's inequality and the bound for $\left\|E_{\alpha,\alpha}(\cdot)\right\|_{L(\mathbb{X})}$ to obtain

$$\mathcal{I}_3(t) \leq 4M_2^2 \frac{T^{2\alpha-1}}{2\alpha-1} \int_0^t \left\| \sigma\left(s, X_v(s), \int_0^s g(s,\tau,X_v(\tau))\mathrm{d}\tau\right)(v^\epsilon(s) - v(s)) \right\|_{\mathbb{X}}^2 \mathrm{d}s. \tag{32}$$

Finally consider the stochastic integral $\mathcal{I}_4(t)$ and taking supremum and then taking expectation on both sides and making use of the Burkholder-Davis-Gundy inequality, we get

$$\begin{aligned}
\mathbb{E}\left[\sup_{t\in J} \mathcal{I}_4(t)\right] &= 4\,\epsilon\,\mathbb{E}\left\{\sup_{t\in J}\left\| \int_0^t (t-s)^{\alpha-1} E_{\alpha,\alpha}(A(t-s)^\alpha)\sigma\left(s, X^\epsilon_{v^\epsilon}(s), \int_0^s g(s,\tau,X_v(\tau))\mathrm{d}\tau\right)\mathrm{d}W(s)\right\|_{\mathbb{X}}^2\right\} \\
&\leq 4\epsilon M_2^2 \mathbb{E}\int_0^T (T-s)^{2\alpha-2} \left\|\sigma\left(s, X^\epsilon_{v^\epsilon}(s), \int_0^s g(s,\tau,X^\epsilon_{v^\epsilon}(\tau))\,\mathrm{d}\tau\right)\right\|_{L_Q}^2 \mathrm{d}s.
\end{aligned}$$

Using the linear growth property of '$\sigma$' and '$g$' and simplifying, we get

$$\begin{aligned}
\mathbb{E}\left[\sup_{t\in J} \mathcal{I}_4(t)\right] &\leq 4\epsilon K_\sigma M_2^2 \mathbb{E}\int_0^T (T-s)^{2\alpha-2}\left[1 + \|X^\epsilon_{v^\epsilon}(s)\|^2_{\mathbb{X}} + K_g T\int_0^s (1 + \|X^\epsilon_{v^\epsilon}(\tau)\|^2_{\mathbb{X}})\mathrm{d}\tau\right]\mathrm{d}s \\
&\leq 4\epsilon K_\sigma M_2^2\left[\mathbb{E}\int_0^T (T-s)^{2\alpha-2}[1 + \left\|X^\epsilon_{v^\epsilon}(s)\right\|^2_{\mathbb{X}}]\mathrm{d}s + K_g T \frac{T^{2\alpha-1}}{2\alpha-1}\mathbb{E}\int_0^T [1 + \left\|X^\epsilon_{v^\epsilon}(s)\right\|^2_{\mathbb{X}}]\mathrm{d}s\right] \\
&\leq 4\epsilon K_\sigma M_2^2 \frac{T^{2\alpha-1}}{2\alpha-1}(1 + K_g T^2)\left\{1 + \mathbb{E}\left[\sup_{t\in J}\left\|X^\epsilon_{v^\epsilon}(t)\right\|^2_{\mathbb{X}}\right]\right\}.
\end{aligned} \tag{33}$$

With all these estimates on the integrals $\mathcal{I}_i(t), i = 1,2,3,4$, given by (30) - (33), equation (25) becomes, after taking supremum over $t \in J$ and then taking expectation,

$$\begin{aligned}
\mathbb{E}\left[\sup_{t\in J}\left\|Y^\epsilon(t)\right\|^2_{\mathbb{X}}\right] &\leq 8L_b^2 M_2^2(1 + L_f^2 T^2)\frac{T^{2\alpha-1}}{2\alpha-1}\mathbb{E}\int_0^T \left\|Y^\epsilon(s)\right\|^2_{\mathbb{X}}\mathrm{d}s \\
&+ 8L_\sigma^2 M_2^2 \frac{T^{2\alpha-1}}{2\alpha-1}\left[\mathbb{E}\int_0^T \left\|Y^\epsilon(s)\right\|^2_{\mathbb{X}}\left\|v^\epsilon(s)\right\|^2_0 \mathrm{d}s + L_g^2 N T\mathbb{E}\int_0^T \left\|Y^\epsilon(\tau)\right\|^2_{\mathbb{X}}\mathrm{d}\tau\right] \\
&+ 4M_2^2\frac{T^{2\alpha-1}}{2\alpha-1}\mathbb{E}\int_0^T\left\|\sigma\left(s, X_v(s), \int_0^s g(s,\tau,X_v(\tau))\mathrm{d}\tau\right)(v^\epsilon(s) - v(s))\right\|^2_{\mathbb{X}}\mathrm{d}s \\
&+ 4\epsilon K_\sigma M_2^2\frac{T^{2\alpha-1}}{2\alpha-1}(1 + K_g T^2)\left\{1 + \mathbb{E}\left[\sup_{t\in J}\left\|X^\epsilon_{v^\epsilon}(t)\right\|^2_{\mathbb{X}}\right]\right\}.
\end{aligned}$$

Applying Gronwall's inequality and further simplifying, we end up with

$$\begin{aligned}
\mathbb{E}\left[\sup_{t\in J}\left\|Y^\epsilon(t)\right\|^2_{\mathbb{X}}\right] &\leq \left\{4M_2^2\frac{T^{2\alpha-1}}{2\alpha-1}\mathbb{E}\int_0^T\left\|\sigma\left(s, X_v(s), \int_0^s g(s,\tau,X_v(\tau))\,\mathrm{d}\tau\right)(v^\epsilon(s) - v(s))\right\|^2_{\mathbb{X}}\mathrm{d}s\right. \\
&\left. + 4\,\epsilon K_\sigma M_2^2\frac{T^{2\alpha-1}}{2\alpha-1}(1 + K_g T^2)\left(1 + \mathbb{E}\left[\sup_{t\in J}\left\|X^\epsilon_{v^\epsilon}(t)\right\|^2_{\mathbb{X}}\right]\right)\right\} \\
&\times\exp\left(8M_2^2\frac{T^{2\alpha-1}}{2\alpha-1}\left[L_b^2(1 + L_f^2T^2)T + L_\sigma^2(1 + L_g^2T^2)N\right]\right).
\end{aligned} \tag{34}$$

Since $\sigma$ is a Hilbert-Schmidt operator and hence compact and since $v^\epsilon \rightharpoonup v$ weakly in $\mathbb{L}^2(J;\mathcal{H}_0)$ as $\epsilon \to 0$, we have that $\sigma v^\epsilon \to \sigma v$ strongly in $\mathbb{L}^2(J;\mathbb{X})$ and so $Y^\epsilon = X^\epsilon_{v^\epsilon} - X_v \to 0$ in probability in the space $\mathbb{L}^2(\Omega;\mathbb{C}(J;\mathbb{X}))$. Since convergence in probability always implies convergence in expectation, we have finally proved the required weak convergence criterion. □

## 4. Examples

**Example 4.1.** Consider the stochastic fractional integrodifferential equation with additive noise given by

$$\left.\begin{array}{rcl} {}^{C}D^{\alpha}X(t) & = & \int_0^t X(s)\,\mathrm{d}s + \sin(X(t)) + \int_0^t \sqrt{1+X^2(s)}\,\mathrm{d}s + \sqrt{\epsilon}\,\frac{\mathrm{d}W(t)}{\mathrm{d}t},\ t \in (0,T], \\ X(0) & = & X_0, \end{array}\right\} \tag{35}$$

with $X_0 \in \mathbb{R}$ and $1/2 < \alpha \leq 1$. The corresponding controlled equation with control $v \in L^2(0,T;\mathbb{R})$ takes the form

$$\begin{array}{rcl} {}^{C}D^{\alpha}X_v(t) & = & \int_0^t X_v(s)\mathrm{d}s + \sin(X_v(t)) + \int_0^t \sqrt{1+X_v^2(s)}\mathrm{d}s + v(t), t \in (0,T], \\ X_v(0) & = & X_0. \end{array}$$

It is observed that if there exists a unique solution $X_v(\cdot)$ for the above mentioned equation, then the control $v \in L^2([0,T],\mathbb{R})$ with which the unique solution $X_v$ is attained is also unique and hence the rate function $\mathcal{I} : C([0,T];\mathbb{R}) \to [0,\infty]$ is given explicitly by

$$\mathcal{I}(\phi) = \frac{1}{2}\int_0^T \left| {}^{C}D^{\alpha}\phi - \sin\phi - \int_0^t \left(\phi(s) + \sqrt{1+\phi^2(s)}\right)\mathrm{d}s \right|^2 \mathrm{d}t, \tag{36}$$

if $\phi$ satisfies (35) for appropriate control $v$, and $\infty$ otherwise.

**Example 4.2.** As an example for (7) with multiplicative type noise, consider the following stochastic equation:

$$\left.\begin{array}{rcl} {}^{C}D^{3/4}X(t) & = & \beta\int_0^t \left[X(s) + \exp\left(\frac{1}{1+X^2(s)}\right)\right]\mathrm{d}s + \sqrt{\epsilon}\eta\int_0^t X(s)\,\mathrm{d}s\,\frac{\mathrm{d}W(t)}{\mathrm{d}t},\ t \in (0,1], \\ X(0) & = & 1, \end{array}\right\} \tag{37}$$

where $\eta, \beta > 0$ are positive constants. Then the rate function $\mathcal{I} : C([0,1];\mathbb{R}) \to [0,\infty]$ is given by

$$\mathcal{I}(\phi) = \inf\left\{\frac{1}{2}\int_0^1 |v(t)|^2\mathrm{d}t : v \in L^2([0,1],\mathbb{R}) \text{ such that } X_v = \phi\right\}, \tag{38}$$

where $\inf\emptyset = \infty$ and $X_v$ is the unique solution of

$$\begin{array}{rcl} X_v(t) & = & 1 + \frac{\beta}{\Gamma\left(\frac{3}{4}\right)}\int_0^t \frac{1}{(t-s)^{1/4}}\int_0^s \left[X_v(r) + \exp\left(\frac{1}{1+X_v^2(r)}\right)\right]\mathrm{d}r\,\mathrm{d}s \\ & & + \frac{\eta}{\Gamma\left(\frac{3}{4}\right)}\int_0^t \frac{v(s)}{(t-s)^{1/4}}\int_0^s X_v(r)\,\mathrm{d}r\,\mathrm{d}s, t \in [0,1]. \end{array} \tag{39}$$

It is evident from (38) that estimating the rate function $\mathcal{I}(\phi)$ is a problem of finding the minimal cost $\frac{1}{2}\int_0^1 |v(t)|^2\mathrm{d}t$, out of all the controls $v$ that steers the desired solution $\phi = X_v$ from (39).

## Acknowledgments

The first author would like to thank the Department of Science and Technology, New Delhi for their financial support under the INSPIRE Fellowship Scheme. The work of the third author is supported by the University Grants Commission [grant number: MANF-2015-17-TAM-50645] from the Government of India.

## Conflict of Interest

All authors declare that there is no conflict of interest.

## References

1. K. Balachandran, S. Divya, M. Rivero and J.J. Trujillo, *Controllability of nonlinear implicit neutral fractional Volterra integrodifferential systems,* Journal of Vibration and Control, **22** (2016), 2165-2172.
2. K. Balachandran, V. Govindaraj, L. Rodrguez-Germa and J.J. Trujillo, *Controllability results for nonlinear fractional-order dynamical systems,* Journal of Optimization Theory and Applications, **156** (2013), 33-44.
3. K. Balachandran, M. Matar and J. J. Trujillo, *Note on controllability of linear fractional dynamical systems,* Journal of Control and Decision, **3** (2016), 267-279.
4. L. Bo and Y. Jiang, *Large deviation for the nonlocal Kuramoto-Sivashinsky SPDE,* Nonlinear Analysis: Theory, Methods and Applications, **82** (2013), 100-114.
5. M. Boue and P. Dupuis, *A variational representation for certain functionals of Brownian motion,* Annals of Probability, **26**(1998), 1641-1659.
6. A. Budhiraja P. Dupuis and V. Maroulas, *Large deviations for infinite dimensional stochastic dynamical systems,* Annals of Probability, **36** (2008), 1390-1420.
7. A. Budhiraja and P. Dupuis, *A variational representation for positive functionals of infinite dimensional Brownian motion,* Probability of Mathematics and Statistics, **20** (2000), 39-61.
8. A. Dembo and O. Zeitouni, *Large Deviations Techniques and Applications*, New York: Springer, 2007.
9. A. Di Crescenzo and A. Meoli, *On a fractional alternating Poisson process,* AIMS Mathematics, **1** (2016), 212-224.
10. N. Dunford and J. Schwartz, *Linear Operators, Part I*, New York: Wiley-Interscience, 1958.
11. P. Dupuis and R.S. Ellis, *A Weak Convergence Approach to the Theory of Large Deviations*, New York: Wiley-Interscience, 1997.
12. W.H. Fleming, *A stochastic control approach to some large deviations problems,* Recent Mathematical Methods in Dynamic Programming, Springer Lecture Notes in Math., **1119** (1985), 52-66.
13. M.I. Freidlin and A.D. Wentzell, *On small random perturbations of dynamical systems,* Russian Mathematical Surveys, **25** (1970), 1-55.

14. M.I. Freidlin and A.D. Wentzell, *Random Perturbations of Dynamical Systems*, New York:Springer, 1984.
15. R. Joice Nirmala and K. Balachandran, *Controllability of nonlinear fractional delay integrodifferential system,* Discontinuity, Nonlinearity, and Complexity, **5** (2016), 59-73.
16. M. Kamrani, *Numerical solution of stochastic fractional differential equations,* Numerical Algorithms, **68** (2015), 81-93.
17. E. Kreyszig, *Introductory Functional Analysis with Applications*, New York: John Wiley and Sons Inc, 1978.
18. W. Liu, *Large deviations for stochastic evolution equations with small multiplicative noise,* Applied Mathematics and Optimization, **61** (2010), 27-56.
19. A. Kilbas, H. M. Srivastava, J. J. Trujillo, *Theory and Applications of Fractional Differential Equations*, Amsterdam: Elsevier, 2006.
20. R. Mabel Lizzy, K. Balachandran and M. Suvinthra, *Controllability of nonlinear stochastic fractional systems with distributed delays in control,* Journal of Control and Decision, **4** (2017), 153-167.
21. K. S. Miller and B. Ross, *An Introduction to the Fractional Calculus and Fractional Differential Equations*, New York: John-Wiley, 1993.
22. C. Mo and J. Luo, *Large deviations for stochastic differential delay equations,* Nonlinear Analysis: Theory, Methods and Applications, **80** (2013), 202-210.
23. S.A. Mohammed and T.S. Zhang, *Large deviations for stochastic systems with memory,* Discrete and Continuous Dynamical Systems Series B, **6** (2006), 881-893.
24. J. C. Pedjeu and G. S. Ladde, *Stochastic fractional differential equations: Modelling, method and analysis,* Chaos, Solitons and Fractals, **45** (2012), 279-293.
25. I. Podlubny, *Fractional Differential Equations*, London: Academic Press, 1999.
26. J. Ren and X. Zhang, *Schilder theorem for the Brownian motion on the diffeomorphism group of the circle,* Journal of Functional Analysis, **224** (2005), 107-133.
27. J. Ren and X. Zhang, *Freidlin-Wentzell large deviations for homeomorphism flows of non-Lipschitz SDE,* Bulletin of Science, **129** (2005), 643-655.
28. S.S. Sritharan and P. Sundar, *Large deviations for two dimensional Navier-Stokes equations with multiplicative noise,* Stochastic Processes and their Applications, **116** (2006), 1636-1659.
29. M. Suvinthra and K. Balachandran, *Large deviations for nonlinear Itô type stochastic integrodifferential equations,* Journal of Applied Nonlinear Dynamics, **6** (2017), 1-15.
30. M. Suvinthra, K. Balachandran and J.K. Kim, *Large deviations for stochastic differential equations with deviating arguments,* Nonlinear Functional Analysis and Applications, **20** (2015), 659-674.
31. S.R.S. Varadhan, *Asymptotic probabilities and differential equations,* Communications on Pure and Applied Mathematics, **19** (1966), 261-286.

# 6

# A Probabilistic Characterization of g-Harmonic Functions

**Liang Cai**[1,*] **Huan-Huan Zhang**[1] **and Li-Yun Pan**[2,3]

[1] School of Mathematics and Statistics, Beijing Institute of Technology, Beijing 100081, China
[2] Department of Basic Science, Beijing Institute of Graphic Communication, Beijing 102600, China
[3] Beijing Institute of Education, Beijing 100120, China

[*] **Correspondence:** Email: cailiang@bit.edu.cn

**Abstract:** Associated with a quasi-linear generator function $g$, we give a definition of g-harmonic functions. The relation between the g-harmonic functions and g-martingales will be delineated. It is direct to construct such relation for smooth case, but for continuous case we need the theory of viscosity solution. Under the nonlinear expectation mechanism, we can also get the similar relation between harmonic functions and martingales. The strict converse problem of mean value property of g-harmonic functions are discussed finally.

**Keywords:** BSDE; g-martingale; g-harmonic function; nonlinear Feynman-Kac formula; viscosity solution

## 1. Introduction and Preliminary

Harmonic function ($\Delta u = 0$) has a probabilistic interpretation as that if $\Delta u = 0$ on $R^n$, then $u(B_t^x)$ is a martingale for any $x \in R^n$ (see for example [6]). This relation between martingale and harmonic function connects probability with potential analysis. It helps us to give probabilistic characterization for harmonic function and more generalized X-harmonic function [6]. In 1997, Peng [9] introduced the notions of g-expectation and conditional g-expectation via backward stochastic differential equations (BSDE) with quasi-linear generator function $g$. Further, Peng [10] introduced the notion of g-martingale. Thanks to these works, we will give a probabilistic characterization of the g-harmonic functions which have quasi-linear generator function $g$.

Now we state our problem in detail. Let $(\Omega, \mathcal{F}, P)$ be a probability space endowed with the natural filtration $\{\mathcal{F}_t\}_{t\geq 0}$ generated by an $n$-dimensional Brownian motion $\{B_t\}_{t\geq 0}$, i.e.

$$\mathcal{F}_t = \sigma\{B_s : s \leq t\}.$$

Then we can define a g-martingale by an $\mathcal{F}_t$-adapted process $\{y_t\}_{t\geq 0}$ which satisfies the following BSDE for any $0 \leq s \leq t$:

$$y_s = y_t + \int_s^t g(y_r, z_r)dr - \int_s^t z_r dB_r. \tag{1.1}$$

Here $g : R \times R^n \longrightarrow R$, satisfies the conditions:
(H1). $g(y, 0) \equiv 0$ and the Lipschitz condition: $\exists C > 0$, for any $(y_1, z_1), (y_2, z_2) \in R \times R^n$ we have

$$|g(y_1, z_1) - g(y_2, z_2)| \leq C(|y_1 - y_2| + |z_1 - z_2|).$$

And the equality (1.1) can also be formulated simply as [11]:

$$\mathcal{E}^g_{s,t}(y_t) := y_s.$$

Then we can also get the definition of g-super(sub)martingale when

$$\mathcal{E}^g_{s,t}(y_t) \leq (\geq)\, y_s.$$

This definition derives from the definition of g-expectation in the beginning paper Peng [9]. When $g(y, z) \equiv 0$ the g-expectation is actually the classical expectation. Except that g-expectation is nonlinear in general, it holds many other important properties as its classical counterpart [2,4,10,12].

Given an $n$-dimensional Itô's diffusion process $\{X^x_t\}_{t\geq 0}$:

$$\begin{aligned} dX^x_t &= b(X^x_t)dt + \sigma(X^x_t)dB_t, \\ X^x_0 &= x \in R^n, \end{aligned} \tag{1.2}$$

where $b(x) : R^n \longrightarrow R^n$, $\sigma(x) : R^n \longrightarrow R^{n\times n}$ satisfy the Lipschitz condition: $\exists C > 0$ s.t.

$$|b(x_1) - b(x_2)| + |\sigma(x_1) - \sigma(x_2)| \leq C|x_1 - x_2|, \quad \forall x_1, x_2 \in R^n,$$

our problem is that: what kind of function $u(x) : R^n \longrightarrow R$ satisfies that $u(X^x_t)$ is a g-martingale for any $x \in R^n$?

This problem also has its classical counterpart:

First if $\{X^x_t\}$ is just the Brownian motion $\{B^x_t\}$, then we have the result that when $u(x)$ is harmonic on $R^n$ i.e.

$$\Delta u = \sum_i \frac{\partial^2 u}{\partial x_i^2} = 0, \quad \text{for any } x \in R^n,$$

the process $u(B^x_t)$ is a martingale for any $x$. And conversely if $u(x)$ satisfies that $u(B^x_t)$ is a martingale for any $x$, then $u(x)$ must be harmonic on $R^n$. The proof may have many editions, here we can give a sketch of one which may induce the extension to g-martingale case.

If $u(x)$ is harmonic on $R^n$, then we use Itô's formula to $u(B^x_t)$ and get

$$du(B^x_t) = \sum_i \frac{\partial u}{\partial x_i}(B^x_t)dB_{i,t} + \frac{1}{2}\sum_i \frac{\partial^2 u}{\partial x_i^2}(B^x_t)dt = \sum_i \frac{\partial u}{\partial x_i}(B^x_t)dB_{i,t}.$$

Then we get $u(B_t^x)$ is a martingale for any $x \in R^n$. Conversely if $u(x)$ is continuous on $R^n$ and for any $x \in R^n$, $u(B_t^x)$ is a martingale, then we have $E[u(B_\tau^x)] = u(x)$ for any stopping time $\tau$. Particularly for any sphere $S(x, r) = \{y \in R^n : |y - x| < r\}$, we have

$$u(x) = E[u(B^x_{\tau_{S(x,r)}})] = \int_{\partial S(x,r)} u(y) d\sigma_y,$$

where $\tau_{S(x,r)}$ is the exit time of $\{B_t^x\}$ from the sphere $S(x, r)$, i.e.

$$\tau_{S(x,r)} = \inf\{t > 0 : |B_t^x - x| \geq r\},$$

and $\sigma_y$ is the harmonic measure on the $\partial S(x, r)$. Then from the familiar converse of the mean value property for harmonic function, we can get $u(x)$ must be harmonic function.

Further we can extend the Brownian motion $\{B_t^x\}$ to the general diffusion process $\{X_t^x\}$:

If $u(x) \in C_0^2(R^n)$ and satisfies

$$\sum_i b_i \frac{\partial u}{\partial x_i}(x) + \frac{1}{2} \sum_{i,j} (\sigma\sigma^T)_{i,j} \frac{\partial^2 u}{\partial x_i \partial x_j}(x) = 0, \tag{1.3}$$

then we have $u(X_t^x)$ is a martingale for any $x$. The proof also uses the Itô's formula. But conversely if $u(X_t^x)$ is a martingale for any $x$, we can't conclude that $u(x)$ is smooth. Then with additional assumption $u(x) \in C_0^2(R^n)$ we can get that $u(x)$ satisfies the PDE (1.3) [6].

Then naturally we will ask that what happens when we substitute the expectation mechanism by the g-expectation mechanism. First we will define the infinitesimal generator:

**Definition 1.** *Let*

$$\mathcal{A}_g^X f(x) := \lim_{t \downarrow 0} \frac{\mathcal{E}_{0,t}^g [f(X_t^x)] - f(x)}{t}, \tag{1.4}$$

*then we call $\mathcal{A}_g^X$ the infinitesimal generator of a diffusion process $\{X_t^x\}$ under g-expectations.*

Thanks to the celebrating nonlinear Feynman-Kac formula [8], we can get the explicit form of $\mathcal{A}_g^X$ when $f \in C_0^2(R^n)$ by considering the following type of quasilinear parabolic PDE:

$$\begin{cases} \frac{\partial u}{\partial t}(t, x) - \mathcal{L}u(t, x) - g(u(t, x), u_x(t, x)\sigma(x)) = 0, \\ u(0, x) = f(x). \end{cases} \tag{1.5}$$

where

$$\mathcal{L}u(t, x) = \sum_i b_i \frac{\partial u}{\partial x_i}(t, x) + \frac{1}{2} \sum_{i,j} (\sigma\sigma^T)_{i,j} \frac{\partial^2 u}{\partial x_i \partial x_j}(t, x). \tag{1.6}$$

When $f \in C_0^2(R^n)$, we assert that

$$u(t, x) = \mathcal{E}_{0,t}^g [f(X_t^x)] \tag{1.7}$$

is the solution of PDE (1.5). Then under the case $t = 0$, we get

$$\mathcal{A}_g^X f(x) = \mathcal{L}f(x) + g(f(x), f_x(x)\sigma(x)). \tag{1.8}$$

Then we finish the preliminary and we can introduce our main results. In section 2, we give a characterization of g-harmonic function under smooth case. In section 3, we characterize it under continuous case, where the differential operator is interpreted as viscosity solution. In section 4, we will investigate the strict converse problem of mean value property of g-harmonic function evoked by its classical counterpart [7].

## 2. Smooth Case

The equality (1.8) implies the relation between the g-martingales and the g-harmonic functions when $f \in C_0^2(R^n)$. In fact, the left side of (1.8) is related to a g-martingale and the right side is related to a harmonic PDE. At first we will give the definition of g-harmonic functions:

**Definition 2.** *Let $f \in C_0^2(R^n)$. We call it a g-(super)harmonic function w.r.t. $\{X_t^x\}$ if it satisfies*

$$\mathcal{A}_g^X f(x)(\leq) = 0, \qquad \textit{for any} \quad x \in R^n. \tag{2.1}$$

Then we suffice to construct the relation between the g-supermartingales and the g-superharmonic functions.

**Theorem 1.** *If $f(x) \in C_0^2(R^n)$, then the following assertions are equivalent:*
*(1)$f(x)$ is a g-superharmonic function.*
*(2)$\{f(X_t^x)\}$ is a g-supermartingale for any $x \in R^n$.*

*Proof.* (i) (1) $\Rightarrow$ (2):

For any $f \in C^2(R^n)$, by Itô's formula, we can get $f(X_t^x)$ is still an Itô's diffusion process:

$$f(X_t^x) = f(X_s^x) + \int_s^t \mathcal{L}f(X_r^x)dr + \int_s^t f_x(X_r^x)\sigma(X_r^x)dB_r, \quad 0 \leq s \leq t.$$

and then we insert the term $g(f(X_r^x), f_x(X_r^x)\sigma(X_x^r))$ and get

$$\begin{aligned} f(X_s^x) &= f(X_t^x) - \int_s^t \mathcal{L}f(X_r^x)dr - \int_s^t f_x(X_r^x)\sigma(X_r^x)dB_r \\ &= f(X_t^x) + \int_s^t g(f(X_r^x), f_x(X_r^x)\sigma(X_x^r))dr - \int_s^t f_x(X_r^x)\sigma(X_r^x)dB_r \\ &\quad - \int_s^t [\mathcal{L}f(X_r^x) + g(f(X_r^x), f_x(X_r^x)\sigma(X_x^r))]dr. \end{aligned}$$

$f(x)$ is a g-superharmonic function, so

$$\mathcal{L}f(X_r^x) + g(f(X_r^x), f_x(X_r^x)\sigma(X_r^x)) = \mathcal{A}_g^X f(X_r^x) \leq 0.$$

And then according to the comparison theory of BSDE [10], we can get $\{f(X_t^x)\}$ is a g-supermartingale.

(ii) (2) $\Rightarrow$ (1):

By the definition of the $\mathcal{A}_g^X$:

$$\mathcal{A}_g^X f(x) = \lim_{t \downarrow 0} \frac{\mathcal{E}_{0,t}^g[f(X_t^x)] - f(x)}{t}.$$

$\{f(X_t^x)\}$ is a g-supermartingale, so

$$\mathcal{E}_{0,t}^g[f(X_t^x)] - f(x) \leq 0,$$

then

$$\mathcal{A}_g^X f(x) \leq 0.$$

So we get $f(x)$ is a g-superharmonic function. □

## 3. Continuous Case

If we generalize the requirement of function $f(x)$ to be only continuous on $R^n$, how we get a function $f$ which satisfies that $f(X_t^x)$ is a g-martingale for any $x \in R^n$? With the help of viscosity solution [3], we can also refer to the quasi-linear second order PDEs. Here we need a lemma due to Peng [8].

**Lemma 1.** *Let* $0 \leq t \leq T$ *and*

$$u(t,x) = \mathcal{E}_{0,T-t}^g[f(X_{T-t}^x)].$$

*Then* $u(t,x)$ *is the viscosity solution of the following PDE on* $(0,T) \times R^n$*:*

$$\begin{cases} \frac{\partial u}{\partial t} + \mathcal{L}u(t,x) + g(u(t,x), u_x(t,x)\sigma(x)) = 0, \\ u(T,x) = f(x). \end{cases} \tag{3.1}$$

*Here* $g(y,z)$ *and* $f(x)$ *satisfy:*
*(H2) Let* $F(u,p) = g(u, p\sigma(x))$*, then* $\exists C > 0$ *s.t.*

$$|F(u,p)| \leq C(1 + |u| + |p|);$$
$$|D_u F(u,p)|, |D_p F(u,p)| \leq C;$$

*and (H3)* $f(x)$ *is a continuous function with a polynomial growth at infinity.*

**Definition 3.** *Let* $u(t,x) \in C(R \times R^n)$*.* $u(t,x)$ *is said to be a viscosity super-solution (resp. sub-solution) of the following PDE (3.2):*

$$\frac{\partial u}{\partial t} + \mathcal{L}u(t,x) + g(u(t,x), u_x(t,x)\sigma(x)) = 0, \tag{3.2}$$

*if for any* $(t,x) \in R \times R^n$ *and* $\varphi \in C^{1,2}(R \times R^n)$ *such that* $\varphi(t,x) = u(t,x)$ *and* $(t,x)$ *is a maximum (resp. minimum) point of* $\varphi - u$*,*

$$\frac{\partial \varphi}{\partial t}(t,x) + \mathcal{L}\varphi(t,x) + g(\varphi(t,x), \varphi_x(t,x)\sigma(x)) \leq 0.$$

$$(resp.\quad \frac{\partial \varphi}{\partial t}(t,x) + \mathcal{L}\varphi(t,x) + g(\varphi(t,x), \varphi_x(t,x)\sigma(x)) \geq 0.)$$

*$u(t,x)$ is said to be a viscosity solution of PDE (3.2) if it is both a viscosity super- and sub-solution of (3.2).*

We also consider the viscosity solution of the following type of quasilinear elliptic PDE (3.3):

$$\mathcal{L}u(x) + g(u(x), u_x(x)\sigma(x)) = 0. \tag{3.3}$$

We can directly get an relation between the two solutions of (3.2) and (3.3):

**Lemma 2.** *Let $\tilde{u}(t,x) = u(x)$ for all $(t,x) \in R \times R^n$, then we have:*
*$\tilde{u}(t,x)$ is the viscosity super-(sub-)solution of PDE (3.2) $\Leftrightarrow$ $u(x)$ is the viscosity super-(sub-)solution of PDE (3.3).*

*Proof.* We suffice to prove the case of viscosity super-solution.

(i) "$\Rightarrow$":

For any $(t_0, x_0) \in R \times R^n$, and a function $\varphi(x) \in C^2(R^n)$ which satisfies $\varphi(x) \leq u(x), \varphi(x_0) = u(x_0)$, we define $\tilde{\varphi}(t,x) = \varphi(x)$ for all $(t,x) \in R \times R^n$. Then

$$\frac{\partial \tilde{\varphi}}{\partial t} = 0, \quad \tilde{\varphi}(t_0, x_0) = \tilde{u}(t_0, x_0), \quad \tilde{\varphi}(t,x) \leq \tilde{u}(t,x),$$

and due to the assumption that $\tilde{u}(t,x)$ is the viscosity super-solution of PDE (3.2), we have

$$\frac{\partial \tilde{\varphi}}{\partial t}(t_0, x_0) + \mathcal{L}\tilde{\varphi}(t_0, x_0) + g(\tilde{\varphi}(t_0, x_0), \tilde{\varphi}_x(t_0, x_0)\sigma(x_0)) \leq 0,$$

i.e.

$$\mathcal{L}\varphi(x_0) + g(\varphi(x_0), \varphi_x(x_0)\sigma(x_0)) \leq 0.$$

So $u(x)$ is the viscosity super-solution of PDE (3.3).

(ii). " $\Leftarrow$ ":

For any $(t_0, x_0) \in R \times R^n$, and a function $\varphi(t,x) \in C^2(R \times R^n)$ which satisfies

$$\varphi(t,x) \leq \tilde{u}(t,x) \quad \text{and} \quad \varphi(t_0, x_0) = \tilde{u}(t_0, x_0),$$

then

$$\frac{\partial \varphi}{\partial t}(t_0, x_0) = 0, \tag{3.4}$$

and due to the assumption that $u(x)$ is the viscosity super-solution of PDE (3.3), we have

$$\mathcal{L}\varphi(t_0, x_0) + g(\varphi(t_0, x_0), \varphi_x(t_0, x_0)\sigma(x_0)) \leq 0.$$

Combined with (3.4), we get

$$\frac{\partial \varphi}{\partial t}(t_0, x_0) + \mathcal{L}\varphi(t_0, x_0) + g(\varphi(t_0, x_0), \varphi_x(t_0, x_0)\sigma(x_0)) \leq 0.$$

So $\tilde{u}(t,x)$ is the viscosity super-solution of PDE (3.2). □

Then we can introduce our main result of this section:

**Theorem 2.** *We have the following two consequences:*
*(i) For any $f(x) \in C(R^n)$, and $g(y,z)$ satisfying (H1), if $\forall x \in R^n$, $f(X_t^x)$ is a g-supermartingale, then $f(x)$ is a viscosity super-solution of PDE (3.3).*
*(ii) For any $f(x)$ satisfying (H3), and $g(y,z)$ satisfying (H1) and (H2), let $f(x)$ is a viscosity super-solution of PDE (3.3), then $\{f(X_t^x)\}$ is a g-supermartingale for all $x \in R^n$.*

Actually, the consequence (ii) is the answer of our main problem and the consequence (i) is the converse of it. But (i) is easier to be proved, so we are going to prove (i) at first.

*Proof.* (i) For any $x \in R^n$, let $\varphi \in C^2(R^n)$, $\varphi(x) = f(x)$ where $x$ is a maximum point of $\varphi - f$. It means $\forall \tilde{x} \in R^n$, we have $\varphi(\tilde{x}) \le f(\tilde{x})$. Then from (1.8), we get

$$\begin{aligned}\mathcal{L}\varphi(x) + g(\varphi(x), \varphi_x(x)\sigma(x)) &= \mathcal{A}_g^X \varphi(x)\\ &= \lim_{t\downarrow 0} \frac{\mathcal{E}_t^g[\varphi(X_t^x)] - \varphi(x)}{t}\\ &= \lim_{t\downarrow 0} \frac{\mathcal{E}_t^g[\varphi(X_t^x)] - f(x)}{t}.\end{aligned}$$

According to the comparison theory of BSDE, we get

$$\mathcal{E}_t^g[\varphi(X_t^x)] \le \mathcal{E}_t^g[f(X_t^x)],$$

and with the assumption $\{f(X_t^x)\}$ is a g-supermartingale, we can get

$$\mathcal{E}_t^g[\varphi(X_t^x)] - f(x) \le \mathcal{E}_t^g[f(X_t^x)] - f(x) \le 0.$$

Then

$$\mathcal{A}_g^X \varphi(x) = \lim_{t\downarrow 0} \frac{\mathcal{E}_t^g[\varphi(X_t^x)] - f(x)}{t} \le 0,$$

i.e.

$$\mathcal{L}\varphi(x) + g(\varphi(x), \varphi_x(x)\sigma(x)) \le 0.$$

By definition, it means $f(x)$ is a viscosity super-solution of PDE (3.3).
(ii) We want to prove $\{f(X_t^x)\}$ is a g-supermartingale for any $x \in R^n$. It means that we need to prove $\forall x \in R^n$ *and* $\forall 0 \le s \le t$, we have

$$\mathcal{E}_{s,t}^g[f(X_t^x)] \le f(X_s^x).$$

Under the assumption, in fact $b(x), \sigma(x)$ and $g(y,z)$ are all independent of time $t$, so we can get the Markovian property of $\mathcal{E}_{s,t}^g$, i.e.

$$\mathcal{E}_{s,t}^g[f(X_t^x)] = \mathcal{E}_{t-s}^g[f(X_{t-s}^y)]|_{y=X_s^x}.$$

Then we get an equivalence relation:

$$\{f(X_t^x)\} \text{ is a g-(super)martingale for any } x \in R^n \Leftrightarrow$$

$$\mathcal{E}_t^g[f(X_t^x)] = (\leq) f(x) \text{ for any } t \geq 0 \text{ and } x \in R^n. \tag{3.5}$$

So we suffice to prove the latter assertion.

According to lemma 2, for any $T \geq 0$, the assumption $f(x)$ is a viscosity super-solution of PDE (3.3) implies that $\tilde{f}(t,x) := f(x)$ is a viscosity super-solution to the following PDE:

$$\begin{cases} \frac{\partial u}{\partial t}(t,x) + \mathcal{L}u(t,x) + g(u(t,x), u_x(t,x)\sigma(x)) = 0, \\ u(T,x) = f(x). \end{cases} \tag{3.6}$$

And with the help of lemma 1,

$$u(t,x) = \mathcal{E}_{0,T-t}^g[f(X_{T-t}^x)]$$

is actually the viscosity solution of PDE (3.6). Moreover by the maximum principle of the viscosity solution [1], we can get

$$u(t,x) \leq \tilde{f}(t,x), \quad \text{for any } 0 \leq t \leq T.$$

Especially, we have

$$u(0,x) \leq \tilde{f}(0,x),$$

i.e.

$$\mathcal{E}_T^g[f(X_T^x)] \leq f(x).$$

□

**Corollary 1.** *(i) For any $f(x) \in C(R^n)$, and $g(y,z)$ satisfying (H1), if $\forall x \in R^n$, $f(X_t^x)$ is a g-martingale, then $f(x)$ is a viscosity solution of PDE (3.3).*
*(ii) For any $f(x)$ satisfying (H3), and $g(y,z)$ satisfying (H1) and (H2), let $f(x)$ is a viscosity solution of PDE (3.3), then $\{f(X_t^x)\}$ is a g-martingale for all $x \in R^n$.*

It is an immediate consequence from the theorem 2.

## 4. Strict Converse of Mean Value Property

For classical harmonic function, many generalized results of the converse problem of mean value property have been investigated [5,7]. In [7], Øksendal and Stroock give a technique to solve a strict converse of the mean value property for harmonic functions. Now we will generalize it to the case of g-harmonic function. Here the strictness means that for each $x \in R^n$ we don't need justify that for any stopping time $\tau$ whether $\mathcal{E}_{0,\tau}^g(f(X_\tau^x))$ equals $f(x)$. We only need to justify one appropriate stopping time of each $x$.

In the sequel we put $\Delta(x,r) = \{y \in R^n; |y-x| < r\}$ for any $x \in R^n$ and $r > 0$. Let $\tau_U = \inf\{t > 0; X_t^x \in U^c\}$ for any open set $U$. And we suppose the operator (1.6) is elliptic on $R^n$.

**Theorem 3.** *$f(x)$ is a local bounded continuous function on $R^n$. If for any $x \in R^n$, there exists a radius $r(x)$, the mean value property holds:*

$$\mathcal{E}_{0,\tau_x}^g[f(X_{\tau_x}^x)] = f(x), \quad \textit{here} \quad \tau_x = \tau_{\Delta(x,r(x))}. \tag{4.1}$$

*And $r(x)$ is a measurable function of $x$ and satisfies that for each $x$, there exists a bounded open set $U_x$, $x \in U_x$ and moreover $r(y)$, $y \in U_x$ should satisfy the following two conditions:*

$$0 \le r(y) \le dist(y, \partial U_x), \tag{4.2}$$

*and*

$$inf\{r(y); y \in K\} > 0 \tag{4.3}$$

*for all closed subsets $K$ of $U_x$ with $dist(K, \partial U_x) > 0$. Then we can get*
*(i) For each $y \in U_x$ the mean value property holds on the boundary:*

$$\mathcal{E}^g_{0,\tau_y}[f(X^y_{\tau_y})] = f(y), \quad here\ \tau_y = inf\{t > 0; X^y_t \in U^c_x\}.$$

*and furthormore*
*(ii) $f(x)$ is the viscosity solution of PDE (12).*

*Proof.* *(i)* $\Rightarrow$ *(ii)* is also based on the nonlinear Feynman-Kac formula for elliptic PDE [8]. So we sufficiently prove the first conclusion.

For each $y \in U_x$, we define a sequence of stopping times $\tau_k$ for $\{X^y_t\}$ by induction as follows:

$$\begin{aligned} &\tau_0 \equiv 0 \\ &\tau_k = \inf\{t \ge \tau_{k-1}; |X^y_t - X^y_{\tau_{k-1}}| \ge r(X^y_{\tau_{k-1}})\}, \quad k \ge 1. \end{aligned}$$

By the mean property (4.1), and the strong markovian property we can get

$$\begin{aligned} \mathcal{E}^g_{0,\tau_k}[f(X^y_{\tau_k})] &= \mathcal{E}^g_{0,\tau_{k-1}}[\mathcal{E}^g_{\tau_{k-1},\tau_k}[f(X^y_{\tau_k})]] \\ &= \mathcal{E}^g_{0,\tau_{k-1}}[\mathcal{E}^g_{0,\tau_k-\tau_{k-1}}[f(X^{X^y_{\tau_{k-1}}}_{\tau_k-\tau_{k-1}})]] \\ &= \mathcal{E}^g_{0,\tau_{k-1}}[f(X^y_{\tau_{k-1}})], \end{aligned}$$

then by induction we get

$$\mathcal{E}^g_{0,\tau_k}[f(X^y_{\tau_k})] = f(y).$$

In the following we will prove $\tau_k \to \tau_y$ *a.e.* when $k \to \infty$. Obviously

$$\tau_k \ge \tau_{k-1},$$

so there exists a stopping time $\tau$ s.t. $\tau_k \uparrow \tau$. If $\tau \ne \tau_y$, then there exists $\epsilon > 0$ s.t.

$$\mathrm{dist}(X^y_{\tau_k}, \partial U_x) \ge \epsilon, \quad \text{for any } k.$$

Let $r_k = r(X^y_{\tau_k})$, according to the condition (4.3), we get there exists $r > 0$,

$$r_k \ge r, \qquad \text{for any } k.$$

It means

$$\mathrm{dist}(X^y_{\tau_k}, X^y_{\tau_{k-1}}) \ge r.$$

And since $X_t^y$ is continuous, then $\tau_k \to \infty$, which implies $\tau_y = \infty$. So

$$P(\tau_k \text{ don't converge to } \tau_y) \leq P(\tau_y = \infty).$$

But for (1.6) is elliptic and $U_x$ is bounded, we have $P(\tau_y < \infty) = 1$. So

$$P(\tau_k \text{ converge to } \tau_y) = 1.$$

Then we get

$$\begin{aligned} f(y) &= \mathcal{E}^g_{0,\tau_k}[f(X^y_{\tau_k})] \\ &= \lim_{k\uparrow\infty} \mathcal{E}^g_{0,\tau_k}[f(X^y_{\tau_k})] \\ &= \mathcal{E}^g_{0,\tau_y}[f(X^y_{\tau_y})]. \end{aligned}$$

So we have finished the proof. □

## Acknowledgement

The first author is supported by the National Natural Science Foundation of China (11026125) ,and the third author is supported by the BIGC Key Project (Ea201606).

## Conflict of Interest

All authors declare no conflicts of interest in this paper.

## References

1. G. Barles and E. Lesigne, *SDE, BSDE and PDE.* Pitman Research Notes in Mathematics Series, **364**, Backward Stochastic Differential Equation, Ed. by N. El Karoui and L.Mazliak (1997), 47-80.
2. Z. Chen and S. Peng, *Continuous properties of g-martingales.* Chin. Ann. of Math, **22** (2001), 115-128.
3. M. G. Crandall, H. Ishii and P. L. Lions, *User's guide to viscosity solutions of second order Partial differential equations.* Bull. Amer. Math. Soc., **27** (1992), 1-67.
4. L. Jiang, *Convexity, translation invariance and subadditivity for g-expectations and related risk measures.* The Annals of Applied Probability, **18** (2008), 245-258.
5. O. D. Kellogg, *Converses of Gauss' theorem on the arithmetic mean.* Tran. Amer. Math. Soc., **36** (1934), 227-242.
6. B. Øksendal, Stochastic differential Equations, Sixth Edition, Springer, Berlin, 2003.
7. B. Øksendal and D. W. Stroock, *A characterization of harmonic measure and markov processs whose hitting distritributions are preserved by rotations, translations and dilatations.* Ann. Inst. Fourier. **32** (1982), 221-232.
8. S. Peng, *A generalized dynamic programming principle and Hamilton-Jacobi-Bellman equation.* Stochastics and Stochastic Reports, **38** (1992), 119-134.

9. S. Peng, *BSDE and related g-expectation*. Pitman Research Notes in Mathematics Series, **364**, Backward Stochastic Differential Equation, Ed. by N. El Karoui and L.Mazliak (1997), 141-159.

10. S. Peng, *Monotonic limit theorem of BSDE and nonlinear decomposition theorem of Doob-Meyer's type*. Prob. Theory Rel. Fields, **113** (1999), 473-499.

11. S. Peng, *Nonlinear expectations, nonlinear evaluations and risk measures.* Stochastic Methods in Finance. Lecture Notes in Mathematics Series. **1856** (2004), 165-253.

12. W. Wang, *Maximal inequalities for g-martingales*. Statist. Probab. Lett., **79** (2009), 1169-1174.

# 7

# The viscosity solutions of a nonlinear equation related to the *p-Laplacian*

**Qitong Ou*and Huashui Zhan**

School of Applied Mathematics, Xiamen University of Technology, Xiamen 361024, P. R. China

* **Correspondence:** ouqitong@xmut.edu.cn

**Abstract:** The viscosity solutions of a nonlinear equation related to the *p-Laplacian* are considered. Besides there is a damping term in the equation, a nonlocal function is added. By considering the regularized problem and using Moser iteration technique, we get the uniformly local bounded properties of the solutions and the $L^p$-norm for the gradients. By the compactness theorem, we prove the existence of the viscosity solution of the equation.

**Keywords:** Nonlinear equation; *p-Laplacian*; moser iteration; viscosity solution
**Mathematics Subject Classification**: 35K55, 35K65, 35B40.

---

## 1. Introduction

The objective of the paper is to study the nonnegative weak solutions of nonlinear parabolic equation with the type

$$u_t = \mathrm{div}(|\nabla u^m|^{p-2}\nabla u^m) - a(x)u^{mq_1}|\nabla u^m|^{p_1} + f_0(u^m)\int_\Omega K(y)|u^m(y,t)|^\beta dy + g(x), \quad \text{in } S = \Omega\times(0,\infty), \tag{1.1}$$

$$u(x,0) = u_0(x), \quad x \in \Omega, \tag{1.2}$$

$$u(x,t) = 0, \quad (x,t) \in \partial\Omega \times (0,\infty), \tag{1.3}$$

where $\Omega \subset \mathbb{R}^N$ is a bounded open domain with smooth boundary $\partial\Omega$, $\int_\Omega K(y)|u(y,t)|^\beta dy$ represents a nonlocal function dependent on spatial domain $\Omega$, $a(x) \geq 0$ is a bounded function, $K(x)$ and $g(x)$ are bounded functions too, and $\nabla$ is the spatial gradient operator. We assume that $p > 1$, $m > 1$, $p_1 \leq 2$, $p > 2p_1$, $N \geq 1$,

$$0 \leq u_0^m(x) \in L^{q-1+\frac{1}{m}}(\Omega), q > 1, \ |f_0(s)| \leq c|s|^{\frac{1}{m}}, \ s \in R^1 = (-\infty, \infty). \tag{1.4}$$

As usual, the here and after, the constants $c$ may be different from one to another. The equation with the type of (1.1) has been suggested as the mathematical model for a variety of problems in mechanics, physics and biology, which can be found in [10, 11, 15, 17] et al. Equation (1.1) has been widely researched, whether it is linear or nonlinear, is uniformly parabolic or degenerate parabolic. In what follows, we only give a very roughly review.

If $a(x) = g(x) = f_0 \equiv 0$, the existence of nonnegative solution of the problem (1.1)-(1.3), defined in weak sense, is well established (see [10], [6] et al.).

If $g(x) = f \equiv 0$, some special cases of equation (1.1) had been researched by Bertsh [3], Zhou [36] and Zhang [34] et al. For examples, the existence and the properties of the viscosity solution to the following equation are obtained in [3, 36]

$$u_t = u\Delta u - \gamma|\nabla u|^2, \tag{1.5}$$

where $\gamma$ is a positive constant. The existence and the properties of the viscosity solution to the following equation are obtained in [34]

$$u_t = \Delta u - b(x)|u|^{q-1}|\nabla u|^2, \tag{1.6}$$

where $b(x)$ is a known function. The most important characteristic of the equation (1.5) or (1.6) lies in that, generally, the uniqueness of the solutions is not true, one can refer to [4,9,29,34,36] for the details. Thus, for the equation with the type of (1.1), one mainly concerns with the existence of the viscosity solution and the related properties such as the large time behavior, one can refer to [8, 20, 33, 35] et al. for some progresses in the direction.

But if $p_1 = 0$, it is well-known the uniqueness of the solutions is true. Aassila [1] studied equation (1.1) when $p = 2, m = 1$ and proved the existence of solution by Schauder fixed point theorem, studied the convergence of the solution towards a steady state by using the point of view in dynamical systems. Cholewa and Dlotko [7], Teman [28] considered the following problem

$$u_t - \operatorname{div}(|\nabla u|^{p-2}\nabla u) + |u|^\alpha u = f_0(u) + g(x), \tag{1.7}$$

and proved the existence of global attractor in $L^2$ which is in fact a bounded set in $W_0^{1,p} \bigcap L^{\alpha+2}$. Chen [20] studied the long time behavior of solutions for following equation

$$u_t - \operatorname{div}(|\nabla u|^{p-2}\nabla u) + a(x)|u|^\alpha u = f_0(u)\int_\Omega K(y)|u(y,t)|^\beta dy + g(x), \tag{1.8}$$

and obtained the existence and $L^p$ estimate of the global attractor.

While the papers, first by Nakao-Chen [25] and later by Chen-Wang [6], had studies the global existence and the gradient estimate for the quasilinear parabolic equation of $m$-Laplacian type with a nonlinear convection term, the typical equations included in [6, 25] are with the form as

$$u_t = \operatorname{div}(u^r|\nabla u|^{p-2}\nabla u) + \nabla A(u). \tag{1.9}$$

In our paper, we will study the global solution of equation (1.1) with the initial value (1.2) and homogeneous boundary value (1.3) by the usual regularized method. The main techniques are inspired by [6, 25]. However, due to the local and the nonlocal nonlinearity of the equation we considered, even to prove the initial value condition, we have to put some restrictions in the exponents of $m, p, p_1, q_1$. In particular, as we have said, instead of the nonlinear convection term $\nabla A(u)$ in equation (1.9), equation (1.1) contains the damping term $-a(x)u^{mq_1}|\nabla u^{mp_1}|$, the uniqueness of the solutions generally is not true.

We can only prove the uniqueness of the solutions under the condition $p_1 = 0$. If $p_1 \neq 0$ we only can prove the uniqueness of the viscosity solutions. At the same time, comparing with [5], since equation (1.1) is more complicated, how to get the estimate in the gradient term of the solution, and how to prove the continuity of the solution etc, become more difficult. A clear promotion lies in that we put not any restrictions in the derivative $f_0'(s)$ of the function $f_0(s)$, while it must satisfy that $|f_0'(s)| \leq c|s|^{r-1}$ in [5]. Other related works on equation (1.1), one can refer to the references [2,14,16,18,19,22,24,27,30–32] et al.

Now we quote the following definition.

**Definition 1.1.** *A nonnegative function $u(x,t)$ is called a weak solution of (1.1)-(1.3) if $u$ satisfies*

*(i)*

$$u \in L^{\infty}_{loc}(0,\infty; L^{\infty}(\Omega)), \tag{1.10}$$

$$u_t \in L^2_{loc}(0,\infty; L^2(\Omega)), \quad u^m \in L^{\infty}_{loc}(0,\infty; W_0^{1,p}(\Omega)), \tag{1.11}$$

*(ii)*

$$\iint_S \left[u\varphi_t - |\nabla u^m|^{p-2}\nabla u^m \cdot \nabla\varphi - a(x)u^{mq_1}|\nabla u^m|^{p_1}\varphi\right]dxdt$$

$$+\iint_S \left[f_0(u^m)\int_\Omega K(y)|u^m(y,t)|^\beta dy + g(x)\right]\varphi dxdt = 0, \ \forall\varphi \in C_0^1(S); \tag{1.12}$$

*(iii)*

$$\lim_{t\to 0}\int_\Omega |u(x,t) - u_0(x)|\, dx = 0. \tag{1.13}$$

We are to get the solution of problem (1.1)-(1.3) by considering the regularized equation

$$u_t = \operatorname{div}((|\nabla u^m|^2 + \frac{1}{k})^{\frac{p-2}{2}}\nabla u^m) - a(x)u^{mq_1}|\nabla u^m|^{p_1} + f_0(u^m)\int_\Omega K(y)|u^m(y,t)|^\beta dy + g(x), \tag{1.14}$$

with the initial value (1.2) and the homogeneous boundary value (1.3). Here $0 \leq u_{0k}(x)$ is a suitable smooth function such that $u_{0k}(x) \in L^\infty(\Omega)$, $\lim_{k\to\infty}\|u_{0k}^m\|_{q-1+\frac{1}{m}} = \|u_0^m\|_{q-1+\frac{1}{m}}$.

**Definition 1.2.** *If $u_k$ is the solution of the initial boundary value problem of (1.14)-(1.2)-(1.3), $\lim_{k\to\infty} u_k = u$, a.e in $S$, $u$ is a weak solution of (1.1)-(1.3), then $u$ is said to be a viscosity solution.*

We need some important lemmas in order to get our results.

**Lemma 1.1.** *If $1 \leq l < N$, $1+\beta \leq q$, $1 \leq r \leq q \leq (1+\beta)Nl/(N-l)$, $u^{1+\beta} \in W^{1,l}(\Omega)$, then*

$$\|u\|_q \leq c^{1/(1+\beta)}\|u\|_r^{1-\theta}\|u^{1+\beta}\|_{1,l}^{\theta/(1+\beta)}, \tag{1.15}$$

*where $\theta = (\beta+1)(r^{-1} - q^{-1})/(N^{-1} - l^{-1} + (\beta+1)r^{-1})$.*

This lemma is a general version of Gagliardo-Nirenberg inequality, it is first proved by M. Nakao [23].

**Lemma 1.2.** *Let $y(t)$ be a nonnegative function on $(0,T]$. If it satisfies*

$$y'(t) + At^{\lambda\theta-1}y^{1+\theta}(t) \leq Bt^{-k}y(t) + Ct^{-\delta}, 0 < t \leq T, \tag{1.16}$$

*where $A, \theta > 0$, $\lambda\theta \geq 1$, $B, C \geq 0, k \leq 1$, then*

$$y(t) \leq A^{-\frac{1}{\theta}}(2\lambda + 2BT^{1-k})^{\frac{1}{\theta}}t^{-\lambda} + 2C(\lambda + BT^{1-k})^{-1}t^{1-\delta}, 0 < t \leq T. \tag{1.17}$$

This lemma can be found in [26].

**Lemma 1.3.** *Suppose* $L_1 \geq 1$, $r, R, M > 0$, $\lambda_1 > 0$. *For* $n = 2, 3, \cdots$, *let*

$$L_n = RL_{n-1} - M,\ \theta_n = NR(1 - L_{n-1}L_n^{-1})(N(R-1)+r)^{-1},$$

$$\beta_n = (L_n + M)\theta_n^{-1} - L_n,\ \lambda_n = (1 + \lambda_{n-1}(\beta_n - M))\beta_n^{-1}.$$

*Then*

$$\lim_{n\to\infty} \lambda_n = \frac{L_1\lambda_1 r + N}{l_1 + MN}. \tag{1.18}$$

This lemma also was first proved in [25], then used in [6].

In our paper, we assume that $p > 1 + \frac{1}{m}$, so equation (1.1) is a doubly degenerate parabolic equation. By considering the solution $u_k$ of the regularized problem (1.14)-(1.2)-(1.3) and using Moser iteration technique, we get $u_k$'s local bounded properties and the local bounded properties of the $L^p$-norm of the gradient $\nabla u_k$. By the compactness theorem, we get the existence of the viscosity solution of the diffusion equation itself. In details, we will prove the following theorems.

**Theorem 1.1.** *It is supposed that* $K$, $g$ *are suitable smooth bounded functions,* $a(x) \in C(\overline{\Omega})$ *and exists* $a_0 > 0$, *such that* $a(x) \geq a_0$ *in* $\Omega$, $f_0$ *satisfies (1.4). If* $p > 1 + \frac{1}{m}$, $u_0(x) \geq 0$,

$$u_0^m(x) \in L^{q-1+\frac{1}{m}}(\Omega), 3 > q > 2 - \frac{1}{m}, \tag{1.19}$$

$$p_1 \leq 2,\ 2p_1 < p,\ \beta < \max\{p - 1 - \frac{1}{m}, q - 1 + \frac{1}{m}\}, \tag{1.20}$$

$$\epsilon = \max\{\frac{mNq_1}{Nm(p-1) - N + mq} + \frac{p_1(m(p-1)+m-2)}{m(p-1)-1}, \frac{(\beta+m)N}{Nm(p-1)-N+mq}\} < 1, \tag{1.21}$$

*then the problem (1.1)-(1.3) has a weak viscosity solution* $u$, *satisfying*

$$u^m \in L^\infty_{\text{loc}}(0, \infty; L^{q+1-\frac{1}{m}}(\Omega)) \bigcap L^\infty_{loc}(0, \infty; W_0^{1,p}(\Omega)), \tag{1.22}$$

*and*

$$\|u^m(t)\|_\infty \leq c(1 + t^{-\lambda})(1+t)^{-1/(p-1-\frac{1}{m})}, t > 0, \tag{1.23}$$

*where* $\lambda = N(pq + (p - 1 - \frac{1}{m})N)^{-1}$. *Moreover, if* $p > 2$, *then*

$$\|\nabla u^m\|_p \leq c(1 + t^{-\delta_1})(1+t)^{-\sigma}, t > 0, \tag{1.24}$$

*where*

$$\delta_1 = \max\{1 + \frac{m-1}{m(p-1)-1}, \delta - 1\}, \delta = \max\{\frac{m+1}{m}, 2\beta\},$$

*and*

$$\sigma = \frac{p[m(2q_1+1)-1] + mp_1}{[m(p-1)-1](p-p_1)}.$$

**Remark 1.1.** *The condition (1.21) is only used to prove (1.13). We conjecture that this condition can be weaken.*

**Theorem 1.2.** *Let* $u$ *be a nonnegative weak solution of problem (1.1)-(1.3). If* $g(x) \leq 0$, $f_0'(s) \geq 0$, *if* $p > 1 + \frac{1}{m}$, $p_1 + q_1 > (p-1)$ *then*

$$\text{supp}u(., s) \subset \text{supp}u(., t), \tag{1.25}$$

*for all* $s, t$ *with* $0 < s < t$.

## 2. The $L^\infty$ estimation of the solution

Instead of considering the regularized problem (1.14)-(1.2)-(1.3) directly as one deals with the case $m = 1$, we have to consider the following approximate problem. For small $s > 0$, we consider

$$u_t = \operatorname{div}((|\nabla u^m|^2 + \frac{1}{k})^{\frac{p-2}{2}}\nabla u^m) - a(x)u^{mq_1}|\nabla u^m|^{p_1} + f_0(u^m)\int_\Omega K(y)|u^m(y,t)|^\beta dy + g(x), \tag{2.1}$$

$$u(x,0) = u_{0k}(x) + s, x \in \Omega, \tag{2.2}$$

$$u(x,t) = s, x \in \partial\Omega, t \geq 0, \tag{2.3}$$

where $0 \leq u_{0k}(x)$ is a suitable smooth function such that $u_{0k}(x) \in L^\infty(\Omega)$, $\lim_{k\to\infty} \|u_{0k}^m\|_{q-1+\frac{1}{m}} = \|u_0^m\|_{q-1+\frac{1}{m}}$.

Similar as the chapter 8 of [13], in which the existence of the initial boundary value problem of the quaslinear equation in the divergent form is obtained, by Leray-Schauder fixed point theory, using the condition $p_1 \leq 2$, we know that problem (2.1)-(2.3) has a nonnegative classical solution $u_{ks}$, we omit the details here.

Let $s \to 0$. In a similar way as [33], we are able to prove that

$$u_{ks} \to u_k, \text{ in } C(S),$$

$$\nabla u_{ks}^m \rightharpoonup \nabla u_k^m, \text{ in } L^p(S),$$

$$u_{kst} \rightharpoonup \nabla u_{kt}, \text{ in } L^2(S),$$

$$|\nabla u_{ks}^m|^{p-2}\nabla u_{ksx_i}^m \rightharpoonup * \ |\nabla u_k^m|^{p-2}\nabla u_{kx_i}^m, \text{weakly star in } L^\infty_{\text{loc}}(0,\infty; L^{\frac{p}{p-1}}(\Omega)),$$

and $u_k$ is the solution of equation (2.1) with the following initial boundary values

$$u(x,0) = u_{0k}(x), x \in \Omega, \tag{2.4}$$

$$u(x,t) = 0, x \in \partial\Omega, t \geq 0. \tag{2.5}$$

**Lemma 2.1.** *Assume that*

*($H_1$) $a(x) \in C(\overline{\Omega})$ and exists $a_0 > 0$, such that $a(x) \geq a_0$ in $\Omega$;*

*($H_2$) $f_0(s) \in C(R^1)$, $|f_0(s)| \leq K_0|s|^{\frac{1}{m}}$, for some $K_0 > 0$.*

*($H_3$) $g(x), K(x) \in L^\infty$.*

*In addition, $\beta + \frac{1}{m} < q_1$, $3 > q \geq 2 - \frac{1}{m}$, then $u_k^m \in L^\infty_{loc}(0,\infty; L^{q-1+\frac{1}{m}}(\Omega))$ and*

$$\|u_k^m\|_{q-1+\frac{1}{m}} \leq c(1+t)^{-\frac{1}{p-1-\frac{1}{m}}}, t \geq 0. \tag{2.6}$$

*Proof.* In the proof what follows, we only denote $u_k$ as $u$ for simplicity. We only give the proof of the case $q > 2 - \frac{1}{m}$, if $q = 2 - \frac{1}{m}$, one can get the conclusion just a minor version. Let $A_n = (q-2)n^{3-q}$, $B_n = (3-q)n^{2-q}$, and

$$f_n(s) = \begin{cases} s^{q-1}, & \text{if } s \geq \frac{1}{n}, \\ A_n s^2 + B_n s, & \text{if } 0 \leq s < \frac{1}{n}. \end{cases}$$

Suppose that $n > k$, multiply (2.1) with $f_n(u^m)$ and integrate it on $\Omega$. Since $f'(s) > 0$,then we have

$$\int_\Omega f_n(u^m)\mathrm{div}(|\nabla u^m|^2 + \frac{1}{k})^{\frac{p-2}{2}}\nabla u^m)dx = -\int_\Omega (|\nabla u^m|^2 + \frac{1}{k})^{\frac{p-2}{2}}|\nabla u^m|^2 f_n'(u^m)dx$$

$$\leq -\int_\Omega |\nabla u^m|^p f_n'(u^m)dx = -\int_\Omega |\nabla \int_0^{u^m} (f_n'(s))^{\frac{1}{p}}ds|^p\,dx, \tag{2.7}$$

$$-\int_\Omega a(x)f_n(u^m)u^{mq_1}|\nabla u^m|^{p_1}dx \leq 0. \tag{2.8}$$

Suppose that $| f_0(s) |\leq K_0 s^r$. Then

$$|\int_{\Omega\bigcap\{u^m\leq\frac{1}{n}\}} f_0(u^m)f_n(u^m)\int_\Omega K(y)|u^m(y,t)|^\beta dydx|$$

$$\leq c(K)\int_{\Omega\bigcap\{u^m\leq\frac{1}{n}\}} u^{mr}(A_n u^{2m} + B_n u^m)dx\int_\Omega |u|^{m\beta}dy$$

$$\leq c(K)n^{1-q-r}\int_\Omega |u|^{m\beta}dy \leq c(K)n^{1-q-r}||u^m||^\beta_{q-1+\frac{1}{m}}.$$

If $r = \frac{1}{m}$,

$$|\int_{\Omega\bigcap\{u^m>\frac{1}{n}\}} f_0(u^m)f_n(u^m)\int_\Omega K(y)|u^m(y,t)|^\beta dydx|$$

$$\leq c(K)\int_\Omega u^{m(r+q-1)}dx\int_\Omega |u|^{m\beta}dy \leq c||u^m||^{q-1+\frac{1}{m}+\beta}_{q-1+\frac{1}{m}},$$

we have

$$|\int_\Omega f_0(u^m)f_n(u^m)\int_\Omega K(y)|u^m(y,t)|^\beta dydx|$$

$$\leq c||u^m||^\beta_{q-1+\frac{1}{m}}[n^{1-q-\frac{1}{m}} + ||u^m||^{q-1+\frac{1}{m}}_{q-1+\frac{1}{m}}]. \tag{2.9}$$

$$|\int_{\Omega\bigcap\{u^m>\frac{1}{n}\}} f_n(u^m)g(x)dx| \leq c(g)\int_\Omega u^{m(q-1)}dx \leq c(g)||u^m||^{q-1}_{q-1+\frac{1}{m}}. \tag{2.10}$$

From the above calculations, we have

$$\int_\Omega f_n(u^m)u_t dx + \int_\Omega |\nabla \int_0^{u^m} (f_n'(s))^{\frac{1}{p}}ds|^p\,dx \leq c||u^m||^{q-1+\frac{1}{m}+\beta}_{q-1+\frac{1}{m}} + O(\frac{1}{n^{q-1}}), \tag{2.11}$$

by Poincare inequality, we have

$$\int_\Omega f_n(u^m)u_t dx + c\int_\Omega |\int_0^{u^m} (f_n'(s))^{\frac{1}{p}}ds|^p\,dx \leq c||u^m||^{q-1+\frac{1}{m}+\beta}_{q-1+\frac{1}{m}} + O(\frac{1}{n^{q-1}}). \tag{2.12}$$

Let $n \to \infty$ in (2.12). We can deduce that

$$\frac{d}{dt}\int_\Omega u^{m(q-1)+1}dx + c\int_\Omega u^{m[q-1+\frac{1}{m}+p-1-\frac{1}{m}]}dx \leq c||u^m||^{q-1+\frac{1}{m}+\beta}_{q-1+\frac{1}{m}}. \tag{2.13}$$

By Jessen inequality, from (2.13) we get

$$\frac{d}{dt}\|u^m\|_{q-1+\frac{1}{m}}^{q-1+\frac{1}{m}} + c\|u^m\|_{q-1+\frac{1}{m}}^{q-1+\frac{1}{m}+p-1-\frac{1}{m}} \le c\|u^m\|_{q-1+\frac{1}{m}}^{q-1+\frac{1}{m}+\beta}.$$

If

$$\beta < p - 1 - \frac{1}{m}$$

by young inequality,

$$\frac{d}{dt}\|u^m\|_{q-1+\frac{1}{m}}^{q-1+\frac{1}{m}} + c\|u^m\|_{q-1+\frac{1}{m}}^{q-1+\frac{1}{m}+p-1-\frac{1}{m}} \le c,$$

then

$$\|u^m\|_{q+1-\frac{1}{m}} \le c(1+t)^{-\frac{1}{p-1-\frac{1}{m}}}.$$

We get the desired result. □

**Lemma 2.2.** *If $p > 1 + \frac{1}{m}$, $u_k$ is the solution of problem (2.1)-(2.4)-(2.5), then*

$$\|u_k^m\|_\infty \le ct^{-\lambda},\ 0 < t \le 1, \tag{2.14}$$

$$\|u_k^m\|_\infty \le c(1+t)^{-\frac{1}{p-1-\frac{1}{m}}}, t \ge 1, \tag{2.15}$$

*where* $\lambda = \frac{N}{(p-1-\frac{1}{m})N+qp}$.

*Proof.* Multiply (2.1) with $u^{m(l-1)}$, and integrate it on $\Omega$, then

$$\begin{aligned}
\int_\Omega u^{m(l-1)}u_t dx &= \int_\Omega \mathrm{div}(|\nabla u^m| + \frac{1}{k})^{\frac{p-2}{2}}\nabla u^m)u^{m(l-1)}dx - \int_\Omega a(x)u^{mq_1}|\nabla u^m|^{p_1}u^{m(l-1)}dx \\
&+ \int_\Omega f_0(u^m)u^{m(l-1)}\int_\Omega K(y)|u^m(y,t)|^\beta dydx + \int_\Omega g(x)u^{m(l-1)}dx \\
&= -(l-1)\int_\Omega (|\nabla u^m| + \frac{1}{k})^{\frac{p-2}{2}}|\nabla u^m|^2 u^{m(l-2)}dx - \int_\Omega a(x)u^{mq_1}|\nabla u^m|^{p_1}u^{m(l-1)}dx \\
&+ \int_\Omega K(y)|u^m(y,t)|^\beta dy\int_\Omega f_0(u^m)u^{m(l-1)}dx + \int_\Omega g(x)u^{m(l-1)}dx \\
&\le -(l-1)\int_\Omega (|\nabla u^m| + \frac{1}{k})^{\frac{p-2}{2}}|\nabla u^m|^2 u^{m(l-2)}dx \\
&+c(K)\int_\Omega |u^m(y,t)|^\beta dy\int_\Omega u^{m(l-1)+1}dx + c(g)\int_\Omega u^{m(l-1)}dx,
\end{aligned}$$

which deduces that

$$\begin{aligned}
\frac{d}{dt}\|u^m\|_{l-1+\frac{1}{m}}^{l-1+\frac{1}{m}} + c(l-1+\frac{1}{m})^{2-p}\int_\Omega |\nabla u^{m\frac{p+l-1+\frac{1}{m}-1-\frac{1}{m}}{p}}|^p dx &\le c\|u^m\|_{l-1+\frac{1}{m}}^{l-1+\frac{1}{m}}\|u^m\|_{q-1+\frac{1}{m}}^{q-1+\frac{1}{m}+\beta} + c\|u^m\|_{l-1+\frac{1}{m}}^{l-1} \\
&\le \|u^m\|_{l-1+\frac{1}{m}}^{l-1+\frac{1}{m}} + c\|u^m\|_{l-1+\frac{1}{m}}^{l-1}, \text{(by (2.6))}.
\end{aligned}$$

Set $L = l - 1 + \frac{1}{m}$. Then

$$\frac{d}{dt}\|u^m\|_L^L + cL^{2-p}\int_\Omega |\nabla u^{m\frac{L+p-1-\frac{1}{m}}{p}}|^p\, dx \le c\|u^m\|_L^{L+\beta} + c\|u^m\|_L^{L-\frac{1}{m}}, \tag{2.16}$$

where $c$ is a constant independent of $l$.

Now, if we choose $L_1 = q - 1 + \frac{1}{m}$, $L_n = rL_{n-1} - (p - 1 - \frac{1}{m})$, $\theta_n = rN(1 - L_{n-1}L_n^{-1})(p + N(r-1))^{-1}$, $\mu_n = (L_n + p - 1 - \frac{1}{m})\theta_n^{-1} - L_n$, $r > 1 + (p - 1 - \frac{1}{m})q^{-1}$, $n = 2, 3, \cdots$, by Lemma 1.3, we have

$$\|u^m\|_{L_n} \le c^{p/(L_n+p-1-\frac{1}{m})}\|u^m\|_{L_{n-1}}^{1-\theta_n}\|\nabla u^{m(L_n+p-1-\frac{1}{m})/p}\|_p^{p\theta_n/(p-1-\frac{1}{m}+L_n)}. \tag{2.17}$$

If we choose $L = L_n$ in (2.16), by (2.17), we have

$$\frac{d}{dt}\|u^m\|_{L_n}^{L_n} + c^{-p/\theta_n}L_n^{2-p}\|u^m\|_{L_n}^{L_n+\mu_n}\|u^m\|_{L_{n-1}}^{p-1-\frac{1}{m}-\mu_n} \le c\|u^m\|_{L_n}^{L_n+\beta} + c\|u^m\|_{L_n}^{L_n-\frac{1}{m}}.\ 0 < t \le 1. \tag{2.18}$$

We will prove that there exist two bounded sequences $\{\xi_n\}$, $\{\lambda_n\}$ such that

$$\|u^m\|_{L_n} \le \xi_n t^{-\lambda_n},\ 0 < t \le 1. \tag{2.19}$$

Without loss of the generality, we may assume that $\|u^m\|_{L_n} \ge 1$. Otherwise, choosing $\xi_n \equiv 1$, (2.17) is true naturally. Thus, by (2.16), we have

$$\frac{d}{dt}\|u^m\|_{L_n}^{L_n} + c^{-p/\theta_n}L_n^{2-p}\|u^m\|_{L_n}^{L_n+\mu_n}\|u^m\|_{L_{n-1}}^{p-1-\frac{1}{m}-\mu_n} \le c\|u^m\|_{L_n}^{L_n+\beta}.\ 0 < t \le 1.$$

If $n = 1$, by Lemma 2.1, $\lambda_1 = 0, \xi_1 = \sup_{t\ge 0}\|u^m(t)\|_{q-1+\frac{1}{m}}$ makes (2.19) sure. If (2.19) is true for $n - 1$, from (2.18),

$$\frac{d}{dt}\|u^m\|_{L_n}^{L_n} + c^{-p/\theta_n}L_n^{2-p}\|u^m\|_{L_n}^{L_n+\mu_n}\xi_{n-1}^{p-1-\frac{1}{m}-\mu_n}t^{-(p-1-\frac{1}{m}-\mu_n)\lambda_{n-1}} \le c\|u^m\|_{L_n}^{L_n+\beta}.\ 0 < t \le 1. \tag{2.20}$$

we can choose

$$\lambda_n = (\lambda_{n-1}(\mu_n - p + 1 + \frac{1}{m}) + 1)\mu_n^{-1},\ \xi_n = \xi_{n-1}(c^{p/\theta_n}L_n^{p-1}\lambda_n)^{1/\mu_n},\ n = 2, 3, \cdots,$$

$$\frac{d}{dt}\|u^m\|_{L_n}^{L_n} + c\|u^m\|_{L_n}^{L_n+\lambda_n} \le c\|u^m\|_{L_n}^{L_n+\beta}.\ 0 < t \le 1. \tag{2.21}$$

Suppose that

$$\beta < \frac{N}{(p - 1 - \frac{1}{m})N + qp}, \tag{2.22}$$

and notice that as $n \to \infty$, $\lambda_n \to \lambda = \frac{N}{(p-1-\frac{1}{m})N+pq}$.

$$\frac{d}{dt}\|u^m\|_{L_n}^{L_n} + c\|u^m\|_{L_n}^{L_n+\lambda_n} \le 0.\ 0 < t \le 1. \tag{2.23}$$

By Lemma 1.2 and (2.23), we know (2.19) is true.

Moreover, it is easy to see that $\{\xi_n\}$ is bounded. Thus, by Lemma 1.2, (2.14) is true.

To prove (2.15), we set $\tau = \log(1 + t), t \ge 1$, $w(\tau) = (1 + t)^{\frac{1}{p-1-\frac{1}{m}}}u^m(t)$. By (2.16), we have

$$\frac{d}{d\tau}\|w(\tau)\|_L^L + cL^{2-p}\|\nabla w^{\frac{L+p-1-\frac{1}{m}}{p}}\|_p^p \le \frac{L}{p - 1 - \frac{1}{m}}\|w(\tau)\|_L^L + c\|w(\tau)\|_L^{L+\beta},\ \tau \ge \log 2. \tag{2.21}$$

By the lemma 3.1 in [24], we can get (2.15), we omit details here. □

## 3. The $L^\infty$ estimation of the gradient

**Lemma 3.1.** *If $p > \max\{2, 1+\frac{1}{m}\}$, $u_k$ is the solution of problem (2.1)-(2.4)-(2.5), then*

$$\|\nabla u_k^m\|_p \le ct^{-(1+\frac{m-1}{m(p-1)-1})} + ct^{1-\delta},\ 0 < t \le 1, \tag{3.1}$$

$$\|\nabla u_k^m\|_p \le c(1+t)^{-\frac{p(m(2q_1+1)-1)+mp_1}{(m(p-1)-1)(p-p_1)}}, t \ge 1. \tag{3.2}$$

*Here* $\delta = \max\{\frac{m-1}{m}, 2\beta\}$.

*Proof.* Multiply (2.1) with $u_t^m$, and integrate it on $\Omega$, then

$$m\int_\Omega u^{m-1}(u_t)^2 dx = \int_\Omega \mathrm{div}((|\nabla u^m|^2 + \frac{1}{k})^{\frac{p-2}{2}}\nabla u^m)u_t^m dx - \int_\Omega a(x)u^{mq_1}|\nabla u^m|^{p_1}u_t^m dx$$

$$+ \int_\Omega f_0(u^m)u_t^m dx \int_\Omega K(y)|u^m(y,t)|^\beta dy + \int_\Omega g(x)u_t^m dx. \tag{3.3}$$

$$\int_\Omega \mathrm{div}((|\nabla u^m|^2 + \frac{1}{k})^{\frac{p-2}{2}}\nabla u^m)u_t^m dx = -\int_\Omega (|\nabla u^m| + \frac{1}{k})^{\frac{p-2}{2}}\nabla u^m \nabla u_t^m dx$$

$$= -\frac{1}{2}\int_\Omega (|\nabla u^m|^2 + \frac{1}{k})^{\frac{p-2}{2}}|\nabla u^m|_t^2 dx$$

$$= -\frac{1}{2}\int_\Omega \frac{d}{dt}\int_0^{|\nabla u^m|^2}(s+\frac{1}{k})^{\frac{p-2}{2}}ds dx = -\frac{1}{2}\frac{d}{dt}\Gamma_k(|\nabla u^m|^2), \tag{3.4}$$

where we define that

$$\Gamma_k(|\nabla u^m|^2) = \int_\Omega\int_0^{|\nabla u^m|^2}(s+\frac{1}{k})^{\frac{p-2}{2}}ds dx.$$

At the same time,

$$|-a(x)u^{mq_1}|\nabla u^m|^{p_1}u_t^m dx| \le \frac{m}{2}\int_\Omega u^{m-1}(u_t)^2 dx + c\int_\Omega |u^m|^{2q_1+\frac{m-1}{m}}|\nabla u^m|^{2p_1}dx. \tag{3.5}$$

By Lemma 2.1, using Young inequality and Hölder inequality,

$$|\int_\Omega f_0(u^m)u_t^m dx \int_\Omega K(y)|u^m(y,t)|^\beta dy|$$

$$\le c(\varepsilon\int_\Omega u^{m-1}(u_t)^2 dx + c\int_\Omega u^{m+1}dx)\|u^m\|^\beta_{q-1+\frac{1}{m}}$$

$$\le c\varepsilon\int_\Omega u^{m-1}(u_t)^2 dx + c\int_\Omega u^{m+1}dx$$

$$|\int_\Omega g(x)u_t^m dx| \le \varepsilon\int_\Omega u^{m-1}(u_t)^2 dx + c\int_\Omega u^{m-1}dx.$$

By (3.3)-(3.5), we have

$$\int_\Omega u^{m-1}(u_t)^2 dx + \frac{1}{m}\frac{d}{dt}\Gamma_k(|\nabla u^m|^2) \le c\int_\Omega |u^m|^{2q_1+\frac{m-1}{m}}|\nabla u^m|^{2p_1}dx + c\int_\Omega u^{m+1}dx + c\int_\Omega u^{m-1}dx. \tag{3.6}$$

Multiply (2.1) with $u^m$, and integrate it on $\Omega$, then

$$\frac{1}{m+1}\int_\Omega \frac{d}{dt}u^{m+1}dx = \int_\Omega \operatorname{div}((|\nabla u^m|^2+\frac{1}{k})^{\frac{p-2}{2}}\nabla u^m)u^m dx - \int_\Omega a(x)u^{mq_1}|\nabla u^m|^{p_1}u^m dx$$
$$+\int_\Omega f_0(u^m)u^m\int_\Omega K(y)|u^m(y,t)|^\beta dydx + \int_\Omega g(x)u^m dx$$
$$= -\int_\Omega (|\nabla u^m|^2+\frac{1}{k})^{\frac{p-2}{2}}|\nabla u^m|^2 dx - \int_\Omega a(x)u^{mq_1}|\nabla u^m|^{p_1}u^m dx$$
$$+\int_\Omega f_0(u^m)u^m\int_\Omega K(y)|u^m(y,t)|^\beta dydx + \int_\Omega g(x)u^m dx$$

and

$$\Gamma_k(|\nabla u^m|^2) \le \int_\Omega (|\nabla u^m|^2+\frac{1}{k})^{\frac{p-2}{2}}|\nabla u^m|^2 dx$$
$$= -\frac{1}{m+1}\int_\Omega \frac{d}{dt}u^{m+1}dx - \int_\Omega a(x)u^{mq_1}|\nabla u^m|^{p_1}u^m dx + \int_\Omega f_0(u^m)u^m\int_\Omega K(y)|u^m(y,t)|^\beta dydx + \int_\Omega g(x)u^m dx$$
$$\le \frac{1}{m+1}\|u^{\frac{m+1}{2}}\|_2\|u^{\frac{m-1}{2}}u_t\|_2 + c(K)\int_\Omega |u^m(y,t)|^\beta dy\int_\Omega u^{m+1}dx + c(g)\int_\Omega u^m dx,$$

so

$$\frac{1}{m}\frac{d}{dt}\Gamma_k(|\nabla u^m|^2) + (m+1)^2\|u^{\frac{m+1}{2}}\|_2^{-2}\Gamma_k^2(|\nabla u^m|^2)$$
$$\le c\int_\Omega |u^m|^{2q_1+\frac{m-1}{m}}|\nabla u^m|^{2p_1}dx + c\int_\Omega u^{m+1}dx + c\int_\Omega u^{m-1}dx$$
$$+c\|u^{\frac{m+1}{2}}\|_2^{-2}(\int_\Omega |u^m(y,t)|^\beta dy\int_\Omega u^{m+1}dx + \int_\Omega u^m dx)^2$$
$$\le c\int_\Omega |u^m|^{2q_1+\frac{m-1}{m}}|\nabla u^m|^{2p_1}dx + c\int_\Omega u^{m+1}dx + c\int_\Omega u^{m-1}dx$$
$$+c(\int_\Omega |u^m(y,t)|^\beta dy)^2\int_\Omega u^{m+1}dx + c\|u^{\frac{m+1}{2}}\|_2^{2\frac{m}{m+1}-2}. \tag{3.7}$$

Setting $2\gamma = 2q_1 + 1 - \frac{1}{m}$, for $\forall a \in [0, 2\gamma]$, if we notice that $p > 2p_1$, then we have

$$\int_\Omega |u^m|^{2a}|\nabla u^m|^{2p_1}dx \le \|u^m(t)\|_\infty^a\left(\int_\Omega |u^m|^{\frac{(2\gamma-a)p}{p-2p_1}}dx\right)^{\frac{p-2p_1}{p}}\|\nabla u^m\|_p^{2p_1}. \tag{3.8}$$

If $2\gamma \ge (p-2p_1)(N+1)/N$, let $a = (2\gamma - (p-2p_1)(1+\frac{q}{N}))^+$. By Lemma 1.3,

$$\left(\int_\Omega |u^m|^{\frac{(2\gamma-a)p}{p-2p_1}}dx\right)^{\frac{p-2p_1}{p}} \le c\|u^m(t)\|_s^{(2\gamma-a)(1-\theta)}\|\nabla u^m\|_p^{p-2p_1}, \tag{3.9}$$

where $\theta = (s^{-1} - (1-\frac{2p_1}{p})(2\gamma-a)^{-1})/(N^{-1}-p^{-1}+s^{-1})$, and $s = (2\gamma - p + 2p_1 - a)N/(p-2p_1)$ when $2\gamma \ge (p-2p_1)(1+q/N)$, $s = q$ when $(p-2p_1)(1+N^{-1}) \le 2\gamma \le (p-2p_1)(1+q/N)$. By Lemma 2.1 and Lemma 2.2, from (3.8), we have

$$\int_\Omega |u^m|^{2a}|\nabla u^m|^{2p_1}dx \le ct^{-\lambda a}\|\nabla u^m\|_p^p \le ct^{-\lambda a}\Gamma_k(|\nabla u^m|^2).\ 0 < t \le 1. \tag{3.10}$$

At the same time, if we choose $q = 2$ in Lemma 2.1, we have

$$\|u^m\|_{1+\frac{1}{m}} = \left(\int_\Omega u^{m+1}dx\right)^{\frac{m}{m+1}} \le c(1+t)^{-(p-1-\frac{m}{m+1})^{-1}} \le c,$$

and

$$\int_\Omega u^{m-1}dx \le ct^{\frac{m-1}{m}\lambda},\ \|u^{\frac{m+1}{2}}\|_2^2 = \int_\Omega u^{m+1}dx \le c. \tag{3.11}$$

By (3.7) and Lemma 2.2, we have

$$\Gamma_k'(t) + ct^{\frac{m+1}{m(p-1)-1}}\Gamma_k^2(t) \le ct^{-\lambda a}\Gamma_k(t) + c(t^{-\lambda\frac{m-1}{m}} + t^{-2\beta\lambda}),\ 0 < t \le 1. \tag{3.12}$$

If $2\gamma < (p-2p_1)(N+1)/N$ and $p - 2p_1 \le 2a \le 2\gamma$,

$$\int_\Omega |u^m|^{2a}|\nabla u^m|^{2p_1}dx \le c\|\nabla u^m\|_1^{2a(1-\theta)}\|\nabla u^m\|_p^{2a\theta+2p_1} \le c\|\nabla u^m\|_p^p \le c\Gamma_k(|\nabla u^m|^2).\ 0 < t \le 1. \tag{3.13}$$

If $2\gamma < (p-2p_1)(N+1)/N$ and $p - 2p_1 \ge 2a \ge 0$,

$$\int_\Omega |u^m|^{2a}|\nabla u^m|^2dx \le c(1 + \|\nabla u^m\|_p^p) \le c(1 + \Gamma_k(|\nabla u^m|^2)).\ 0 < t \le 1. \tag{3.14}$$

(3.13) and (3.14) imply that (3.12) is still true when $2\gamma < (p-2p_1)(N+1)/N$. Using Lemma 1.2,

$$\Gamma_k(t) \le ct^{-(1+\frac{m-1}{m(p-1)-1})} + ct^{1-\delta},\ 0 < t \le 1,$$

where $\delta = \max\{\frac{m+1}{m}, 2\beta\}$. Then (3.1) is true. Now, we will prove (3.2). For $t \ge 1$, by (2.15)

$$\int_\Omega |u^m|^{2a}|\nabla u^m|^{2p_1}dx \le c\|\nabla u^m\|_p^2\|u^m(t)\|_{2\gamma p/p-2p_1}^{2\gamma} \le c(1+t)^{-2\gamma/(p-1-\frac{1}{m})}\|\nabla u^m\|_p^{2p_1}.\ t \ge 1. \tag{3.15}$$

$$\Gamma_k(|\nabla u^m|^2) = \int_0^{|\nabla u^m|^2}(s^2 + \frac{1}{k})^{\frac{p-2}{2}}ds \le c\|\nabla u^m\|_p^p = c(\|\nabla u^m\|_p^{2p_1})^{\frac{p}{2p_1}},\ t \ge 1. \tag{3.16}$$

$$\|u^{\frac{m+1}{2}}\|_2^2 = \left(\int_\Omega u^{m+1}dx\right)^2 \le c(1+t)^{-(p-1-\frac{1}{m})^{-1}},\ t \ge 1. \tag{3.17}$$

by (3.7), using (3.15)-(3.17)

$$\Gamma_k'(t) + c(1+t)^{-(p-1-\frac{1}{m})^{-1}}\Gamma_k^2(t) \le c(1+t)^{2\gamma/(p-1-\frac{1}{m})}(\Gamma_k(t))^{\frac{2p_1}{p}}$$

$$+c\int_\Omega u^{m+1}dx + c\int_\Omega u^{m-1}dx + c(\int_\Omega |u^m(y,t)|^\beta dy)^2\int_\Omega u^{m+1}dx + c\|u^{\frac{m+1}{2}}\|_2^{2(m-1)},$$

by Young inequality,

$$\Gamma_k'(t) + c(1+t)^{-(p-1-\frac{1}{m})^{-1}}\Gamma_k^2(t) \le c(1+t)^{\frac{-m(2\gamma p+p_1)}{(m(p-1)-1)(p-p_1)}} + c(1+t)^{-\frac{m(m+1)}{m(p-1)-1}}$$

$$= c(1+t)^{-\frac{p(m(2q_1+1)-1)+mp_1}{(m(p-1)-1)(p-p_1)}} + c(1+t)^{-\frac{m(m+1)}{m(p-1)-1}},$$

which means (3.2) is true. □

**Lemma 3.2.** *If $p > 1 + \frac{1}{m}$, $u_k$ is the solution of problem (2.1) -(2.4)-(2.5), then*

$$\int_t^T \int_\Omega u_k^{m-1}(u_{kt})^2 dxds \le ct^{-(1+\frac{m-1}{m(p-1)-1})} + ct^{-(\lambda\gamma+\frac{m-1}{m(p-1)-1})} + ct^{-\frac{1+m}{m}\lambda},\ 0 < t \le T. \tag{3.18}$$

*Proof.* From (3.6), (3.10) and (2.14), we have

$$\int_\Omega u^{m-1}(u_t)^2 dx + \frac{1}{m}\frac{d}{dt}\Gamma_k(|\nabla u^m|^2) \le c\int_\Omega |u^m|^{2q_1+\frac{m-1}{m}}|\nabla u^m|^{2p_1}dx + c\int_\Omega u^{m+1}dx + c\int_\Omega u^{m-1}dx$$

$$\begin{aligned}\int_t^T \int_\Omega u^{m-1}(u_t)^2 dxds &\le \Gamma_k(t) + c\int_t^T \int_\Omega |u^m|^{2q_1+\frac{m-1}{m}}|\nabla u^m|^{2p_1}dxds + c\int_t^T \int_\Omega u^{m+1}dx \\ &\le \Gamma_k(t) + c\int_t^T s^{-\lambda(2q_1+\frac{m-1}{m})}\Gamma_k(s)ds + c\int_t^T \int_\Omega u^{m+1}dx \\ &\le ct^{-(1+\frac{m-1}{m(p-1)-1})} + ct^{-(\lambda\gamma+\frac{m-1}{m(p-1)-1})} + ct^{-\frac{1+m}{m}\lambda}.\end{aligned} \tag{3.19}$$

□

## 4. The proof of Theorem 1.1

The proof of Theorem 1.1 from Lemma 2.1, Lemma 2.2, Lemma 3.1 and Lemma 3.2, using the compactness theory (cf [21]), there is a sequence (still denoted it as $\{u_k\}$) of $\{u_k\}$ such that when $k \to \infty$, $u_k \to u$, *a.e.* in $S$ and so

$$\lim_{k\to\infty} f_0(u_k^m)\int_\Omega K(y)|u_k^m(y,t)|^\beta dy = f_0(u^m)\int_\Omega K(y)|u^m(y,t)|^\beta dy.$$

Moreover, we have

$$u_k \rightharpoonup u, \text{weakly* star in } L^\infty_{\rm loc}(0,\infty; L^{m(q-1)+1}(\Omega)), \tag{4.1}$$

$$u_{kt} \rightharpoonup u_t,\ \text{weakly in } L^2(0,\infty; L^2(\Omega)), \nabla u_k^m \rightharpoonup \nabla u^m,\ \text{weakly in } L^p_{\rm loc}(0,\infty; L^p(\Omega)) \tag{4.2}$$

$$|\nabla u_k^m|^{p-2}u_{kx_i}^m \rightharpoonup \chi_i, \text{weakly* in } L^\infty_{\rm loc}(0,\infty; L^{\frac{p}{p-1}}(\Omega)), \tag{4.3}$$

$$a(x)u_k^{mq_1}|\nabla u_k^m|^{p_1} \rightharpoonup \nu, \text{weakly* in } L^\infty_{\rm loc}(0,\infty; L^{\frac{p}{p_1}}(\Omega)), \tag{4.4}$$

where $\chi = \{\chi_i : 1 \le i \le N\}$ and every $\chi_i$ is a function in $L^\infty_{loc}(0,T; L^{\frac{p}{p-1}}(\Omega))$, $\nu \in L^\infty_{loc}(0,\infty; L^{\frac{p}{p_1}}(\Omega))$. (4.1) and (4.2) are clearly true.

In what follows, we only need to prove that

$$\chi = |\nabla u^m|^{p-2}\nabla u^m,\ \text{ in } L^\infty_{\rm loc}(0,\infty; L^{\frac{p}{p-1}}(\Omega)). \tag{4.5}$$

and

$$\nu = a(x)u^{mq_1}|\nabla u^m|^{p_1},\ \text{ in } L^\infty_{\rm loc}(0,\infty; L^{\frac{p}{p_1}}(\Omega)). \tag{4.6}$$

It is easy to know that

$$\iint_S \left(u\varphi_t - \chi\cdot\nabla\varphi - \nu\varphi + f_0(u^m)\int_\Omega K(y)|u^m(y,t)|^\beta dy\varphi + g(x)\varphi\right)dxdt = 0,\ \forall\varphi \in C_0^\infty(S), \tag{4.7}$$

so, if we can prove that

$$\iint_S |\nabla u^m|^{p-2}\nabla u^m\cdot\nabla\varphi dxdt=\iint_S \chi\cdot\nabla\varphi dxdt,\ \forall\varphi\in C_0^1(S); \tag{4.8}$$

$$\iint_S a(x)u^{mq_1}|\nabla u^m|^{p_1}\varphi dxdt=\iint_S \nu\varphi dxdt,\ \ \forall\varphi\in C_0^1(S); \tag{4.9}$$

then (4.5),(4.6) and (1.12) are true.

First, for any $\psi\in C_0^\infty(S)$, $0\le\psi\le 1$; $v^m\in L^p_{\text{loc}}(0,T;W_0^{1,p}(\Omega))$, we have

$$\iint_S \psi(|\nabla u_k^m|^{p-2}\nabla u_k^m-|\nabla v^m|^{p-2}\nabla v^m)\cdot\nabla(u_k^m-v^m)dxdt\ge 0, \tag{4.10}$$

If we multiply with $u_k^m\psi$ on two sides of (2.1), then we have

$$\iint_S \psi\left(|\nabla u_k^m|^2+\frac{1}{k}\right)^{\frac{p-2}{2}}|\nabla u_k^m|^2dxdt=\frac{1}{m+1}\iint_S\psi_t u_k^{m+1}dxdt-\iint_S u_k^m\left(|\nabla u_k^m|^2+\frac{1}{k}\right)^{\frac{p-2}{2}}\nabla u_k^m\cdot\nabla\psi dxdt$$

$$-\iint_S a(x)u_k^{m(q_1+1)}|\nabla u_k^m|^{p_1}\psi dxdt+\iint_S[f_0(u_k^m)\int_\Omega K(y)|u_k^m(y,t)|^\beta dy+g(x)]u_k^m\psi dxdt. \tag{4.11}$$

Noticing that when $1<p<2$,

$$|\nabla u_k^m|^2\ge(|\nabla u_k^m|^2+\frac{1}{k})^{\frac{p}{2}}-(\frac{1}{k})^{\frac{p}{2}},$$

$$(|\nabla u_k^m|^2+\frac{1}{k})^{\frac{p-2}{2}}|\nabla u_k^m|\le(|\nabla u_k^m|^2+\frac{1}{k})^{\frac{p-1}{2}},$$

and when $p\ge 2$,

$$(|\nabla u_k^m|^2+\frac{1}{k})^{\frac{p-2}{2}}|\nabla u_k^m|^2\ge|\nabla u_k^m|^p,$$

$$(|\nabla u_k^m|^2+\frac{1}{k})^{\frac{p-2}{2}}|\nabla u_k^m|\le(|\nabla u_k^m|^{p-1}+1),$$

by (4.10), (4.11), we have

$$\frac{1}{m+1}\iint_S\psi_t u_k^{m+1}dxdt-\iint_S u_k^m\left(|\nabla u_k^m|^2+\frac{1}{k}\right)^{\frac{p-2}{2}}\nabla u_k^m\cdot\nabla\psi dxdt$$

$$-\iint_S a(x)u_k^{m(q_1+1)}|\nabla u_k^m|^{p_1}\psi dxdt+(\frac{1}{k})^{\frac{p-2}{2}}\text{mes}\Omega$$

$$+\iint_S[f_0(u_k^m)\int_\Omega K(y)|u_k^m(y,t)|^\beta dy+g(x)]u_k^m\psi dxdt$$

$$-\iint_S\psi|\nabla u_k^m|^{p-2}\nabla u_k^m\cdot\nabla v^m dxdt-\iint_S\psi|\nabla v^m|^{p-2}\nabla v^m\cdot\nabla(u_k^m-v^m)dxdt\ge 0. \tag{4.12}$$

Since

$$\left(|\nabla u_k^m|^2+\frac{1}{k}\right)^{\frac{p-2}{2}}\nabla u_k^m=|\nabla u_k^m|^{p-2}\nabla u_k^m+\frac{p-2}{2k}\int_0^1(|\nabla u_k^m|^2+\frac{s}{k})^{\frac{p-4}{2}}ds\nabla u_k^m,$$

and

$$\lim_{k\to\infty}\iint_S\int_0^1(|\nabla u_k^m|^2+\frac{s}{k})^{\frac{p-4}{2}}ds\nabla u_k^m\cdot\nabla\psi u_k^m dxdt=0,$$

if we let $k\to\infty$ in (4.12), we have

$$\frac{1}{m+1}\iint_S\psi_t u^{m+1}dxdt-\iint_S u^m v\psi dxdt-\iint_S u^m\chi\nabla\psi dxdt$$
$$-\iint_S\psi\chi\cdot\nabla v^m dxdt-\iint_S\psi|\nabla v^m|^{p-2}\nabla v^m\cdot\nabla(u^m-v^m)dxdt$$
$$+\iint_S[f_0(u^m)\int_\Omega K(y)|u^m(y,t)|^\beta dy+g(x)]u^m\psi dxdt\geq 0. \tag{4.13}$$

Now, we choose $\varphi=\psi u^m$ in (4.7),

$$\frac{1}{m+1}\iint_S\psi_t u^{m+1}dxdt-\iint_S u^m v\psi dxdt-\iint_S\chi\cdot\nabla\psi u^m dxdt$$
$$+\iint_S[f_0(u^m)\int_\Omega K(y)|u^m(y,t)|^\beta dy+g(x)]\psi u^m dxdt=\iint_S\psi\chi\cdot\nabla u^m dxdt.$$

From this formula and (4.13), we have

$$\iint_S\psi(\chi-|\nabla v^m|^{p-2}\nabla v^m)\cdot\nabla(u^m-v^m)dxdt\geq 0. \tag{4.14}$$

Let $v^m=u^m-\lambda\varphi$, $\lambda\geq 0,\varphi\in C_0^\infty(S)$. Then

$$\iint_S\psi(\chi_i-|\nabla(u^m-\lambda\varphi)|^{p-2}(u^m-\lambda\varphi)_{x_i})dxdt\geq 0.$$

Let $\lambda\to 0$. We obtain

$$\iint_S\psi(\chi_i-|\nabla u^m|^{p-2}u^m_{x_i})dxdt\geq 0,\ \forall\varphi\in C_0^\infty(S).$$

Moreover, if we choose $\lambda\leq 0$, we are able to get

$$\iint_S\psi(\chi_i-|\nabla u^m|^{p-2}u^m_{x_i})dxdt\leq 0,\ \forall\varphi\in C_0^\infty(S).$$

Now, if we choose $\psi$ such that $\mathrm{supp}\varphi\subset\mathrm{supp}\psi$, and on $\mathrm{supp}\varphi$, $\psi=1$, then we can get (4.8).

By a process of limitation, we can choose the test function $\varphi$ in (4.8) as $u^m$, then we have

$$\lim_{k\to 0}\iint_S|\nabla u_k^m|^p dxdt=\iint_S\chi\cdot\nabla u^m dxdt=\iint_S|\nabla u^m|^p dxdt. \tag{4.15}$$

Due to (1.20), $2p_1<p$, then by Hölder inequality, we have

$$\lim_{k\to 0}\iint_S|\nabla u_k^m|^{p_1}dxdt=\iint_S\chi\cdot\nabla u^m dxdt=\iint_S|\nabla u^m|^{p_1}dxdt. \tag{4.16}$$

By a refinement of Fatou's lemma, the theorem 1.4.1 in [12], we are easy to prove (4.9), and so (1.12) is true.

Secondly, we are to prove (1.13).

For small $r > 0$, denote $\Omega_r = \{x \in \Omega : \mathrm{dist}(x, \partial\Omega) \leq r\}$. For any $\eta > 0$, let

$$\mathrm{sgn}_\eta(s) = \begin{cases} 1, & \text{if } s > \eta, \\ \frac{s}{\eta}, & \text{if } |s| \leq \eta, \\ -1, & \text{if } s < -\eta. \end{cases}$$

For any given small $r > 0$, large enough $k, l$, we declare that

$$\int_{\Omega_{2r}} |u_k(x,t) - u_l(x,t)|dx \leq \int_{\Omega_r} |u_k(x,0) - u_l(x,0)|dx + c_r(t), \tag{4.17}$$

where $c_r(t)$ is independent of $k, l$, and $\lim_{t\to 0} c_r(t) = 0$. By (2.1)

$$\int_0^t \int_{\Omega_r} \varphi(u_{kt} - u_{lt})dxd\tau + \int_0^t \int_{\Omega_r} \nabla\varphi[(|\nabla u_k^m|^2 + \frac{1}{k})^{\frac{p-2}{2}}\nabla u_k^m - (|\nabla u_l^m|^2 + \frac{1}{l})^{\frac{p-2}{2}}\nabla u_l^m]dxd\tau$$

$$+ \int_0^t \int_{\Omega_r} a(x)(u_k^{mq_1}|\nabla u_k^m|^{p_1} - u_l^{mq_1}|\nabla u_l^m|^{p_1})\varphi dxd\tau$$

$$+ \int_0^t \int_{\Omega_r} [f_0(u_k^m)\int_\Omega K(y)|u_k^m(y,t)|^\beta dy - f_0(u_l^m)\int_\Omega K(y)|u_l^m(y,t)|^\beta dy]\varphi dxd\tau = 0, \tag{4.18}$$

for $\forall\varphi \in L^p(0,T; W_0^{1,p}(\Omega))$. Suppose that $\xi(x) \in C_0^1(\Omega_r)$ such that

$$0 \leq \xi \leq 1; \ \ \xi\,|_{\Omega_{2r}} = 1,$$

and choose $\varphi = \xi\mathrm{sgn}_\eta(u_k^m - u_l^m)$ in (4.18), then

$$\int_0^t \int_{\Omega_r} \xi\mathrm{sgn}_\eta(u_k^m - u_l^m)(u_{kt} - u_{lt})dxd\tau$$

$$+ \int_0^t \int_{\Omega_r} [(|\nabla u_k^m|^2 + \frac{1}{k})^{\frac{p-2}{2}}\nabla u_k^m - (x|\nabla u_l^m|^2 + \frac{1}{l})^{\frac{p-2}{2}}\nabla u_l^m]\nabla\xi\mathrm{sgn}_\eta(u_k^m - u_l^m)dxd\tau$$

$$+ \int_0^t \int_{\Omega_r} [(|\nabla u_k^m|^2 + \frac{1}{k})^{\frac{p-2}{2}}\nabla u_k^m - (x|\nabla u_l^m|^2 + \frac{1}{l})^{\frac{p-2}{2}}\nabla u_l^m]\nabla(u_k^m - u_l^m)\xi\mathrm{sgn}'_\eta(u_k^m - u_l^m)dxd\tau$$

$$+ \int_0^t \int_{\Omega_r} a(x)(u_k^{mq_1}|\nabla u_k^m|^{p_1} - u_l^{mq_1}|\nabla u_l^m|^{p_1})\xi\mathrm{sgn}_\eta(u_k^m - u_l^m)dxd\tau$$

$$+ \int_0^t \int_{\Omega_r} [f_0(u_k^m)\int_\Omega K(y)|u_k^m(y,t)|^\beta dy - f_0(u_l^m)\int_\Omega K(y)|u_l^m(y,t)|^\beta dy]\xi\mathrm{sgn}_\eta(u_k^m - u_l^m)dxd\tau \leq 0. \tag{4.19}$$

If we notice that the third term in the left hand side on (4.19) is nonnegative when $\eta \to 0$, then we have

$$\lim_{\eta\to 0} \int_0^t \int_{\Omega_r} \xi\mathrm{sgn}_\eta(u_k^m - u_l^m)(u_{kt} - u_{lt})dxd\tau$$

$$+\lim_{\eta\to 0}\int_0^t\int_{\Omega_r}[(|\nabla u_k^m|^2+\frac{1}{k})^{\frac{p-2}{2}}\nabla u_k^m-(|\nabla u_l^m|^2+\frac{1}{l})^{\frac{p-2}{2}}\nabla u_l^m]\nabla\xi \mathrm{sgn}_\eta(u_k^m-u_l^m)dxd\tau$$

$$+\lim_{\eta\to 0}\int_0^t\int_{\Omega_r}a(x)(u_k^{mq_1}|\nabla u_k^m|^{p_1}-u_l^{mq_1}|\nabla u_l^m|^{p_1})\xi \mathrm{sgn}_\eta(u_k^m-u_l^m)dxd\tau$$

$$+\lim_{\eta\to 0}\int_0^t\int_{\Omega_r}[f_0(u_k^m)\int_\Omega K(y)|u_k^m(y,t)|^\beta dy-f_0(u_l^m)\int_\Omega K(y)|u_l^m(y,t)|^\beta dy]\xi \mathrm{sgn}_\eta(u_k^m-u_l^m)dxd\tau=0. \quad (4.20)$$

At the same time,

$$\lim_{\eta\to 0}\int_0^t\int_{\Omega_r}\xi \mathrm{sgn}_\eta(u_k^m-u_l^m)(u_{kt}-u_{lt})dxd\tau=\int_0^t\int_{\Omega_r}\xi \mathrm{sgn}(u_k^m-u_l^m)(u_{kt}-u_{lt})dxd\tau$$

$$=\int_0^t\int_{\Omega_r}\xi \mathrm{sgn}(u_k-u_l)(u_{kt}-u_{lt})dxd\tau$$

$$\lim_{\eta\to 0}\int_0^t\int_{\Omega_r}\xi \mathrm{sgn}_\eta(u_k-u_l)(u_{kt}-u_{lt})dxd\tau=\lim_{\eta\to 0}\int_0^t\int_{\Omega_r}\xi(\int_0^{u_k-u_l}\mathrm{sgn}_\eta(s)ds)_\tau dxd\tau$$

$$=\lim_{\eta\to 0}\int_0^t\int_{\Omega_r}\xi\int_0^{u_k-u_l}\mathrm{sgn}_\eta(s)ds\,|_0^t\,dx=\int_{\Omega_r}\xi|u_k-u_l|dx-\int_{\Omega_r}\xi|u_{0k}-u_{0l}|dx. \quad (4.21)$$

By (4.20) (4.21), we have

$$\int_{\Omega_{2r}}\xi|u_k-u_l|dx\le\int_{\Omega_r}|u_{0k}-u_{0l}|dx+c\int_0^t\int_{\Omega_r}[(|\nabla u_k^m|^2+\frac{1}{k})^{\frac{p-1}{2}}+(|\nabla u_l^m|^2+\frac{1}{l})^{\frac{p-1}{2}}]dxd\tau$$

$$+\int_0^t\int_{\Omega_r}a(x)|u_k^{mq_1}|\nabla u_k^m|^{p_1}-u_l^{mq_1}|\nabla u_l^m|^{p_1}|dxd\tau$$

$$\int_0^t\int_{\Omega_r}|f_0(u_k^m)\int_\Omega K(y)|u_k^m(y,t)|^\beta dy-f_0(u_l^m)\int_\Omega K(y)|u_l^m(y,t)|^\beta dy|dxd\tau. \quad (4.22)$$

By Lemma 2.2 and Lemma 3.1, if $0<t\le 1$,

$$\int_0^t\int_{\Omega_r}a(x)|u_k^{mq_1}|\nabla u_k^m|^{p_1}-u_1^{mq_1}|\nabla u_1^m|^{p_1}|dxd\tau\le c\int_0^t\int_{\Omega_r}t^{-\epsilon}dxd\tau,$$

which means (4.17) is true. Here

$$\epsilon=\max\{\frac{mNq_1}{Nm(p-1)-N+mq}+\frac{p_1(m(p-1)+m-2)}{m(p-1)-1},\frac{(\beta+m)N}{Nm(p-1)-N+mq}\}<1$$

Now, for any given small $r$, if $k,l$ are large enough, by (4.17), we have

$$\int_{\Omega_{2r}}|u(x,t)-u_0(x)|dx\le\int_{\Omega_r}|u(x,t)-u_k(x,t)|dx+\int_{\Omega_{2r}}|u_{0k}(x)-u_{0l}(x)|dx$$

$$+\int_{\Omega_{2r}}|u_l(x,t)-u_{0l}(x)|dx+\int_{\Omega_{2r}}|u_{0l}(x)-u_0(x)|dx$$

letting $t\to 0$, we get (1.13).

## 5. The uniqueness of the solutions and the proof of Theorem 1.2

As we have said in the introduction, the uniqueness of the solutions of problem (1.1)-(1.3) is not true generally. But it is not difficult to prove the following theorems.

**Theorem 5.1.** *Let* $u_1, u_2$ *be the two solutions of the problem (1.1)-(1.3) with the different initial values* $u_{01}(x), u_{02}(x)$ *respectively. If* $(\frac{1}{m}+\beta-2)\setminus q_1 < 1$ *and*

$$p_1 = 0, \tag{5.1}$$

*then*

$$\int_\Omega |u_1(x,t)-u_2(x,t)|dx \le \int_\Omega |u_{01}(x)-u_{02}(x)|dx,\ \forall t\ge 0. \tag{5.2}$$

*Proof.* Let $u_1(t), u_2(t)$ be two solutions of equation (1.1). Let $v_1 = u_1^m(t), v_2 = u_2^m(t)$. Denote $w(t) = v_1^{\frac{1}{m}}(t) - u_2^{\frac{1}{m}}(t)$, $v(t) = v_1(t) - v_2(t)$. Then $w(t), v_1(t), v_2(t)$ satisfy that

$$w'(t) - [\mathrm{div}(\|\nabla v_1\|^{p-2}\nabla v_1) - \mathrm{div}(\|\nabla v_2\|^{p-2}\nabla v_2) + a(x)(v_1^{q_1} - v_2^{q_1})$$

$$= f_0(v_1)\int_\Omega K(y)|v_1|^\beta dy - f_0(v_2)\int_\Omega K(y)|v_2|^\beta dy. \tag{5.3}$$

For any positive integer $n$, let $g_n(s)$ be an odd function and

$$g_n(s) = \begin{cases} 1, & \text{if } s > \frac{1}{n}, \\ n^2s^2e^{1-n^2s^2}, & \text{if } s \le \frac{1}{n}. \end{cases}$$

Clearly, when $|s| \ge n^{-1}$, $g_n'(s) = 0$; when $|s| \le n^{-1}$, $0 \le g_n'(s) = 6s^{-1}$.

Multiplying (5.3) with $g_n(v_1 - v_2)$ and integrating on $\Omega$, we have

$$\int_\Omega g_n(v)w'(t)dx + \int_\Omega [|\nabla|v_1|^{p-2}|\nabla v_1 - |\nabla|v_2|^{p-2}|\nabla v_2]\nabla(v_1-v_2)g_n'(v)dx + \int_\Omega a(x)(v_1^{q_1} - v_2^{q_1})g_n(v)dx$$

$$= \int_\Omega g_n(v)[f_0(v_1)\int_\Omega K(y)|v_1|^\beta dy - f_0(v_2)\int_\Omega K(y)|v_2|^\beta dy]dx. \tag{5.4}$$

Moreover,

$$\lim_{n\to\infty}\int_\Omega g_n(v)w'(t)dx = \frac{d}{dt}\|w(t)\|_1,$$

$$\int_\Omega [|\nabla|v_1|^{p-2}|\nabla v_1 - |\nabla|v_2|^{p-2}|\nabla v_2]\nabla(v_1-v_2)g_n'(v)dx \ge 0,$$

$$\int_\Omega a(x)(v_1^{q_1} - v_2^{q_1})g_n(v)dx \ge 0,$$

$$|\int_\Omega g_n(v)[f_0(v_1)\int_\Omega K(y)|v_1|^\beta dy - f_0(v_2)\int_\Omega K(y)|v_2|^\beta dy]dx|$$

$$\le |\int_\Omega K(y)|v_1|^\beta dy\int_\Omega [f_0(v_1) - f_0(v_2)]dx| + c|\int_\Omega f_0(v_2)dx||\int_\Omega\int_{v_2}^{v_1} s^{\beta-1}dsdy|$$

$$\leq c\|w(t)\|_1\|v_1\|_\beta^\beta + c\|v_2\|_1\int_\Omega |\xi|^{\beta-1}|v(t)|dx,$$

where $\xi \in [v_1, v_2]$.

So

$$\frac{d}{dt}\|w(t)\|_1 \leq c\|w(t)\|_1\|v_1\|_\beta^\beta + c\|v_2\|_1(\|v_1\|_\beta^\beta + \|v_2\|_\beta^\beta), \tag{5.5}$$

By using (1.23)-(1.24) of Theorem 1.1 to (5.5), letting $n \to \infty$. By Gronwall's inequality, for any given $T > 0$, we can deduce that

$$\|w(t)\|_1 \equiv 0, 0 \leq t \leq T. \tag{5.6}$$

□

Another aim of the section is to prove the uniqueness of the viscosity solution of problem (1.1)-(1.3)

**Theorem 5.2.** *Suppose that $a(x)$ and $K(x)$ are bounded functions. If $u(x,t) \in L^\infty(S)$, $|\nabla u| \leq c$ in addition, $2 \geq p_1 \geq 1$, then the viscosity solution of (1.1)-(1.3) is unique.*

*Proof.* Let $u, v$ be the two viscosity solutions of (1.1)-(1.3). Then there are two sequences $\{u_k\}$ and $\{v_l\}$, which are the solutions of problem (1.14)-(1.2)-(1.3), such that

$$\lim_{k\to\infty} u_k = u, \quad \lim_{l\to\infty} v_l = v, \text{ a.e.in } S. \tag{5.7}$$

Clearly, since $u(x,t), v(x,t) \in L^\infty(S)$, we may assume

$$\|u_k\|_\infty \leq c, \ \|v_l\|_\infty \leq c. \tag{5.8}$$

Let

$$w = u_k - v_l, \quad w_1 = u_k^m - v_l^m.$$

Then

$$w_t = \left(a_{ij}(x,t)w_{1x_j}\right)_{x_i} + b(x,t,w,\nabla w), (x,t) \in \Omega \times (0,\infty) \tag{5.9}$$

$$w(x,0) = u_{0k}(x) - v_{0l}(x), \ x \in \Omega \tag{5.10}$$

$$w(x,t) = 0, \ (x,t) \in \partial\Omega \times (0,\infty), \tag{5.11}$$

where

$$a_{ij}(x,t) = \int_0^1 \left|s\nabla u_k^m + (1-s)\nabla v_l^m\right|^{p-2} ds \cdot \delta_{ij}$$

$$+\int_0^1 (p-2)\left|s\nabla u_k^m + (1-s)\nabla v_l^m\right|^{p-4}(su_{kx_i}^m + (1-s)v_{lx_i}^m)(su_{kx_j}^m + (1-s)v_{lx_j}^m)ds,$$

and since $p_1 \geq 1$, using the convexity of the function $s^{p_1}$, by (5.8), we have

$$b(x,t,w,\nabla w) = a(x)[u_k^{mq_1}|\nabla u_k^m|^{p_1} - v_l^{mq_1}|\nabla v_l^m|^{p_1}]$$

$$+f_0(u_k^m)\int_\Omega K(y)|u_k^m(y,t)|^\beta dy - f_0(v_l^m)\int_\Omega K(y)|v_l^m(y,t)|^\beta dy,$$

$$|b(x,t,w,\nabla w)| \leq c|\nabla(u_k^m - v_l^m)|^{p_1} \leq c|\nabla w|^{p_1} \leq c|\nabla w|^2 + c.$$

By the chapter 8 of [13], we know that

$$\|u_k(x,t) - v_l(x,t)\|_\infty \leq c\|u_{0k} - v_{0l}\|_\infty.$$

Let $k, l \to \infty$, we know that the uniqueness of the viscosity solution (1.1)-(1.3) is true. □

Suppose that the viscosity solution of problem (1.1)-(1.3) is unique in what follows. Then, by considering the regularized problem (1.14)-(1.2)-(1.3), we easily get the following Theorem 5.3, and Theorem 1.2 is a simple corollary of Theorem 5.3.

**Theorem 5.3.** *Let u be a weak solution of problem (1.1)-(1.3). If v satisfies*

$$v_t \geq div(|\nabla v^m|^{p-2} \nabla v^m) - a(x)v^{mq_1}|\nabla v^m|^{p_1}$$

$$+f_0(v^m)\int_\Omega K(y)|v^m(y,t)|^\beta dy + g(x) \quad in\ S = \Omega \times (0,\infty), \tag{5.12}$$

$$v(x,0) \geq u_0(x), \quad x \in \Omega, \tag{5.13}$$

$$v(x,t) = 0, \quad (x,t) \in \partial\Omega \times (0,\infty), \tag{5.14}$$

*then*

$$u(x,t) \geq v(x,t), \ \forall (x,t) \in S. \tag{5.15}$$

Now, let

$$v(x,t) = u_{kr}(x,t) = ru_k(x, r^{m(p-1)-1}t), \quad r \in (0,1).$$

Then

$$v_t(x,t) = \text{div}(|Dv^m|^{p-2} Dv^m) - a(x)r^{m(p-1-q_1-p_1)}v^{mq_1} | Dv^m |^{p_1}$$

$$+r^{m[p-1-\beta]}f_0(r^{-m}v^m)\int_\Omega K(y)|v^m|^\beta dy + r^{m(p-1)}g(x), (x,t) \in \Omega \times (0,\infty) \tag{5.16}$$

$$v(x,0) = ru_k(x,0), x \in \Omega, \tag{5.17}$$

$$v(x,t) = 0, \ (x,t) \in \partial\Omega \times (0,\infty). \tag{5.18}$$

Noticing that $g(x) \leq 0$, $f_0(r^{-m}v^m) \geq f_0(v^m)$, and

$$p_1 + q_1 < p-1, p-1-\beta < 0, \ 0 < r < 1,$$

which implies that

$$r^{m(p-1-q_1-p_1)} < 1, \ r^{m[p-1-\beta]} > 1,$$

$$v_t(x,t) \geq \text{div}(|Dv^m|^{p-2} Dv^m) - a(x)v^{q_1 m} | Dv^m |^{p_1} + f_0(v^m)\int_\Omega K(y)|v^m|^\beta dy + g(x),$$

using the argument similar to that in the proof Lemma 3.5 of [35], we can prove

$$u_k \geq u_{kr}.$$

It follows that

$$\frac{u_k(x, r^{m(p-1)-1}t) - u_k(x,t)}{(r^{m(p-1)-1} - 1)t}$$

$$\geq \frac{r-1}{(1-r^{m(p-1)-1})t}u_k(x, r^{m(p-1)-1}t).$$

Letting $r \to 1$, we get

$$u_{kt} \geq -\frac{u_k}{(m(p-1)-1)t}. \tag{5.19}$$

By (5.19), we can easily get Theorem 1.2.

## Acknowledgement

The paper is supported by Natural Science Foundation of Fujian province in China (No: 2015J01592), supported by Science Foundation of Xiamen University of Technology.

## Conflict of Interest

All authors declare no conflicts of interest in this paper.

## References

1. M. Aassila, *The influence of nonlocal nonlinearities on the long time behavior of solutions of diffusion problems,* J. of Diff. Equ., **192** (2003), 47-69.
2. A. V. Babin and M. I. Vishik, *Attractors of evolution equations,* North-Holland, Amsterdam, 1992.
3. M. Bertsch, R. Dal Passo and M. Ughi, *Discontinuous viscosity solutions of a degenerate parabolic equation,* Trans. Amer. Math. Soc., **320** (1990), 779-798.
4. M. Bertsch, R. Dal Passo and M. Ughi, *Non-uniqueness of solutions of a degenerate parabolic equation,* Ann. Math. Pura Appl., **161** (1992), 57-81.
5. C. Chen, *On global attractor for m-Laplacian parabolic equation with local and nonlocal nonlinearity,* J. Math. Anal. Appl., **337** (2008), 318-332.
6. C. Chen and R. Wang, *Global existence and $L^\infty$ estimates of solution for doubly degenerate parabolic equation (in Chinese),* ACTA Math. Sinica, **44** (2001), 1089-1098.
7. J. W. Cholewa and T. Dlotko, *Global attractors in abstract parabolic problems,* London Math. Soc. Lecture Note Ser., Cambridge Univ. Press, **278** (2000) .
8. A. Dall'Aglioa, D. Giachetti, I. Peral and S. *León, Global existence for some slightly super-linear parabolic equations with measure data,* J. Math. Anal. Appl., **345** (2008), 892-902.
9. R. Dal Passo and S. Luckhaus, *A degenerate diffusion problem not in divergence form,* J. Diff. Equa., **69** (1987), 1-14.
10. E. DiBenedetto, *Degenerate parabolic equations,* Universitext, Springer Verlag, 1993.
11. J. R. Esteban and J. L. Vazquez, *Homogeneous diffusion in R with power-like nonlinear diffusivity,* Arch. Rational Mech. Anal., **103** (1988), 39-88.
12. L. C. Evans, *Weak convergence methods for nonlinear partial differential equations, Conference Board of the Mathematical Sciences,* Regional Conferences Series in Mathematics Number **74**, 1998.
13. L. Gu, *Second order parabolic partial differential equations,* The Publishing Company of Xiamen University, China, 2002.
14. A. V. Ivanov, *Hölder estimates for quasilinear parabolic equations,* J.Soviet Mat., **56** (1991), 2320-2347.
15. A. S. Kalashnikkov, *Some problems of nonlinear parabolic equations of second order,* USSR. Math, Nauk, T., **42** (1987), 135-176.

16. S. Kamin and J. L. Vázquez, *Fundamental solutions and asymptotic behavior for the p-Laplacian equation,* Rev. Mat. Iberoamericana, **4** (1988), 339-354.

17. O. A. Ladyzenskaja, *New equations for the description of incompressible fluids and solvability in the large boundary value problem for them,* Proc. Steldov Inst. Math., **102** (1976), 95-118.

18. K. Lee, A. Petrosyan and J.L. Vázquez, *Large time geometric properties of solutions of the evolution p-Laplacian equation,* J. Diff. Equ., **229** (2006), 389-411.

19. K. Lee and J. L. Vázque, *Geometrical properties of solutions of the Porous Medium Equation for large times,* Indiana Univ. Math. J., **52** (2003), 991-1016.

20. P. Lei, Y. Li and P. Lin, *Null controllability for a semilinear parabolic equation with gradient quadratic growth,* Nonlinear Anal. T.M.A., **68** (2008), 73-82.

21. J. L. Lions, *Quelques méthodes de resolution des problè mes aux limites non linear,* Dound Gauthier-Villars, Paris, 1969.

22. J. Manfredi and V. Vespri, *Large time behavior of solutions to a class of doubly nonlinear parabolic equations,* Electronic J. Diff. Equ., **1994** (1994), 1-16.

23. M. Nakao, *Global solutions for some nonlinear parabolic equations with non-monotonic perturbations,* Non. Anal. TMA, **10** (1986), 299-314.

24. M. Nakao, *$L^p$ estimates of solutions of some nonlinear degenerate diffusion equation,* J. Math. Soc. Japan, **37** (1985), 41-63.

25. M. Nakao and C. Chen, *Global existence and gradient estimate for the quasilinear parabolic equation of m-Laplacian type with a nonlinear convection term,* J. Diff. Equ., 162 (2000), 224-250.

26. Y. Ohara, *$L^\infty$ estimates of solutions of some nonlinear degenerate parabolic equations,* Nonlinear Anal. TMA, **18** (1992), 413-426.

27. M. Pierre, *Uniqueness of solution of $u_t - \triangle\varphi(u) = 0$ with initial datum a measure,* Nonlinear Analysis TMA., **6** (1982), 175-187.

28. R. Temam, *Infinite-dimensional dynamical in mechanics and physics,* Springer-Verlag, New York, 1997.

29. M. Ughi, *A degenerate parabolic equation modelling the spread of an epidemic,* Ann. Math. Pura Appl., **143** (1986), 385-400.

30. M. Winkler, *Large time behavior of solutions to degenerate parabolic equations with absorption,* NoDEA. **8** (2001), 343-361.

31. Z. Wu, J. Zhao, J. Yin and H. Li, *Nonlinear diffusion equations,* Word Scientific Publishing, 2001.

32. J. Yuan, Z. Lian, L. Cao, J. Gao and J. Xu, *Extinction and positivity for a doubly nonlinear degenerate parabolic equation,* Acta Math. Sinica, Eng. Ser., **23** (2007), 1751-1756.

33. H. Zhan, *The Asymptotic Behavior of Solutions for a Class of Doubly Nonlinear Parabolic Equations,* J. Math. Anal. Appl., **370** (2010), 1-10.

34. Q. Zhang and P. Shi, *Global solutions and self-similar solutions of semilinear parabolic equations with nonlinear gradient terms,* Nonlinear Anal. T.M.A., **72** (2010), 2744-2752.

35. J. Zhao and H. Yuan, *The Cauchy problem of some doubly nonlinear degenerate parabolic equations (in Chinese),* Chinese Ann. Math., A, **16** (1995), 179-194.

# 8

# Logarithmically improved regularity criteria for the Boussinesq equations

**Sadek Gala**[1,2,*], **Mohamed Mechdene**[1] **and Maria Alessandra Ragusa**[2,*]

[1] Department of Mathematics, University of Mostaganem, Box 227, Mostaganem, 27000, Algeria
[2] Dipartimento di Mathematicae Informatica, Universit à di Catania Viale Andrea Doria, 6, 95125 Catania, Italy

[*] **Correspondence:** sadek.gala@gmail.com; maragusa@dmi.unict.it

**Abstract:** In this paper, logarithmically improved regularity criteria for the Boussinesq equations are established under the framework of Besov space $\dot{B}^{-r}_{\infty,\infty}$. We prove the solution $(u,\theta)$ is smooth up to time $T > 0$ provided that

$$\int_0^T \frac{\|u(\cdot,t)\|^{\frac{2}{1-r}}_{\dot{B}^{-r}_{\infty,\infty}}}{\log(e + \|u(t,.)\|_{\dot{B}^{-r}_{\infty,\infty}})}dt < \infty$$

for some $0 \le r < 1$ or

$$\|u(\cdot,t)\|_{L^\infty(0,T;\dot{B}^{-1}_{\infty,\infty}(\mathbb{R}^3))} << 1.$$

This result improves some previous works.

**Keywords:** Regularity criterion; Boussinesq equations; A priori estimates
**2010 Mathematics Subject Classification:** 35Q35; 76D03

## 1. Introduction and main result

This paper is concerned with the regularity criterion of the 3D Boussinesq equations with the incompressibility condition :

$$\begin{cases} \partial_t u + u\cdot\nabla u - \Delta u + \nabla\pi = \theta e_3, \\ \partial_t\theta + u\cdot\nabla\theta - \Delta\theta = 0, \\ \nabla\cdot u = 0, \\ (u,\theta)(x,0) = (u_0,\theta_0)(x), \quad x\in\mathbb{R}^3, \end{cases} \tag{1.1}$$

where $u = u(x,t)$ and $\theta = \theta(x,t)$ denote the unknown velocity vector field and the scalar function temperature, while $u_0$, $\theta_0$ with $\nabla\cdot u_0 = 0$ in the sense of distribution are given initial data. $e_3 = (0,0,1)^T$. $\pi = \pi(x,t)$ the pressure of fluid at the point $(x,t)\in\mathbb{R}^3\times(0,\infty)$. The Boussinesq equation is one of important subjects for researches in nonlinear sciences [14]. There are a huge literatures on the incompressible Boussinesq equations such as [1–4, 6, 8–10, 17, 19–22] and the references therein.

When $\theta = 0$, (1.1) reduces to the well-known incompressible Navier-Stokes equations and many results are available. Besides their physical applications, the Navier-Stokes equations are also mathematically significant. From that time on, much effort has been devoted to establish the global existence and uniqueness of smooth solutions to the Navier-Stokes equations.

However, similar to the classic Navier-Stokes equations, the question of global regularity of the weak solutions of the 3D Boussinesq equations still remains a big open problem and the system (1.1) has received many studies. Based on some analysis technique, some regularity criteria via the velocity of weak solutions in the Lebesgue spaces, multiplier spaces and Besov spaces have been obtained in [5, 17, 19, 20, 22, 23].

More recently, the authors of the present paper [7] showed that the weak solution becomes regular if

$$\int_0^T \frac{\|u(\cdot,t)\|_{\dot{B}^{-r}_{\infty,\infty}}^{\frac{2}{1-r}} + \|\theta(\cdot,t)\|_{\dot{B}^{-r}_{\infty,\infty}}^{\frac{2}{1-r}}}{1+\log(e+\|u(\cdot,t)\|_{H^s}+\|\theta(\cdot,t)\|_{H^s})}dt < \infty \quad \text{for some } 0 \le r < 1 \quad \text{and} \quad s \ge \frac{1}{2}, \tag{1.2}$$

where $\dot{B}^{-r}_{\infty,\infty}$ denotes the homogeneous Besov space. Definitions and basic properties of the Sobolev spaces and the Besov spaces can be find in [18]. For concision, we omit them here.

The purpose of this paper is to improve the regularity criterion (1.2) in the following form.

Let $(u, \theta)$ be a smooth solution to (1.1) in $[0, T)$ with the initial data $(u_0, \theta_0) \in H^3(\mathbb{R}^3) \times H^3(\mathbb{R}^3)$ with $\operatorname{div} u_0 = 0$ in $\mathbb{R}^3$. Suppose that the solution $(u, \theta)$ satisfies

$$\int_0^T \frac{\|u(\cdot,t)\|_{\dot{B}^{-r}_{\infty,\infty}}^{\frac{2}{1-r}}}{\log(e+\|u(\cdot,t)\|_{\dot{B}^{-r}_{\infty,\infty}})}dt < \infty \quad \text{for some } r \text{ with } 0 \le r < 1. \tag{1.3}$$

Then it holds

$$\sup_{0\le t\le T}\left(\|u(\cdot,t)\|_{H^3}^2 + \|\theta(\cdot,t)\|_{H^3}^2\right) < \infty.$$

That is, the solution $(u, \theta)$ can be smoothly extended after time $t = T$. In other word, if $T_*$ is the maximal time existence of the solution, then

$$\int_0^{T_*} \frac{\|u(\cdot,t)\|_{\dot{B}^{-r}_{\infty,\infty}}^{\frac{2}{1-r}}}{\log(e+\|u(\cdot,t)\|_{\dot{B}^{-r}_{\infty,\infty}})}dt < \infty.$$

Then the solution can be smoothly extended after $t = T$.

The condition (1.3) can be regarded as a logarithmically improved version of the assumption

$$\int_0^T \|u(\cdot,t)\|_{\dot{B}^{-r}_{\infty,\infty}}^{\frac{2}{1-r}} dt < \infty \quad \text{for some } r \text{ with } 0 \le r < 1.$$

For the case $r = 1$, we have the following result.

Let $(u, \theta)$ be a smooth solution to (1.1) in $[0, T)$ with the initial data $(u_0, \theta_0) \in H^3(\mathbb{R}^3) \times H^3(\mathbb{R}^3)$ with $\operatorname{div} u_0 = 0$ in $\mathbb{R}^3$. Suppose that there exists a small positive constant $\eta$ such that

$$||u(\cdot, t)||_{L^\infty(0,T;\dot{B}^{-1}_{\infty,\infty}(\mathbb{R}^3))} \leq \eta, \tag{1.4}$$

then solution $(u, \theta)$ can be smoothly extended after time $t = T$.

Theorem 1 can be regarded as improvements and limiting cases of those in [7]. It is worth to point out all conditions are valid for the usual Navier-Stokes equations. We refer to a recent work [7] and references therein.

For the case $r = 0$, see [23].

## 2. Proof of Theorem 1

In this section, we will prove Theorem 1 by the standard energy method.

Let $T > 0$ be a given fixed time. The existence and uniqueness of local smooth solutions can be obtained as in the case of the Navier-Stokes equations. Hence, for all $T > 0$ we assume that $(u, \theta)$ is a smooth solution to (1.1) on $[0, T)$ and we will establish a priori bounds that will allow us to extend $(u, \theta)$ beyond time $T$ under the condition (1.3).

Owing to (1.3) holds, one can deduce that for any small $\epsilon > 0$, there exists $T_0 = T_0(\epsilon) < T$ such that

$$\int_{T_0}^{T} \frac{||u(\cdot, t)||_{\dot{B}^{-r}_{\infty,\infty}}^{\frac{2}{1-r}}}{\log(e + ||u(\cdot, t)||_{\dot{B}^{-r}_{\infty,\infty}})} dt \leq \epsilon << 1. \tag{2.1}$$

Thanks to the divergence-free condition $\nabla \cdot u = 0$, from $(1.1)_2$, we get immediately the global a priori bound for $\theta$ in any Lebesgue space

$$||\theta(\cdot, t)||_{L^q} \leq C\, ||\theta_0||_{L^q} \quad \text{for all } \ q \in [2, \infty] \ \text{ and all } \ t \in [0, T].$$

Now, multiplying $(1.1)_2$ by $\theta$ and using integration by parts, we get

$$\frac{1}{2}\frac{d}{dt} ||\theta||_{L^2}^2 + ||\nabla\theta||_{L^2}^2 = 0.$$

Hence, we obtain

$$\theta \in L^\infty\left(0, T; L^2(\mathbb{R}^3)\right) \cap L^2\left(0, T; H^1(\mathbb{R}^3)\right). \tag{2.2}$$

Next, multiplying $(1.1)_1$ by $u$, we have after integration by part,

$$\frac{1}{2}\frac{d}{dt} ||u||_{L^2}^2 + ||\nabla u||_{L^2}^2 = \int_{\mathbb{R}^3} (\theta e_3) \cdot u dx \leq ||\theta||_{L^2} ||u||_{L^2} \leq C\, ||u||_{L^2},$$

which yields

$$u \in L^\infty\left(0, T; L^2(\mathbb{R}^3)\right) \cap L^2\left(0, T; H^1(\mathbb{R}^3)\right), \tag{2.3}$$

where we used (2.2) and

$$\int_{\mathbb{R}^3} (u \cdot \nabla u) \cdot u dx = \frac{1}{2} \int_{\mathbb{R}^3} (u \cdot \nabla)\, u^2 dx = -\frac{1}{2} \int_{\mathbb{R}^3} (\nabla \cdot u)\, u^2 dx = 0$$

by incompressibility of $u$, that is, $\nabla \cdot u = 0$.

Now, apply $\nabla$ operator to the equation of $(1.1)_1$ and $(1.1)_2$, then taking the inner product with $\nabla u$ and $\nabla\theta$, respectively and using integration by parts, we get

$$\begin{aligned}
&\frac{1}{2}\frac{d}{dt}(\|\nabla u\|_{L^2}^2 + \|\nabla\theta\|_{L^2}^2) + \|\Delta u\|_{L^2}^2 + \|\Delta\theta\|_{L^2}^2 \\
= &-\int_{\mathbb{R}^3} \nabla(u\cdot\nabla)u\cdot\nabla u dx + \int_{\mathbb{R}^3}\nabla(\theta e_3)\cdot\nabla u dx - \int_{\mathbb{R}^3}\nabla(u\cdot\nabla)\theta\cdot\nabla\theta dx \\
= &\ \mathcal{I}_1 + \mathcal{I}_2 + \mathcal{I}_3.
\end{aligned} \tag{2.4}$$

Employing the Hölder and Young inequalities, we derive the estimation of the first term $\mathcal{I}_1$ as

$$\begin{aligned}
\mathcal{I}_1 &= \int_{\mathbb{R}^3}(u\cdot\nabla)u\cdot\Delta u dx \le \|\nabla\cdot(u\otimes u)\|_{L^2}\|\Delta u\|_{L^2} \\
&\le C\|u\|_{\dot{B}^{-r}_{\infty,\infty}}\|\nabla u\|_{\dot{H}^r}\|\Delta u\|_{L^2} \\
&\le C\|u\|_{\dot{B}^{-r}_{\infty,\infty}}\|\nabla u\|_{L^2}^{1-r}\|\Delta u\|_{L^2}^{1+r} \\
&\le \frac{1}{2}\|\Delta u\|_{L^2}^2 + C\|u\|_{\dot{B}^{-r}_{\infty,\infty}}^{\frac{2}{1-r}}\|\nabla u\|_{L^2}^2 \\
&\le \frac{1}{2}\|\Delta u\|_{L^2}^2 + C\|u\|_{\dot{B}^{-r}_{\infty,\infty}}^{\frac{2}{1-r}}(\|\nabla u\|_{L^2}^2 + \|\nabla\theta\|_{L^2}^2),
\end{aligned}$$

where we have used the inequality due to [16] :

$$\|u\otimes u\|_{\dot{H}^1} \le C\|u\|_{\dot{B}^{-r}_{\infty,\infty}}\|\nabla u\|_{\dot{H}^r}$$

and the interpolation inequality

$$\|w\|_{\dot{H}^s} = \left\||\xi|^s\widehat{w}\right\|_{L^2} \le \|w\|_{L^2}^{1-s}\|\nabla w\|_{L^2}^s \quad \text{for all } 0\le s\le 1.$$

The term $\mathcal{I}_3$ can be estimated as

$$\begin{aligned}
\mathcal{I}_3 &\le C\|\nabla u\|_{L^2}\|\nabla\theta\|_{L^4}^2 \\
&\le C\|\nabla u\|_{L^2}\|\nabla\theta\|_{\dot{B}^{-1}_{\infty,\infty}}\|\Delta\theta\|_{L^2} \\
&\le C\|\nabla u\|_{L^2}\|\theta\|_{\dot{B}^{0}_{\infty,\infty}}\|\Delta\theta\|_{L^2} \\
&\le \frac{1}{2}\|\Delta\theta\|_{L^2}^2 + C\|\theta\|_{L^\infty}^2\|\nabla u\|_{L^2}^2 \\
&\le \frac{1}{2}\|\Delta\theta\|_{L^2}^2 + C\|\theta\|_{L^\infty}^2(\|\nabla u\|_{L^2}^2 + \|\nabla\theta\|_{L^2}^2),
\end{aligned}$$

where we have used

$$\|\nabla\theta\|_{\dot{B}^{-1}_{\infty,\infty}} \le C\|\theta\|_{\dot{B}^{0}_{\infty,\infty}} \le C\|\theta\|_{L^\infty}.$$

The term $\mathcal{I}_2$ can be estimated as

$$\mathcal{I}_2 \le \|\nabla u\|_{L^2}\|\nabla\theta\|_{L^2} \le \frac{1}{2}(\|\nabla u\|_{L^2}^2 + \|\nabla\theta\|_{L^2}^2).$$

Plugging all the estimates into (2.4) yields that

$$\begin{aligned}&\frac{d}{dt}(\|\nabla u\|_{L^2}^2+\|\nabla\theta\|_{L^2}^2)+\|\Delta u\|_{L^2}^2+\|\Delta\theta\|_{L^2}^2\\ \le\ & C(\frac{1}{2}+\|u\|_{\dot{B}^{-r}_{\infty,\infty}}^{\frac{2}{1-r}}+\|\theta\|_{L^\infty}^2)(\|\nabla u\|_{L^2}^2+\|\nabla\theta\|_{L^2}^2).\end{aligned}$$

Hence, we obtain

$$\begin{aligned}&\frac{d}{dt}(\|\nabla u(\cdot,t)\|_{L^2}^2+\|\nabla\theta(\cdot,t)\|_{L^2}^2)+\|\Delta u\|_{L^2}^2+\|\Delta\theta\|_{L^2}^2\\
\le\ & C\left[\frac{\frac{1}{2}+\|u\|_{\dot{B}^{-r}_{\infty,\infty}}^{\frac{2}{1-r}}+\|\theta\|_{L^\infty}^2}{\log(e+\|u\|_{\dot{B}^{-r}_{\infty,\infty}})}\right](\|\nabla u\|_{L^2}^2+\|\nabla\theta\|_{L^2}^2)\log(e+\|u\|_{\dot{B}^{-r}_{\infty,\infty}})\\
\le\ & C\left[\frac{\frac{1}{2}+\|u\|_{\dot{B}^{-r}_{\infty,\infty}}^{\frac{2}{1-r}}+\|\theta\|_{L^\infty}^2}{\log(e+\|u\|_{\dot{B}^{-r}_{\infty,\infty}})}\right](\|\nabla u\|_{L^2}^2+\|\nabla\theta\|_{L^2}^2)\log(e+\|u\|_{H^3}+\|\theta\|_{H^3})\\
\le\ & C\left[\frac{\frac{1}{2}+\|u\|_{\dot{B}^{-r}_{\infty,\infty}}^{\frac{2}{1-r}}+\|\theta\|_{L^\infty}^2}{\log(e+\|u\|_{\dot{B}^{-r}_{\infty,\infty}})}\right](\|\nabla u\|_{L^2}^2+\|\nabla\theta\|_{L^2}^2)\log(e+\kappa(t))\end{aligned}$$

where $\kappa(t)$ is defined by

$$\kappa(t)=\sup_{T_0\le\tau\le t}(\|u(\cdot,\tau)\|_{H^3}+\|\theta(\cdot,\tau)\|_{H^3})\quad \text{for all } T_0<t<T.$$

It should be noted that the function $\kappa(t)$ is nondecreasing. Moreover, we have used the following fact :

$$\|u\|_{\dot{B}^{-r}_{\infty,\infty}}\le C\|u\|_{H^3}.$$

Integrating the above inequality over $[T_0,t]$ and applying Gronwall's inequality, we have

$$\begin{aligned}&\|\nabla u(\cdot,t)\|_{L^2}^2+\|\nabla\theta(\cdot,t)\|_{L^2}^2+\int_{T_*}^t\|\Delta u(\cdot,\tau)\|_{L^2}^2+\|\Delta\theta(\cdot,\tau)\|_{L^2}^2\,d\tau\\
\le\ & (\|\nabla u(\cdot,T_0)\|_{L^2}^2+\|\nabla\theta(\cdot,T_0)\|_{L^2}^2)\\
&\times\exp\left(C\int_{T_0}^t\frac{\|u\|_{\dot{B}^{-r}_{\infty,\infty}}^{\frac{2}{1-r}}}{\log(e+\|u(\cdot,\tau)\|_{\dot{B}^{-r}_{\infty,\infty}})}\log(e+\kappa(\tau))\,d\tau\right)\\
\le\ & (\|\nabla u(\cdot,T_0)\|_{L^2}^2+\|\nabla\theta(\cdot,T_0)\|_{L^2}^2)\\
&\times\exp\left(C\log(e+\kappa(t))\int_{T_0}^t\frac{\|u\|_{\dot{B}^{-r}_{\infty,\infty}}^{\frac{2}{1-r}}}{\log(e+\|u(\cdot,\tau)\|_{\dot{B}^{-r}_{\infty,\infty}})}d\tau\right)\\
\le\ & \widetilde{C}\exp(C\epsilon\log(e+\kappa(t)))=\widetilde{C}(e+\kappa(t))^{C\epsilon}\end{aligned}\tag{2.5}$$

where $\widetilde{C}$ is a positive constant depending on $\|\nabla u(\cdot, T_0)\|_{L^2}^2$, $\|\nabla \theta(\cdot, T_0)\|_{L^2}^2$, $T_0$, $T$ and $\theta_0$.

$H^3$**−norm.** Next, we start to obtain the $H^3$−estimates under the above estimate (2.5). Applying $\Lambda^3 = (-\Delta)^{\frac{3}{2}}$ to $(1.1)_1$, then taking $L^2$ inner product of the resulting equation with $\Lambda^3 u$, and using integration by parts, we obtain

$$\frac{1}{2}\frac{d}{dt}\left\|\Lambda^3 u(\cdot, t)\right\|_{L^2}^2 + \left\|\Lambda^4 u(\cdot, t)\right\|_{L^2}^2 = -\int_{\mathbb{R}^3} \Lambda^3 (u \cdot \nabla u) \cdot \Lambda^3 u dx + \int_{\mathbb{R}^3} \Lambda^3(\theta e_3) \cdot \Lambda^3 u dx \tag{2.6}$$

Similarly, applying $\Lambda^3 = (-\Delta)^{\frac{3}{2}}$ to $(1.1)_2$, then taking $L^2$ inner product of the resulting equation with $\Lambda^3\theta$, and using integration by parts, we obtain

$$\frac{1}{2}\frac{d}{dt}\left\|\Lambda^3 \theta(\cdot, t)\right\|_{L^2}^2 + \left\|\Lambda^4 \theta(\cdot, t)\right\|_{L^2}^2 = -\int_{\mathbb{R}^3} \Lambda^3 (u \cdot \nabla \theta) \cdot \Lambda^3 \theta dx, \tag{2.7}$$

Using $\nabla \cdot u = 0$, we deduce that

$$\begin{aligned}
&\frac{1}{2}\frac{d}{dt}(\left\|\Lambda^3 u(\cdot, t)\right\|_{L^2}^2 + \left\|\Lambda^3 \theta(\cdot, t)\right\|_{L^2}^2) + \left\|\Lambda^4 u(\cdot, t)\right\|_{L^2}^2 + \left\|\Lambda^4 \theta(\cdot, t)\right\|_{L^2}^2 \\
= &-\int_{\mathbb{R}^3} \left[\Lambda^3(u \cdot \nabla u) - u \cdot \Lambda^3 \nabla u\right] \cdot \Lambda^3 u dx + \int_{\mathbb{R}^3} \Lambda^3(\theta e_3) \cdot \Lambda^3 u dx \\
&-\int_{\mathbb{R}^3}^{3} \left[\Lambda^3(u \cdot \nabla \theta) - u \cdot \Lambda^3 \nabla \theta\right] \cdot \Lambda^3 \theta dx \\
= &\ \Pi_1 + \Pi_2 + \Pi_3.
\end{aligned} \tag{2.8}$$

To bound $\Pi_1$, we recall the following commutator estimate due to [12]:

$$\|\Lambda^\alpha (fg) - f\Lambda^\alpha g\|_{L^p} \leq C\left(\left\|\Lambda^{\alpha-1} g\right\|_{L^{q_1}} \|\nabla f\|_{L^{p_1}} + \|\Lambda^\alpha f\|_{L^{p_2}} \|g\|_{L^{q_2}}\right), \tag{2.9}$$

for $\alpha > 1$, and $\frac{1}{p} = \frac{1}{p_1} + \frac{1}{q_1} = \frac{1}{p_2} + \frac{1}{q_2}$. Hence $\Pi_1$ can be estimated as

$$\begin{aligned}
\Pi_1 &\leq C\|\nabla u\|_{L^3}\|\Lambda^3 u\|_{L^3}^2 \\
&\leq C\|\nabla u\|_{L^2}^{\frac{3}{4}}\|\Lambda^3 u\|_{L^2}^{\frac{1}{4}}\|\nabla u\|_{L^2}^{\frac{1}{3}}\|\Lambda^4 u\|_{L^2}^{\frac{5}{3}} \\
&\leq \frac{1}{6}\|\Lambda^4 u\|_{L^2}^2 + C\|\nabla u\|_{L^2}^{\frac{13}{2}}\|\Lambda^3 u\|_{L^2}^{\frac{3}{2}},
\end{aligned} \tag{2.10}$$

where we used (2.9) with $\alpha = 3$, $p = \frac{3}{2}$, $p_1 = q_1 = p_2 = q_2 = 3$, and the following Gagliardo-Nirenberg inequalities

$$\begin{cases}
\|\nabla u\|_{L^3} \leq C\|\nabla u\|_{L^2}^{\frac{3}{4}}\|\Lambda^3 u\|_{L^2}^{\frac{1}{4}}, \\
\|\Lambda^3 u\|_{L^3} \leq C\|\nabla u\|_{L^2}^{\frac{1}{6}}\|\Lambda^4 u\|_{L^2}^{\frac{5}{6}}.
\end{cases} \tag{2.11}$$

If we use the existing estimate (2.1) for $T_0 \leq t < T$, (2.10) reduces to

$$\Pi_1 \leq \frac{1}{2}\|\Lambda^4 u\|_{L^2}^2 + \widetilde{C}(e + \kappa(t))^{\frac{3}{2}+\frac{13}{2}C\epsilon}. \tag{2.12}$$

Using (2.11) again, we get

$$\Pi_3 \leq C(\|\nabla u\|_{L^3}\|\Lambda^3 \theta\|_{L^3} + \|\nabla \theta\|_{L^3}\|\Lambda^3 u\|_{L^3}) \left\|\Lambda^3 \theta\right\|_{L^3}$$

$$\leq C(\|\nabla u\|_{L^3} + \|\nabla\theta\|_{L^3})(\left\|\Lambda^3 u\right\|_{L^3}^2 + \left\|\Lambda^3\theta\right\|_{L^3}^2)$$
$$\leq \frac{1}{6}(\left\|\Lambda^4 u\right\|_{L^2}^2 + \left\|\Lambda^4\theta\right\|_{L^2}^2) + \widetilde{C}(e+\kappa(t))^{\frac{3}{2}+\frac{13}{2}C\epsilon}.$$

For $\Pi_2$, we have

$$\Pi_2 \leq \frac{1}{2}(\left\|\Lambda^3 u\right\|_{L^2}^2 + \left\|\Lambda^3\theta\right\|_{L^2}^2) \leq \widetilde{C}(e+\kappa(t))^2.$$

Inserting all the inequalities into (2.8) and absorbing the dissipative terms, one finds

$$\frac{d}{dt}(\left\|\Lambda^3 u(\cdot,t)\right\|_{L^2}^2 + \left\|\Lambda^3\theta(\cdot,t)\right\|_{L^2}^2) \leq \widetilde{C}(e+\kappa(t))^{\frac{3}{2}+\frac{13}{2}C\epsilon} + \widetilde{C}(e+\kappa(t))^2, \tag{2.13}$$

with together with the basic energy (2.2)-( [?]) yields

$$\frac{d}{dt}(\|u(\cdot,t)\|_{H^3}^2 + \|\theta(\cdot,t)\|_{H^3}^2) \leq \widetilde{C}(e+\kappa(t))^{\frac{3}{2}+\frac{13}{2}C\epsilon} + \widetilde{C}(e+\kappa(t))^2, \tag{2.14}$$

Choosing $\epsilon$ sufficiently small provided that $\frac{13}{2}C\epsilon < \frac{1}{2}$ and applying the Gronwall inequality to (2.14), we derive that

$$\sup_{T_0\leq\tau\leq t}\left(\|u(\cdot,\tau)\|_{H^3}^2 + \|\theta(\cdot,\tau)\|_{H^3}^2\right) \leq \widetilde{C} < \infty, \tag{2.15}$$

where $\widetilde{C}$ depends on $\|\nabla u(\cdot,T_0)\|_{L^2}^2$ and $\|\nabla\theta(\cdot,T_0)\|_{L^2}^2$.

Noting that the right-hand side of (2.15) is independent of $t$ for $T_0 \leq t < T$ , we know that $(u(\cdot,T),\theta(\cdot,T)) \in H^3(\mathbb{R}^3)\times H^3(\mathbb{R}^3)$. Consequently, $(u,\theta)$ can be extended smoothly beyond $t = T$. This completes the proof of Theorem 1.

## 3. Proof of Theorem 1.

In order to prove Theorem 1, we first recall the following local existence theorem of the three-dimensional Boussinesq equations.

Suppose $(u,\theta) \in L^\alpha\left(\mathbb{R}^3\right)$, for some $\alpha \geq 3$ and $\nabla\cdot u = 0$. Then, there exists $T_0 > 0$ and a unique solution of (1.1) on $[0,T_0)$ such that

$$(u,\theta) \in BC\left([0,T_0); L^\alpha\left(\mathbb{R}^3\right)\right) \cap L^s\left([0,T_0); L^r\left(\mathbb{R}^3\right)\right),\quad t^{\frac{1}{s}}u \in BC\left([0,T_0); L^\alpha\left(\mathbb{R}^3\right)\right) \tag{3.1}$$

Moreover, let $(0,T^*)$ be the maximal interval such that $(u,\theta)$ solves (1.1) in $C\left((0,T^*); L^\alpha\left(\mathbb{R}^3\right)\right)$, $\alpha > 3$. Then for any $t \in (0,T^*)$

$$\|u(\cdot,t)\|_{L^\alpha} \geq \frac{C}{(T^*-t)^{\frac{\alpha-3}{2\alpha}}} \quad\text{and}\quad \|\theta(\cdot,t)\|_{L^\alpha} \geq \frac{C}{(T^*-t)^{\frac{\alpha-3}{2\alpha}}},$$

with the constant $C$ independent of $T^*$ and $\alpha$.

Let $(u,\theta)$ be a strong solution satisfying

$$(u,\theta) \in L^\alpha\left((0,T); L^\beta\left(\mathbb{R}^3\right)\right) \quad\text{for}\quad \frac{2}{\alpha}+\frac{3}{\beta} = 1 \quad\text{and}\quad \beta > 3.$$

Then $(u,\theta)$ belongs to $C^\infty\left(\mathbb{R}^3\times(0,T)\right)$.

For all $T > 0$, we assume that $(u, \theta)$ is a smooth solution to (1.1) on $[0, T)$ and we will establish a priori bounds that will allow us to extend $(u, \theta)$ beyond time $T$ under the condition (1.4).

Similar to the proof of Theorem 1, we can show that

$$(u,\theta) \in L^{\infty}\left(0,T;L^{2}(\mathbb{R}^{3})\right) \cap L^{2}\left(0,T;H^{1}(\mathbb{R}^{3})\right). \tag{3.2}$$

The proof of Theorem 1 is divided into steps.

**Step I.** $H^1$–estimation. In order to get the $H^1$–estimates, we apply $\nabla$ operator to the equation of $(1.1)_1$ and $(1.1)_2$, multiply by $\nabla u$ and $\nabla\theta$, respectively to obtain

$$\begin{aligned}
&\frac{1}{2}\frac{d}{dt}(\|\nabla u(\cdot,t)\|_{L^2}^2 + \|\nabla\theta(\cdot,t)\|_{L^2}^2) + \|\Delta u(\cdot,t)\|_{L^2}^2 + \|\Delta\theta(\cdot,t)\|_{L^2}^2 \\
= &-\int_{\mathbb{R}^3} \nabla(u\cdot\nabla)u\cdot\nabla u dx + \int_{\mathbb{R}^3}\nabla(\theta e_3)\cdot\nabla u dx - \int_{\mathbb{R}^3}\nabla(u\cdot\nabla)\theta\cdot\nabla\theta dx \\
= &\ \mathcal{I}_1 + \mathcal{I}_2 + \mathcal{I}_3.
\end{aligned} \tag{3.3}$$

Next we estimate $\mathcal{I}_1, \mathcal{I}_2$ and $\mathcal{I}_3$ in another way. Hence,

$$\begin{aligned}
\mathcal{I}_1 &\le \|\nabla u\|_{L^3}^3 \le C\,\|\nabla u\|_{\dot{B}^{-2}_{\infty,\infty}}\,\|\Delta u\|_{L^2}^2 \\
&\le C\,\|u\|_{\dot{B}^{-1}_{\infty,\infty}}\,\|\Delta u\|_{L^2}^2,
\end{aligned}$$

where we have used the following interpolation inequality due to [16] :

$$\|w\|_{L^3} \le C\,\|\nabla w\|_{L^2}^{\frac{2}{3}}\,\|w\|_{\dot{B}^{-2}_{\infty,\infty}}^{\frac{1}{3}}.$$

By means of the Hölder and Young inequalities, the term $\mathcal{I}_3$ can be estimated as

$$\begin{aligned}
\mathcal{I}_3 &\le C\,\|\nabla u\|_{L^2}\,\|\nabla\theta\|_{L^4}^2 \\
&\le C\,\|\nabla u\|_{L^2}\,\|\nabla\theta\|_{\dot{B}^{-1}_{\infty,\infty}}\,\|\Delta\theta\|_{L^2} \\
&\le C\,\|\theta\|_{\dot{B}^{0}_{\infty,\infty}}^2\,\|\Delta\theta\|_{L^2}^2 + C\,\|\nabla u\|_{L^2}^2 \\
&\le C\,\|\theta\|_{L^\infty}^2\,\|\Delta\theta\|_{L^2}^2 + C\,\|\nabla u\|_{L^2}^2,
\end{aligned}$$

where we have used the following interpolation inequality due to [16] :

$$\|\nabla\theta\|_{L^4}^2 \le C\,\|\nabla\theta\|_{\dot{B}^{-1}_{\infty,\infty}}\,\|\Delta\theta\|_{L^2}.$$

The term $\mathcal{I}_2$ can be estimated as

$$\mathcal{I}_2 \le \|\nabla u\|_{L^2}\,\|\nabla\theta\|_{L^2} \le \frac{1}{2}(\|\nabla u\|_{L^2}^2 + \|\nabla\theta\|_{L^2}^2).$$

Plugging all the estimates into (3.3) yields that

$$\frac{1}{2}\frac{d}{dt}(\|\nabla u(\cdot,t)\|_{L^2}^2 + \|\nabla\theta(\cdot,t)\|_{L^2}^2) + \|\Delta u(\cdot,t)\|_{L^2}^2 + \|\Delta\theta(\cdot,t)\|_{L^2}^2$$

$$\leq C\|u\|_{\dot{B}^{-1}_{\infty,\infty}}\|\Delta u\|^2_{L^2} + C\|\theta\|^2_{L^\infty}\|\Delta\theta\|^2_{L^2} + C(\|\nabla u\|^2_{L^2} + \|\nabla\theta\|^2_{L^2}).$$

Under the assumption (1.4), we choose $\eta$ small enough so that

$$C\|u\|_{\dot{B}^{-1}_{\infty,\infty}} \leq \frac{1}{2}.$$

Hence, we find that

$$\frac{d}{dt}(\|\nabla u\|^2_{L^2} + \|\nabla\theta\|^2_{L^2}) + \|\Delta u\|^2_{L^2} + \|\Delta\theta\|^2_{L^2} \leq C(\|\nabla u\|^2_{L^2} + \|\nabla\theta\|^2_{L^2}).$$

Integrating in time and applying the Gronwall inequality, we infer that

$$\|\nabla u(\cdot,t)\|^2_{L^2} + \|\nabla\theta(\cdot,t)\|^2_{L^2} + \int_0^T (\|\Delta u(\cdot,\tau)\|^2_{L^2} + \|\Delta\theta(\cdot,\tau)\|^2_{L^2})d\tau \leq C. \tag{3.4}$$

**Step II.** $H^2$**–estimation.** Next, we start to obtain the $H^2$–estimates under the above estimate (3.4). Applying $\Delta$ to $(1.1)_1$, then taking $L^2$ inner product of the resulting equation with $\Delta u$, and using integration by parts, we obtain

$$\frac{1}{2}\frac{d}{dt}\|\Delta u(\cdot,t)\|^2_{L^2} + \left\|\Lambda^3 u(\cdot,t)\right\|^2_{L^2} = -\int_{\mathbb{R}^3}\Delta(u\cdot\nabla u)\cdot\Delta u dx + \int_{\mathbb{R}^3}\Delta(\theta e_3)\cdot\Delta u dx \tag{3.5}$$

Similarly, applying $\Delta$ to $(1.1)_2$, then taking $L^2$ inner product of the resulting equation with $\Delta\theta$, and using integration by parts, we obtain

$$\frac{1}{2}\frac{d}{dt}\|\Delta\theta(\cdot,t)\|^2_{L^2} + \left\|\Lambda^3\theta(\cdot,t)\right\|^2_{L^2} = -\int_{\mathbb{R}^3}\Delta(u\cdot\nabla\theta)\cdot\Delta\theta dx. \tag{3.6}$$

Adding (3.5) and (3.6), we deduce that

$$\begin{aligned}
&\frac{1}{2}\frac{d}{dt}(\|\Delta u(\cdot,t)\|^2_{L^2} + \|\Delta\theta(\cdot,t)\|^2_{L^2}) + \left\|\Lambda^3 u(\cdot,t)\right\|^2_{L^2} + \left\|\Lambda^3\theta(\cdot,t)\right\|^2_{L^2}\\
&= -\int_{\mathbb{R}^3}\Delta(u\cdot\nabla u)\cdot\Delta u dx + \int_{\mathbb{R}^3}\Delta(\theta e_3)\cdot\Delta u dx - \int_{\mathbb{R}^3}\Delta(u\cdot\nabla\theta)\cdot\Delta\theta dx\\
&= \mathcal{K}_1 + \mathcal{K}_2 + \mathcal{K}_3.
\end{aligned} \tag{3.7}$$

Using Hölder's inequality and Young's inequality, $\mathcal{K}_1$ can be estimated as

$$\begin{aligned}
\mathcal{K}_1 &= \int_{\mathbb{R}^3}\Delta(u\otimes u)\cdot\Delta\nabla u dx \leq \|\Delta(u\otimes u)\|_{L^2}\|\Delta\nabla u\|_{L^2}\\
&\leq C\|u\|_{L^\infty}\|\Delta u\|_{L^2}\left\|\Lambda^3 u\right\|_{L^2}\\
&\leq \frac{1}{2}\left\|\Lambda^3 u\right\|^2_{L^2} + C\|u\|^2_{L^\infty}\|\Delta u\|^2_{L^2}.
\end{aligned}$$

Here we have used the bilinear estimates due to Kato-Ponce [12] and Kenig-Ponce-Vega [13] :

$$\|\Lambda^\alpha(fg)\|_{L^p} \leq C\left(\|\Lambda^\alpha g\|_{L^{q_1}}\|f\|_{L^{p_1}} + \|\Lambda^\alpha f\|_{L^{p_2}}\|g\|_{L^{q_2}}\right),$$

for $\alpha > 0$, and $\frac{1}{p} = \frac{1}{p_1} + \frac{1}{q_1} = \frac{1}{p_2} + \frac{1}{q_2}$.

From the incompressibility condition, Hölder's inequality and Young's inequality, one has

$$\begin{aligned} \mathcal{K}_3 &= \int_{\mathbb{R}^3} \Delta(u\theta) \cdot \Delta\nabla\theta dx \leq \|\Delta(u\theta)\|_{L^2} \|\Delta\nabla\theta\|_{L^2} \\ &\leq C(\|u\|_{L^\infty} \|\Delta\theta\|_{L^2} + \|\theta\|_{L^\infty} \|\Delta u\|_{L^2}) \left\|\Lambda^3\theta\right\|_{L^2} \\ &\leq \frac{1}{2}\left\|\Lambda^3\theta\right\|_{L^2}^2 + C(\|u\|_{L^\infty}^2 + \|\theta\|_{L^\infty}^2)(\|\Delta u\|_{L^2}^2 + \|\Delta\theta\|_{L^2}^2). \end{aligned}$$

For $\mathcal{K}_2$, we have

$$\mathcal{K}_2 \leq \frac{1}{2}(\|\Delta u\|_{L^2}^2 + \|\Delta\theta\|_{L^2}^2).$$

Inserting all the inequalities into (3.7) and absorbing the dissipative terms, one finds

$$\begin{aligned} &\frac{d}{dt}(\|\Delta u(\cdot,t)\|_{L^2}^2 + \|\Delta\theta(\cdot,t)\|_{L^2}^2) + \left\|\Lambda^3 u(\cdot,t)\right\|_{L^2}^2 + \left\|\Lambda^3\theta(\cdot,t)\right\|_{L^2}^2 \\ \leq\ & C(\|u\|_{L^\infty}^2 + \|\theta\|_{L^\infty}^2)(\|\Delta u\|_{L^2}^2 + \|\Delta\theta\|_{L^2}^2). \end{aligned} \tag{3.8}$$

Using the following interpolation inequality

$$\|f\|_{L^\infty} \leq C\|f\|_{L^2}^{\frac{1}{4}} \|\Delta f\|_{L^2}^{\frac{3}{4}},$$

together with the key estimate (3.4) yield that

$$\int_0^T (\|u(\cdot,\tau)\|_{L^\infty}^2 + \|\theta(\cdot,\tau)\|_{L^\infty}^2)d\tau \leq C < \infty.$$

Applying the Gronwall inequality to (3.8), we derive that

$$\|\Delta u(\cdot,t)\|_{L^2}^2 + \|\Delta\theta(\cdot,t)\|_{L^2}^2 + \int_0^T (\left\|\Lambda^3 u(\cdot,t)\right\|_{L^2}^2 + \left\|\Lambda^3\theta(\cdot,t)\right\|_{L^2}^2)dt \leq C. \tag{3.9}$$

By estimates (3.4) and (3.9) as well as the following Gagliardo-Nirenberg's inequality

$$\|f\|_{L^6} \leq C\|f\|_{L^2}^{\frac{1}{2}} \|\Delta f\|_{L^2}^{\frac{1}{2}},$$

it is easy to see that

$$(u,\theta) \in L^4(0,T;L^6(\mathbb{R}^3)),$$

from which and Lemma 3 the smoothness of $(u,\theta)$ follows immediately. This completes the proof of Theorem 1.


## Acknowledgments

Part of the work was carried out while the first author was long term visitor at University of Catania. The hospitality and support of Catania University are graciously acknowledged.

All authors would like to thank Professor Bo-Qing Dong for helpful discussion and constant encouragement. They also would like to thank the anonymous reviewers for their valuable comments and suggestions to improve the quality of the paper.

## Conflict of Interest

All authors declare no conflicts of interest in this paper.

## References


1. J. R. Cannon and E. Dibenedetto, *The initial problem for the Boussinesq equation with data in $L^p$*, in: Lecture Notes in Mathematics, Springer, Berlin, **771** (1980), 129-144.
2. D. Chae and H.-S. Nam, *Local existence and blow-up criterion for the Boussinesq equations*, Proc. Roy. Soc. Edinburgh, Sect. A, **127** (1997), 935-946.
3. B. Q. Dong, J. Song and W. Zhang, *Blow-up criterion via pressure of three-dimensional Boussinesq equations with partial viscosity (in Chinese)*, Sci. Sin. Math., **40** (2010), 1225-1236.
4. J. Fan and Y. Zhou, *A note on regularity criterion for the 3D Boussinesq system with partial viscosity,* Appl. Math. Lett., **22** (2009), 802-805.
5. J. Fan and T. Ozawa, *Regularity criteria for the 3D density-dependent Boussinesq equations*, Nonlinearity, **22** (2009), 553-568.
6. S. Gala, *On the regularity criterion of strong solutions to the 3D Boussinesq equations* , Applicable Analysis, **90** (2011), 1829-1835.
7. S. Gala and M.A. Ragusa, *Logarithmically improved regularity criterion for the Boussinesq equations in Besov spaces with negative indices,* Applicable Analysis, **95** (2016), 1271-1279.
8. S. Gala, Z. Guo and M. A. Ragusa, *A remark on the regularity criterion of Boussinesq equations with zero heat conductivity,* Appl. Math. Lett., **27** (2014), 70-73.
9. Z. Guo and S. Gala, *Regularity criterion of the Newton-Boussinesq equations in $\mathbb{R}^3$*, Commun. Pure Appl. Anal., **11** (2012), 443-451.
10. J. Geng and J. Fan, *A note on regularity criterion for the 3D Boussinesq system with zero thermal conductivity,* Appl. Math. Lett., **25** (2012), 63-66.
11. Y. Jia, X. Zhang and B. Dong, *Remarks on the blow-up criterion for smooth solutions of the Boussinesq equations with zero diffusion*, C.P.A.A., **12** (2013), 923-937.
12. T. Kato and G. Ponce, *Commutator estimates and the Euler and Navier-Stokes equations,* Commun. Pure Appl. Math., **41** (1988), 891-907.
13. C. Kenig, G. Ponce and L. Vega, *Well-posedness of the initial value problem for the Korteweg-de-Vries equation,* J. Amer. Math. Soc., **4** (1991), 323-347.
14. A. Majda, *Introduction to PDEs and Waves for the Atmosphere and Ocean,* Courant Lecture Notes in Mathematics, **9** (2003).
15. M. Mechdene, S. Gala, Z. Guo and M.A. Ragusa, *Logarithmical regularity criterion of the three-dimensional Boussinesq equations in terms of the pressure,* Z. Angew. Math. Phys., **67** (2016), 67-120.
16. Y. Meyer, P. Gerard and F. Oru, *Inégalités de Sobolev précisées, Séminaire équations aux dérivées partielles (Polytechnique)*, **4**, 1996-1997.

17. N. Ishimura and H. Morimoto, *Remarks on the blow-up criterion for the 3D Boussinesq equations*, Math. Meth. Appl. Sci., **9** (1999), 1323-1332.

18. H. Triebel, *Theory of Function Spaces, Birkhäuser Verlag,* Basel, 1983.

19. H. Qiu, Y. Du and Z. Yao, *Blow-up criteria for 3D Boussinesq equations in the multiplier space,* Communications in Nonlinear Science and Numerical Simulation, **16** (2011), 1820-1824.

20. H. Qiu, Y. Du and Z. Yao, *A blow-up criterion for 3D Boussinesq equations in Besov spaces,* Nonlinear Analysis TMA, **73** (2010), 806-815.

21. Z. Xiang, *The regularity criterion of the weak solution to the 3D viscous Boussinesq equations in Besov spaces*, Mathematical Methods in the Applied Sciences, **34** (2011), 360-372.

22. F. Xu, Q. Zhang and X. Zheng, *Regularity Criteria of the 3D Boussinesq Equations in the Morrey-Campanato Space*, Acta Appl. Math., **121** (2012), 231-240.

23. Z. Ye, *A Logarithmically improved regularity criterion of smooth solutions for the 3D Boussinesq equations,* Osaka J. Math., **53** (2016), 417-423.

# 9

# On the upper semicontinuity of global attractors for damped wave equations

**Joseph L. Shomberg**

Department of Mathematics and Computer Science, Providence College, 1 Cunningham Square, Providence, Rhode Island 02918, USA

* **Correspondence:** jshomber@providence.edu

**Abstract:** We provide a new proof of the upper-semicontinuity property for the global attractors admitted by the solution operators associated with some strongly damped wave equations. In particular, we demonstrate an explicit control over semidistances between trajectories in the weak energy phase space in terms of the perturbation parameter. This result strengthens the recent work by Y. Wang and C. Zhong [7].

**Keywords:** Upper-semicontinuity; global attractor; strongly damped wave equation
**Mathematics Subject Classification:** 35B41, 35L71, 35Q74, 35L20

## 1. Introduction

In this short article, we revisit the recent work of [7] who examine the upper-semicontinuity properties of the family of global attractors associated with the strong damping perturbation of weakly damped wave equations. Such equations are used in modeling non-Hookean viscoelastic materials. Here, the strong damping term $-\varepsilon\Delta u_t$ present in such equations indicates that we are accounting for the strain *rate* in the material, in addition to other forces. The upper-semicontinuity result in [7] shows that the global attractors do not "blow-up" as the perturbation parameter vanishes. Hence, the asymptotic behavior of the solutions is stable. What we offer here improves this result by communicating that the difference of trajectories corresponding to the perturbation problem and the limit problem, emanating from the same initial data, can be estimated in terms of the perturbation parameter $\varepsilon$ in the topology associated with the weak energy phase space of the model problems.

Let $\Omega$ be a bounded domain in $\mathbb{R}^3$ with boundary $\partial\Omega$ of class $C^2$. We consider the semilinear strongly damped wave equation,

$$u_{tt} - \varepsilon\Delta u_t + u_t - \Delta u + f(u) = 0 \quad \text{in} \quad (0,\infty)\times\Omega, \tag{1.1}$$

where $0 \leq \varepsilon \leq 1$ represents the diffusivity of the momentum. The equation is endowed with Dirichlet boundary condition,

$$u_{|\partial\Omega} = 0 \quad \text{on} \quad (0,\infty) \times \partial\Omega, \tag{1.2}$$

and with the initial conditions

$$u(0,x) = u_0(x), \quad u_t(0,x) = u_1(x) \quad \text{at} \quad \{0\} \times \Omega. \tag{1.3}$$

For the nonlinear term, we assume $f \in C^2(\mathbb{R})$ satisfies the sign condition

$$\liminf_{|s|\to\infty} f'(s) > -\lambda_1, \tag{1.4}$$

where $\lambda_1 > 0$ denotes the first eigenvalue of the Dirichlet–Laplacian, and we assume the growth assumption holds, for all $s \in \mathbb{R}$,

$$|f''(s)| \leq \ell\,(1 + |s|)\,, \tag{1.5}$$

for some positive constant $\ell$. We will refer to equations (1.1)–(1.3) under assumptions (1.4)–(1.5) as Problem $\mathbf{P}_\varepsilon$, for $\varepsilon \in [0,1]$.

It is now well-known that the model problems admit globally defined weak-solutions in the (weak) energy phase space

$$\mathcal{H}_0 := H^1_0(\Omega) \times L^2(\Omega)$$

and, for each $\varepsilon \in [0,1]$, a global attractor $\mathcal{A}_\varepsilon$ is compact in $\mathcal{H}_0$ and bounded in

$$\mathcal{H}_1 := \left(H^2(\Omega) \cap H^1_0(\Omega)\right) \times L^2(\Omega).$$

Furthermore, when $\varepsilon > 0$, the operator associated with the linear part of the abstract Cauchy problem generates an *analytic* semigroup on $\mathcal{H}_0$. On these results we mention the following references [1–5].

The main result in this paper is the following:

**Theorem 1.1.** *The family of global attractors $\{\mathcal{A}_\varepsilon\}_{\varepsilon\in[0,1]}$ is upper-semicontinuous in the topology of $\mathcal{H}_0$ in the following explicit sense: there is a constant $C > 0$ independent of $\varepsilon$ in which*

$$\operatorname{dist}_{\mathcal{H}_0}(\mathcal{A}_\varepsilon, \mathcal{A}_0) := \sup_{a\in\mathcal{A}_\varepsilon} \inf_{b\in\mathcal{A}_0} \|a - b\|_{\mathcal{H}_0} \leq C\sqrt{\varepsilon}.$$

*A word about notation:* we will often drop the dependence on $x$ and even $t$ from the unknown $u(x,t)$ writing only $u$ instead. The norm in the space $L^p(\Omega)$ is denoted $\|\cdot\|_p$ except in the common occurrence when $p = 2$ where we simply write the $L^2(\Omega)$ norm as $\|\cdot\|$. The $L^2(\Omega)$ product is simply denoted $(\cdot,\cdot)$. Other Sobolev norms are denoted by occurrence; in particular, since we are working with the homogeneous Dirichlet boundary conditions (1.2), in $H^1_0(\Omega)$, we will use the equivalent norm

$$\|u\|_{H^1_0} = \|\nabla u\|.$$

Given a subset $B$ of a Banach space $X$, denote by $\|B\|_X$ the quantity $\sup_{x\in B}\|x\|_X$. In many calculations $C$ denotes a *generic* positive constant which may or may not depend on several of the parameters involved in the formulation of the problem. Finally, for each $\varepsilon \in [0,1]$, and $t \geq 0$, we denote by $S_\varepsilon(t)$ the semigroup of solution operators acting on $\mathcal{H}_0$ defined through the weak solution,

$$S_\varepsilon(t)(u_0(x), u_1(x)) := (u_\varepsilon(t,x;u_0,u_1), \partial_t u_\varepsilon(t,x;u_0,u_1)),$$

where $u_\varepsilon$ here denotes the weak solution to Problem $\mathbf{P}_\varepsilon$.

The next section contains a proof of Theorem 1.1.

## 2. Continuity properties of the global attractors

Following [6, Section 10.8], the type of perturbation examined in this article is called *regular* because both classes of Problem $\mathbf{P}_\varepsilon$ ($\varepsilon > 0$ and $\varepsilon = 0$) lie in the same phase space; in particular, the family of global attractors, $\{\mathcal{A}_\varepsilon\}_{\varepsilon\in[0,1]}$, lies in $\mathcal{H}_0$. Hence, we will utilize [6, Theorem 10.16].

**Proposition 2.1.** *Assume that for $\varepsilon \in [0, \varepsilon_0)$ the semigroups $S_\varepsilon$ each admit a global attractor $\mathcal{A}_\varepsilon$ and that there exists a bounded set $X$ such that*

$$\bigcup_{\varepsilon\in[0,\varepsilon_0)} \mathcal{A}_\varepsilon \subset X.$$

*If in addition the semigroup $S_\varepsilon$ converges to $S_0$ in the sense that, for each $t > 0$, $S_\varepsilon(t)x \to S_0(t)x$ uniformly on bounded subsets $Y$ of the phase space $H$, i.e.,*

$$\sup_{x\in Y} \|S_\varepsilon(t)x - S_0(t)x\|_H \to 0 \quad as \quad \varepsilon \to 0,$$

*then*

$$\operatorname{dist}(\mathcal{A}_\varepsilon, \mathcal{A}_0) \to 0 \quad as \quad \varepsilon \to 0.$$

We now arrive at our first result.

**Lemma 2.2.** *Let $T > 0$. There exists a constant $C = C(\|\mathcal{A}_\varepsilon\|_{\mathcal{H}_1}, T) > 0$ such that for all $\zeta_0 \in \mathcal{A}_\varepsilon$ and for all $t \in [0, T]$, there holds, for all $\varepsilon \in (0, 1]$,*

$$\|S_\varepsilon(t)\zeta_0 - S_0(t)\zeta_0\|_{\mathcal{H}_0} \le C\sqrt{\varepsilon}. \tag{2.1}$$

*Proof.* Let $B$ be a bounded set on $\mathcal{H}_0$ and $T > 0$. Let $\zeta_0 = (u_0, u_1) \in \mathcal{A}_\varepsilon$. For $t > 0$, let

$$\zeta^+(t) = (u^+(t), u_t^+(t)) \quad \text{and} \quad \zeta^0(t) = (u^0(t), u_t^0(t)),$$

denote the corresponding global solutions of Problem $\mathbf{P}_\varepsilon$ and Problem $\mathbf{P}_0$, respectively, on $[0, T]$, both with the (same) initial data $\zeta_0$. For all $t \in (0, T]$, set

$$\begin{aligned}\bar{\zeta}(t) &:= \zeta^+(t) - \zeta^0(t)\\ &= (u^+(t), u_t^+(t)) - \left(u^0(t), u_t^0(t)\right)\\ &=: (\bar{u}(t), \bar{u}_t(t))\,.\end{aligned}$$

Then $\bar{\zeta}$ and $\bar{u}$ satisfy the equations

$$\begin{cases}\bar{u}_{tt} - \varepsilon\Delta\bar{u}_t + \bar{u}_t - \Delta\bar{u} + f(u^+) - f(u^0) = -\varepsilon\Delta u_t^0 & \text{in} \quad (0,\infty)\times\Omega\\ \bar{u}_{|\partial\Omega} = 0 & \text{on} \quad (0,\infty)\times\partial\Omega\\ \bar{\zeta}(0) = \mathbf{0} & \text{at} \quad \{0\}\times\Omega.\end{cases} \tag{2.2}$$

After multiplying the equation $(2.2)_1$ by $2\bar{u}_t$ in $L^2(\Omega)$, we estimate the new product to arrive at the differential inequality,

$$\frac{d}{dt}\left\{\|\bar{u}_t\|^2 + \|\nabla\bar{u}\|^2\right\} + 2\varepsilon\|\nabla\bar{u}_t\|^2 + 2\|\bar{u}_t\|^2$$

$$= -2(f(u^1) - f(u^0), \bar{u}_t) - 2\varepsilon(\nabla u_t^0, \nabla \bar{u}_t)$$
$$\leq C||\nabla \bar{u}||^2 + ||\bar{u}_t||^2 + \varepsilon||\nabla u_t^0||^2 + \varepsilon||\nabla \bar{u}_t||^2. \tag{2.3}$$

The constant $C = C(L, \Omega) > 0$ is due to the local Lipschitz condition of $f : H_0^1 \to L^2$ following assumptions (1.4) and (1.5), as well as the embedding $H_0^1 \hookrightarrow L^2$.

It suffices to find an appropriate bound for $||\nabla u_t^0(t)||^2$. Indeed, since the global attractor for Problem $\mathbf{P}_0$ consists of strong solutions ($\mathcal{A}_0$ is bounded in $\mathcal{H}_1$), we are allowed to test/multiply the *weakly* damped wave equation in $L^2(\Omega)$ by $-2\Delta u_t^0(t)$. To this end we obtain,

$$\begin{aligned}\frac{d}{dt}\left\{||\nabla u_t^0||^2 + ||\Delta u^0||^2\right\} + 2||\nabla u_t^0||^2 &\leq 2|(f'(u^0)\nabla u^0, \nabla u_t^0)| \\ &\leq ||f'(u^0)\nabla u^0||^2 + ||\nabla u_t^0||^2 \\ &\leq ||f'(u^0)||_{L^3}^2||\nabla u^0||_{L^6}^2 + ||\nabla u_t^0||^2 \\ &\leq ||u^0||_{H^1}^4||u^0||_{H^2}^2 + ||\nabla u_t^0||^2.\end{aligned}$$

Integrating this inequality over $[0, T]$ yields the desired bound,

$$\int_0^t ||\nabla u_t^0(s)||^2 ds \leq C, \tag{2.4}$$

where the constant $C = C(||\mathcal{A}_0||_{\mathcal{H}_1}, T) > 0$, depends on the bound on $\mathcal{A}_0$ in $\mathcal{H}_1$ (through the initial condition) and on $T > 0$.

Now returning to inequality (2.3), we integrate

$$\frac{d}{dt}\left\{||\bar{u}_t||^2 + ||\nabla \bar{u}||^2\right\} \leq ||\bar{u}_t||^2 + C||\nabla \bar{u}||^2 + \varepsilon||\nabla u_t^0||^2 \tag{2.5}$$

over $[0, T]$ and apply the bound (2.4) to the last term on the right-hand side to produce the claim (2.1). This completes the proof. □

*Remark* 2.3. The above result (2.1) establishes that, on compact time intervals, the difference between trajectories of Problem $\mathbf{P}_\varepsilon$, $\varepsilon \in (0, 1]$, and Problem $\mathbf{P}_0$, originating from the same initial data on $\mathcal{A}_\varepsilon \subset \mathcal{H}_1$, can be controlled, explicitly, in terms of the perturbation parameter $\varepsilon$ in the topology of $\mathcal{H}_0$.

The well-known upper-semicontinuity result in Proposition 2.1 now follows for our family of global attractors.

## Conflict of Interest

The author declares no conflicts of interest in this paper.

## References

1. A. V. Babin and M. I. Vishik, *Attractors of evolution equations*, North-Holland, Amsterdam, 1992.
2. Alexandre N. Carvalho and Jan W. Cholewa, *Attractors for strongly damped wave equations with critical nonlinearities*, Pacific J. Math. **207** (2002), 287-310.

3. Alexandre N. Carvalho and Jan W. Cholewa, *Local well posedness for strongly damped wave equations with critical nonlinearities*, Bull. Austral. Math. Soc. **66** (2002), 443-463.
4. V. Pata and M. Squassina, *On the strongly damped wave equation*, Comm. Math. Phys. **253** (2005), 511-533.
5. Vittorino Pata and Sergey Zelik, *A remark on the damped wave equation*, Commun. Pure Appl. Anal. **5** (2006), 609-614.
6. James C. Robinson, *Infinite–dimensional dynamical systems*, Cambridge Texts in Applied Mathematics, Cambridge University Press, Cambridge, 2001.
7. Yonghai Wang and Chengkui Zhong, *Upper semicontinuity of global attractors for damped wave equations*, Asymptot. Anal. **91** (2015), 1-10.

# 10

# Some Convolution Properties of Multivalent Analytic Functions

**Nazar Khan, Bilal Khan, Qazi Zahoor Ahmad*and Sarfraz Ahmad**

Department of Mathematics Abbottabad University of Science and Technology Abbottabad, Pakistan.

* **Correspondence:** zahoorqazi5@gmail.com

**Abstract:** In this paper, we introduce a new subclass of multivalent functions associated with conic domain in an open unit disk. We study some convolution properties, sufficient condition for the functions belonging to this new class.

**Keywords:** Multivalent functions; Hadamard product; Conic domain; Analytic functions; Sufficient condition

---

## 1. Introduction

Let $A(p)$ denote the class of all functions

$$f(z) = z^p + \sum_{n=1}^{\infty} a_{n+p} z^{n+p}, \qquad (p \in N = \{1, 2, 3.....\}) \tag{1.1}$$

which are analytic and $p$-valent in the open unit disk $E = \{z : |z| < 1\}$. For $p = 1$, $A(1) = A$. Let $f$, $g \in A(p)$, where $f$ is given by (1.1) and $g$ is defined by

$$g(z) = z^p + \sum_{n=1}^{\infty} b_{n+p} z^{n+p}, \qquad (z \in E).$$

Then the Hadamard product (or convolution) $f * g$ of the functions $f$ and $g$ is defined by

$$(f * g)(z) = z^p + \sum_{n=1}^{\infty} a_{n+p} b_{n+p} z^{n+p} = (g * f)(z).$$

Let $UCV$ and $UST$ denote the usual classes of uniformly convex and uniformly starlike functions and are defined by

$$UCV = \left\{ f(z) \in A : Re\left(1 + \frac{zf''(z)}{f'(z)}\right) > \left|\frac{zf''(z)}{f'(z)}\right| \right\}, \quad z \in E,$$

$$UST = \left\{ f(z) \in A : Re\left(\frac{zf'(z)}{f(z)}\right) > \left|\frac{zf'(z)}{f(z)} - 1\right| \right\}, \quad z \in E.$$

These classes were first introduced by Goodman [2, 3] and further investigated by [14] and [6]. Kanas and Wiśniowska [4, 5] introduced the conic domain $\Omega_k$, $k \geq 0$ as

$$\Omega_k = \left\{ u + iv : u > k\sqrt{(u-1)^2 + v^2} \right\}.$$

For fixed $k$ this domain represents the right half plane ($k = 0$), a parabola ($k = 1$), the right branch of hyperbola ($0 < k < 1$) and an ellipse ($k > 1$). For detail study about $\Omega_k$ and its generalizations, see [8, 9, 10]. The extremal functions for these conic regions are

$$p_k(z) = \begin{cases} \frac{1+z}{1-z}, & k = 0, \\ 1 + \frac{2}{\pi^2}\left(\log\frac{1+\sqrt{z}}{1-\sqrt{z}}\right)^2, & k = 1, \\ \frac{1}{1-k^2}\cosh\left\{\left(\frac{2}{\pi}\arccos k\right)\log\frac{1+\sqrt{z}}{1-\sqrt{z}}\right\} - \frac{k^2}{1-k^2}, & 0 < k < 1, \\ \frac{1}{k^2-1}\sin\left(\frac{\pi}{2K(\kappa)}\int_0^{\frac{u(z)}{\sqrt{\kappa}}}\frac{dt}{\sqrt{1-t^2}\sqrt{1-\kappa^2t^2}}\right) + \frac{k^2}{k^2-1}, & k > 1, \end{cases} \tag{1.2}$$

where

$$u(z) = \frac{z - \sqrt{\kappa}}{1 - \sqrt{\kappa z}}, \quad z \in \mathbb{E},$$

and $\kappa \in (0, 1)$ is chosen such that $k = \cosh(\pi K'(\kappa)/(4K(\kappa)))$. Here $K(\kappa)$ is Legendre's complete elliptic integral of first kind and $K'(\kappa) = K(\sqrt{1-\kappa^2})$ and $K'(t)$ is the complementary integral of $K(t)$.
Now we define the following:

**Definition.** *Let $f \in A(p)$ given by (1.1) is said to belong to $k - UR_p$, $k \geq 0$ if it satisfies the following condition*

$$Re\left(\frac{f^{(p)}(z) + zf^{(p+1)}(z)}{p!}\right) > k\left|\frac{f^{(p)}(z) + zf^{(p+1)}(z)}{p!} - 1\right|, \quad z \in E,$$

*where $f^{(p)}(z)$ is the pth derivative of $f(z)$.*

**Special Cases:**

i) For $k = 0$, we have $0 - UR_p = R_p$, introduced and studied by Noor et-al. [7].

ii) For $k = 0$, $p = 1$, we have $0 - UR_1 = R$, introduced and studied by Singh et-al. [15].

## 2. Preliminary Results

**Lemma 2.1.** [12]. For $\alpha \leq 1$ and $\beta \leq 1$

$$p(\alpha) * p(\beta) \subset p(\delta), \qquad \delta = 1 - 2(1-\alpha)(1-\beta).$$

The result is sharp.

**Lemma 2.2.** [1]. Let $\{d_n\}_0^\infty$ be a convex null sequence. Then the function

$$q(z) = \frac{d_0}{2} + \sum_{n=1}^{\infty} d_n z^n$$

is analytic in $E$ and $Req(z) > 0 \quad z \in E$.

**Lemma 2.3.** [13]. For $0 \leq \theta \leq \pi$,

$$\frac{1}{2} + \sum_{n=1}^{m} \frac{\cos n\theta}{n+1} \geq 0.$$

**Lemma 2.4.** [7]. If $f$ and $g$ belong to the class $R_p$ and

$$h^{(p-1)}(z) = f^{(p-1)}(z) * g^{(p-1)}(z).$$

Then $h$ also belong to the class $R_p$.

## 3. Main Result

**Theorem 3.1.** Let $f \in k - UR_P$ then

$$Re\left(\frac{f^{(p)}(z)}{p!}\right) > \frac{k - 1 + 2\log 2}{k+1}.$$

**Proof.** Let $f \in k - UR_p$ then by definition, we have

$$Re\left(\frac{f^{(p)}(z) + zf^{(p+1)}(z)}{p!}\right) > k\left|\frac{f^{(p)}(z) + zf^{(p+1)}(z)}{p!} - 1\right|.$$

After some simple computations, we have

$$Re\left(\frac{f^{(p)}(z) + zf^{(p+1)}(z)}{p!}\right) > \frac{k}{k+1}, \tag{3.1}$$

This can be written as

$$Re\left(1 + \sum_{n=1}^{\infty} \frac{(p+n)!(n+1)}{n!} a_{n+p} z^n\right) > \frac{k}{k+1}, \tag{3.2}$$

or

$$Re\left(1 + \frac{1}{2}\sum_{n=1}^{\infty} \frac{(p+n)!(n+1)}{n!} a_{n+p} z^n\right) > \frac{2k+1}{2k+2}. \tag{3.3}$$

Consider the function

$$h(z) = 1 + 2\sum_{n=1}^{\infty} \frac{z^n}{n+1}. \tag{3.4}$$

Clearly $h$ is analytic, $h(0) = 1$ in $E$ and

$$Reh(z) = Re\left(1 - \frac{2}{z}[z + \log(1-z)]\right) > -1 + 2\log 2. \tag{3.5}$$

From (3.3) and (3.4), we have

$$\left(\frac{f^{(p)}(z)}{p!}\right) = \left(1 + \frac{1}{2}\sum_{n=1}^{\infty} \frac{(p+n)!(n+1)}{n!} a_{n+p} z^n\right) * \left(1 + 2\sum_{n=1}^{\infty} \frac{z^n}{n+1}\right). \tag{3.6}$$

Now using (3.3) , (3.5) and Lemma 2.2 with $\alpha = \frac{2k+1}{2k+2}, \beta = -1 + 2\log 2$ and $\delta = \frac{k-1+2\log 2}{k+1}$, we have

$$Re\left(\frac{f^{(p)}(z)}{p!}\right) > \frac{k-1+2\log 2}{k+1}. \tag{3.7}$$

This completes the result.
For some spacial value of $k$ and $p$ we obtain the following known result.

**Corollary 3.2.** [7]. *Let $f \in R_p$ then*

$$Re\left(\frac{f^{(p)}(z)}{p!}\right) > -1 + 2\log 2.$$

**Theorem 3.3.** *Let $f \in k - UR_p$ then*

$$Re\left(\frac{f^{(p-1)}(z)}{z}\right) > \frac{p!(2k+1)}{2k+2}. \tag{3.8}$$

**Proof.** From (3.3), we have

$$Re\left(1 + \frac{1}{2}\sum_{n=1}^{\infty}\frac{(p+n)!(n+1)}{n!}a_{n+p}z^n\right) > \frac{(2k+1)}{2k+2}.$$

Now consider the convex null sequence $\{d_n\}_0^\infty$ define by $d_0 = 0$, $d_n = \frac{2}{(n+1)^2}$, $n \geq 1$, using Lemma 2.2, we have

$$Re\left(\frac{1}{2} + \sum_{n=1}^{\infty}\frac{2}{(n+1)^2}z^n\right) > 0,$$

or equivalently

$$Re\left(1 + 2\sum_{n=1}^{\infty}\frac{1}{(n+1)^2}z^n\right) > \frac{1}{2}. \tag{3.9}$$

From (3.3) and (3.9), we have

$$\frac{f^{(p-1)}(z)}{p!z} = \left(1 + \frac{1}{2}\sum_{n=1}^{\infty}\frac{(p+n)!(n+1)}{n!}a_{n+p}z^n\right) * \left(1 + 2\sum_{n=1}^{\infty}\frac{1}{(n+1)^2}z^n\right). \tag{3.10}$$

From (3.10) and Lemma (2.1) with $\alpha = \frac{2k+1}{2k+2}$ and $\beta = \frac{1}{2}$, we have

$$Re\left(\frac{f^{(p-1)}(z)}{z}\right) > \frac{p!(2k+1)}{2k+2}. \tag{3.11}$$

Which is the required result.

**Corollary 3.4.** [7]. *Let $f \in R_p$ then*

$$Re\left(\frac{f^{(p-1)}(z)}{z}\right) > \frac{p!}{2}, \quad z \in E.$$

**Corollary 3.5.** [15]. *Let $f \in R$ then*

$$Re\left(\frac{f(z)}{z}\right) > \frac{1}{2}, \quad z \in E.$$

**Theorem 3.6.** Let $f \in k - UR_p$ then for every $n \geq 1$, the $n$th partial sum of $f$ satisfies

$$ReS_n^{(p)}(z, f) > \frac{p!k}{k+1}, \quad z \in E.$$

and hence $S_n(z, f)$ is $p-$valent in $E$.

**Proof.** From (3.2) and (3.4), we have

$$\frac{s_n^{(p)}(z, f)}{p!} = \left(1 + \sum_{n=1}^{\infty} \frac{(p+n)!(n+1)}{p!n} a_{n+p} z^n\right) * \left(1 + \sum_{n=1}^{\infty} \frac{z^n}{n+1}\right). \tag{3.12}$$

Putting $z = re^{i\theta}$, $0 \leq r \leq 1$, $0 \leq \theta \leq \pi$ and the minimum principle for harmonic functions with Lemma 2.3, we have

$$\begin{aligned} Re\left(1 + \sum_{n=1}^{k} \frac{z^n}{n+1}\right) &= Re\left(1 + \sum_{n=1}^{k} \frac{r^n e^{in\theta}}{n+1}\right), \quad 0 \leq \theta \leq \pi \\ &= Re\left(1 + \sum_{n=1}^{k} \frac{r^n}{n+1}(\cos n\theta + i \sin n\theta)\right) \\ &= \left(1 + \sum_{n=1}^{k} \frac{r^n \cos n\theta}{n+1}\right) \\ &= \left(1 + \sum_{n=1}^{k} \frac{r^n \cos n\theta}{n+1}\right) \geq \frac{1}{2}. \end{aligned} \tag{3.13}$$

Using (3.2), (3.12), (3.13) and Lemma 2.1 with $\alpha = \frac{k}{k+1}$ and $\beta = \frac{1}{2}$, we have

$$Re\left(s_n^{(p)}(z, f)\right) > \frac{p!k}{k+1}. \tag{3.14}$$

This completes the proof. From the result given by [11], we see that $s_n(z, f)$ is $p-$valent in $E$ for every $n \geq 1$.

**Corollary 3.7.** [7]. *Let $f \in R_p$, then for every $n \geq 1$, the nth partial sum of $f$ satisfies*

$$ReS_n^{(p)}(z, f) > 0, \quad z \in E$$

*and hence $s_n(z, f)$ is $p-$valent in $E$.*

For $k = 1$ we have the following corollary.

**Corollary 3.8.** [15]. *Let $f \in 1 - UR_p$, then for every $n \geq 1$, the nth partial sum of $f$ satisfies*

$$ReS_n'(z, f) > \frac{p!}{2}, \quad z \in E$$

*and hence $s_n(z, f)$ is univalent in $E$.*

**Theorem 3.9.** *Let $f \in k - UR_p$, $g \in R_p$ and*

$$h^{(p-1)}(z) = f^{(p-1)}(z) * g^{(p-1)}(z).$$

*Then h belong to the class $k - UR_p$.*

**Proof.** Since

$$h^{(p-1)}(z) = f^{(p-1)}(z) * g^{(p-1)}(z). \tag{3.15}$$

It follows that

$$zh^{(p)}(z) = f^{(p)}(z) * g^{(p-1)}(z). \tag{3.16}$$

After simple computations, (3.16) can be written as

$$Re\left(\frac{h^{(p)}(z) + zh^{(p+1)}(z)}{p!}\right) = Re\left(\left(\frac{f^{(p)}(z) + zf^{(p+1)}(z)}{p!}\right) * \left(\frac{g^{(p-1)}(z)}{zp!}\right)\right). \tag{3.17}$$

From (3.17), (3.1), Corollary 3.4 and Lemma 2.1 with $\alpha = \frac{k}{k+1}$ and $\beta = \frac{1}{2}$, we get the required proof.

**Corollary 3.10.** *[15]. If $f(z) = z + \sum_{n=2}^{\infty} a_n z^n$,and $g(z) = z + \sum_{n=2}^{\infty} b_n z^n$ belong to R then so does their Hadamard product*

$$h(z) = f(z) * g(z).$$

**Theorem 3.11.** *If $f, g \in R_p$, $h \in k - UR_p$ and*

$$\varphi^{(p-1)}(z) = h^{(p-1)}(z) * f^{(p-1)}(z) * g^{(p-1)}(z).$$

*Then $\varphi \in k - UR_p$.*

**Proof.** Suppose that

$$m^{(p-1)}(z) = f^{(p-1)}(z) * g^{(p-1)}(z), \tag{3.18}$$

and it is clear from Lemma 2.4 that, $m \in R_p$. From the hypothesis and (3.18), we have

$$\varphi^{(p-1)}(z) = h^{(p-1)}(z) * m^{(p-1)}(z). \tag{3.19}$$

From (3.19) and Theorem 3.9, we get the required result.

**Theorem 3.12.** *If $f_1, f_2, f_3, ..., f_n$ belong to $R_p$, $h \in k - UR_p$ and*

$$g^{(p-1)}(z) = f_1^{(p-1)}(z) * f_2^{(p-1)}(z) * f_3^{(p-1)}(z) * ... * f_n^{(p-1)}(z) * h^{(p-1)}(z). \tag{3.20}$$

*Then $g \in k - UR_p$.*

**Proof.** For proving the above Theorem, we use the principle of mathematical induction. For $n = 2$, we have proved Theorem 3.11, thus (3.20) hold true for $n = 2$. Suppose that (3.20) hold true for $n = k$; that is,

$$g^{(p-1)}(z) = f_1^{(p-1)}(z) * f_2^{(p-1)}(z) * f_3^{(p-1)}(z) * ... * f_k^{(p-1)}(z) * h^{(p-1)}(z). \tag{3.21}$$

Then $g \in k - UR_p$.

We have to prove that (3.20) hold true for $n = k + 1$, for this, consider

$$g^{(p-1)}(z) = f_1^{(p-1)}(z) * f_2^{(p-1)}(z) * f_3^{(p-1)}(z) * ... * f_{k+1}^{(p-1)}(z) * h^{(p-1)}(z). \tag{3.22}$$

Let

$$M^{(p-1)} = f_1^{(p-1)} * f_2^{(p-1)} * f_3^{(p-1)} * \dots\dots\dots * f_k^{(p-1)} * h^{(p-1)}$$

Then by hypothesis $M \in k - UR_p$. Now (3.22) becomes

$$g^{(p-1)}(z) = (M^{(p-1)} * f_{k+1}^{(p-1)})(z). \tag{3.23}$$

Using Theorem 3.9, from (3.23), we have

$$Re\left(\frac{g^{(p)}(z) + zg^{(p+1)}(z)}{p!}\right) > \frac{k}{k+1}. \tag{3.24}$$

(3.24) now implies that $g \in k - UR_p$. Therefore, the result is true for $n = k + 1$ and hence by using mathematical induction, (3.20) holds true for all $n \geq 2$. This completes the proof.

**Theorem 3.13.** *If $f$, $g \in k - UR_p$ and*

$$h^{(p-1)}(z) = f^{(p-1)}(z) * g^{(p-1)}(z).$$

*Then h belong to the class $k - UR_p$.*

**Proof.** Since

$$h^{(p-1)}(z) = f^{(p-1)}(z) * g^{(p-1)}(z). \tag{3.25}$$

Differentiation yields

$$zh^{(p)}(z) = f^{(p)}(z) * g^{(p-1)}(z). \tag{3.26}$$

After simplification, we have

$$Re\left(\frac{h^{(p)}(z) + zh^{(p+1)}(z)}{p!}\right) = Re\left(\left(\frac{f^{(p)}(z) + zf^{(p+1)}(z)}{p!}\right) * \left(\frac{g^{(p-1)}(z)}{zp!}\right)\right). \tag{3.27}$$

From (3.27), (3.1), (3.11) and Lemma 2.1 with $\alpha = \frac{k}{k+1}$ and $\beta = \frac{2k+1}{2k+2}$, we have

$$Re\left(\frac{h^{(p)}(z) + zh^{(p+1)}(z)}{p!}\right) > \frac{k}{k+1}. \tag{3.28}$$

(3.28) implies that $h$ belong to $k - UR_p$.

Our next result give us a sufficient condition for the class $k - UR_p$.

**Theorem 3.14.** *Let $f \in A(p)$ satisfies*

$$\sum_{n=1}^{\infty} \frac{(k-1)(n+1)(p+n)!}{p!n!} \left|a_{n+p}\right| < 1. \tag{3.29}$$

*Then $f \in k - UR_p$.*

**Proof.** To prove the required result it is sufficient to show that

$$k\left|\frac{f^{(p)}(z)+zf^{(p+1)}(z)}{p!}-1\right|-Re\left(\frac{f^{(p)}(z)+zf^{(p+1)}(z)}{p!}-1\right)<1 \tag{3.30}$$

Now

$$\begin{aligned}
&k\left|\frac{f^{(p)}(z)+zf^{(p+1)}(z)}{p!}-1\right|-Re\left(\frac{f^{(p)}(z)+zf^{(p+1)}(z)}{p!}-1\right)\\
\leq\ &(k-1)\left|\frac{f^{(p)}(z)+zf^{(p+1)}(z)}{p!}-1\right|\\
=\ &(k-1)\left|\frac{f^{(p)}(z)+zf^{(p+1)}(z)-p!}{p!}\right|\\
=\ &(k-1)\left|\sum_{n=1}^{\infty}\frac{(n+1)(p+n)!}{p!n!}a_{n+p}z^{n}\right|.
\end{aligned}$$

This can be written as

$$\begin{aligned}
&k\left|\frac{f^{(p)}(z)+zf^{(p+1)}(z)}{p!}-1\right|-Re\left(\frac{f^{(p)}(z)+zf^{(p+1)}(z)}{p!}-1\right)\\
\leq\ &(k-1)\left|\sum_{n=1}^{\infty}\frac{(n+1)(p+n)!}{p!n!}a_{n+p}\right||z^{n}|
\end{aligned} \tag{3.31}$$

(3.31) is bounded above by 1 if (3.29) is satisfied. This completes the proof.

## Conflicts of Interest

All authors declare no conflicts of interest in this paper.

## References


1. L. Fejer, *Uber die positivitat von summen, die nach trigonometrischen order Legendreschen funktionen fortschreiten*, Acta Litt. Ac. Sci. Szeged., **2** (1925), 75-86.
2. A. W. Goodman, *On uniformly convex functions*, Ann. Polon. Math., **56** (1991), 87-92.
3. A. W. Goodman, *On uniformly starlike functions*, J. Math. Anal. Appl., **155** (1991), 364-370.
4. S. Kanas and A. Wiśniowska, *Conic regions and k-uniform convexity*, J. Comput. Appl. Math., **105** (1999), 327-336.
5. S. Kanas and A. Wiśniowska, *Conic domains and starlike functions*, Rev. Roumaine Math. Pures Appl., **45** (2000), 647-657.
6. D. W. Minda, *A unified treatment of some special classes of univalent functions*, Proceedings of the conference on complex analysis, Conf. Proc. Lecture Notes Anal., International Press, Massachusetts, 1994, 157-169.

7. K. I. Noor and N. Khan, *Some convolution properties of a subclass of p-valent functions*, Maejo Int. J. Sci. Technol., **9** (2015), 181-192.

8. K. I. Noor, Q. Z. Ahmad and M. A. Noor, On some subclasses of analytic functions defined by fractional derivative in the conic regions, Appl. Math. Inf. Sci., **9** (2015), 819-824.

9. K. I. Noor, J. Sokol and Q. Z. Ahmad, Applications of conic type regions to subclasses of meromorphic univalent functions with respect to symmetric points, RACSAM, 2016, 1-14.

10. M. Nunokawa, S. Hussain, N. Khan and Q. Z. Ahmad, A subclass of analytic functions related with conic domain, J. Clas. Anal., **9** (2016), 137-149.

11. S. Ozaki, *On the theory of multivalent functions*, Sci. Rep. Tokyo Bunrika Daigaku A., **40** (1935), 167-188.

12. S. Ponnusamy and V. Singh, *Convolution properties of some classes of analytic functions*, J. Math. Sci., **89** (1998), 1008-1020.

13. W. Rogosinski and G. Szego, *Uber die abschimlte von potenzreihen die in ernein kreise beschrankt bleiben.* Math. Z., **28** (1928), 73-94.

14. F. Ronning, *On starlike functions associated with parabolic regions*, Ann. Univ. Mariae Curie-Sklodowska, Sect A., **45** (1991), 117-122.

15. R. Singh and S. Singh, *Convolution properties of a class of starlike functions*, Proc.Amer. Math. Soc., **106** (1989), 145-152.

# 11

# A logarithmically improved regularity criterion for the 3D MHD equations in Morrey-Campanato space

**Sadek Gala**[1,2,*] **Maria Alessandra Ragusa**[2]

[1] Department of Mathematics, University of Mostaganem, Box 227, Mostaganem, 27000, Algeria

[2] Dipartimento di Matematicae Informatica, Università di Catania Viale Andrea Doria, 6, 95125 Catania, Italy

* **Correspondence:** Email: sadek.gala@gmail.com; maragusa@dmi.unict.it.

**Abstract:** In this paper, we will establish a sufficient condition for the regularity criterion to the 3D MHD equation in terms of the derivative of the pressure in one direction. It is shown that if the partial derivative of the pressure $\partial_3\pi$ satisfies the logarithmical Serrin type condition $\partial_3\pi$ satisfies the logarithmical Serrin type condition

$$\int_0^T \frac{\|\partial_3\pi(s)\|_{\mathcal{M}_{2,\frac{3}{r}}}^{\frac{2}{2-r}}}{1+\ln(1+\|b(s)\|_{L^4})}ds < \infty \quad \text{for } 0 < r < 1,$$

then the solution $(u, b)$ remains smooth on $[0, T]$. Compared to the Navier-Stokes result, there is a logarithmic correction involving $b$ in the denominator.

**Keywords:** MHD equations; regularity criteria

---

## 1. Introduction

The MHD equation plays a significant role of mathematical model in fluid dynamics, which can be stated as follows :

$$\begin{cases} \partial_t u - \Delta u + u\cdot\nabla u + \nabla\pi - b\cdot\nabla b = 0, \\ \partial_t b - \Delta b + u\cdot\nabla b - b\cdot\nabla u = 0, \\ \nabla\cdot u = \nabla\cdot b = 0, \\ u(x,0) = u_0(x), \qquad b(x,0) = b_0(x). \end{cases} \tag{1.1}$$

Here $u = u(x,t) \in \mathbb{R}^3$ is the velocity field, $\pi = \pi(x,t) \in \mathbb{R}$, $b = b(x,t) \in \mathbb{R}^3$ denote the velocity vector, scalar pressure and the magnetic field of the fluid, respectively, while $u_0(x)$ and $b_0(x)$ are given initial velocity and initial magnetic fields with $\nabla \cdot u_0 = \nabla \cdot b_0 = 0$ in the sense of distribution.

In their work, Sermange and Temam [19] (see also Duvaut and Lions [6]) proved that the MHD equations admit at least one global weak solution for any divergence-free initial data $(u_0, b_0) \in L^2(\mathbb{R}^3)$ and it has a (unique) local strong solution, if additionally, $(u_0, b_0)$ belongs to some Sobolev space $H^s(\mathbb{R}^3)$ with $s \geq 3$. However, whether a local strong solution can exist globally, or equivalently, whether global weak solutions are smooth is an open and challenge problem.

There are many known mathematical results on the three-dimentional MHD equations (see [4, 5, 10, 14, 20, 21, 22, 25, 26] and the references therein). Realizing the dominant role played by the velocity field, He and Xin [10] were able to derive criteria in terms of the velocity field $u$ alone. In particular, a scaling invariant regularity criterion in terms of $u$ was established (also by Zhou [25] independently) which shows that a weak solution $(u, b)$ is smooth on a time interval $(0, T]$ if

$$\nabla u \in L^\alpha(0, T; L^\gamma(\mathbb{R}^3)) \quad \text{with} \quad 1 \leq \alpha < \infty, \ \ 3/2 < \gamma \leq \infty \ \text{and} \ \frac{2}{\alpha} + \frac{3}{\gamma} = 2.$$

Moreover, the problem of so-called "regularity criteria via partial components" was shown in [3, 9, 11, 12, 13, 15, 17, 23, 24, 27].

Recently, Cao and Wu in [3] presented the regularity criteria on the derivatives of the pressure in one direction. More precisely, they proved that if

$$\frac{\partial \pi}{\partial x_3}(x, t) \in L^\alpha\left(0, T; L^\gamma(\mathbb{R}^3)\right) \quad \text{with} \ \frac{2}{\alpha} + \frac{3}{\gamma} \leq \frac{7}{4} \ \text{and} \ \frac{12}{7} \leq \gamma \leq \infty, \tag{1.2}$$

then $(u, b)$ is smooth on $\mathbb{R}^3 \times [0, T]$. Later, [13] and [24] improve condition (1.2) as:

$$\frac{\partial \pi}{\partial x_3}(x, t) \in L^\alpha\left(0, T; L^\gamma(\mathbb{R}^3)\right) \quad \text{with} \ \frac{2}{\alpha} + \frac{3}{\gamma} \leq 2 \ \text{and} \ \frac{3}{2} \leq \gamma \leq \infty. \tag{1.3}$$

Very recently, Benbernou et al. [2] extend (1.3) to the homogeneous Morrey-Campanato space $\dot{\mathcal{M}}_{2,\frac{3}{r}}(\mathbb{R}^3)$. to obtain the regularity of weak solutions. This space has been used successfully in the study of the uniqueness of weak solutions for the Navier-Stokes equations in [16] where it is pointed out that

$$L^{\frac{3}{r}}\left(\mathbb{R}^3\right) \subset L^{\frac{3}{r},\infty}\left(\mathbb{R}^3\right) \subset \dot{\mathcal{M}}_{2,\frac{3}{r}}\left(\mathbb{R}^3\right).$$

The purpose of this manuscript is to establish a logarithmically improved regularity criterion in terms of the derivatives of the pressure in one direction of the systems (1.1). Our result can be stated as follows.

**Theorem 1.1.** *(regularity criterion) Let* $(u_0, b_0) \in L^2(\mathbb{R}^3) \cap L^4(\mathbb{R}^3)$ *with* $\nabla \cdot u_0 = \nabla \cdot b_0 = 0$. *Suppose that* $(u, b)$ *is a weak solution to the MHD equations (1.1) in the time interval* $[0, T)$ *for some* $0 < T < \infty$. *If the pressure* $\pi(x, t)$ *satisfies the condition :*

$$\int_0^T \frac{\|\partial_3 \pi(s)\|_{\dot{\mathcal{M}}_{2,\frac{3}{r}}}^{\frac{2}{2-r}}}{1 + \ln(1 + \|b\|_{L^4})} ds < \infty \quad \textit{for } 0 < r < 1,$$

*then* $(u, b)$ *is a regular solution on* $\mathbb{R}^3 \times [0, T]$.

Theorem 1.1 is also true for the 3-D incompressible Navier-Stokes equations, so it gives extensions for previous results in [2, 1, 3, 12, 17, 23]. Definitions and basic properties of the Morrey-Campanato spaces can be find in [28] and the references therein. For concision, we omit them here.

Now we are in the position to prove Theorem 1.1.

## 2. Proof of Theorem 1.1

Throughout this paper, $C$ denotes a generic positive constant (generally large), it may be different from line to line. In order to prove regularity, we need to establish the $L^4$ bound of $(u, b)$ and the desired regularity then follows from the standard Serrin-type criteria on the 3D MHD equations.

Instead of considering the equations in the form (1.1), we rewrite it in the following form as that in [7, 8]:

$$\begin{cases} \partial_t w^+ + w^- \cdot \nabla w^+ = \Delta w^+ - \nabla \pi, \\ \partial_t w^- + w^+ \cdot \nabla w^- = \Delta w^- - \nabla \pi, \\ \nabla \cdot w^+ = \nabla \cdot w^- = 0, \\ w^+(x,0) = u_0 + b_0, \quad w^-(x,0) = u_0 - b_0, \end{cases} \tag{2.1}$$

with $w^\pm := u \pm b$.

First, taking the inner product of $(2.1)_1$ with $(0, 0, w_3^+ |w_3^+|^2)$, we have

$$\frac{1}{4}\frac{d}{dt}\int_{\mathbb{R}^3} |w_3^+|^4 \, dx + \int_{\mathbb{R}^3} (w^- \cdot \nabla) w_3^+ . |w_3^+|^2 \, dx$$
$$= \int_{\mathbb{R}^3} \Delta w_3^+ |w_3^+|^2 \, dx - \int_{\mathbb{R}^3} \frac{\partial \pi}{\partial x_3} w_3^+ |w_3^+|^2 \, dx.$$

Integrating by parts over $\mathbb{R}^3$ and using the divergence free property $\nabla \cdot w^+ = 0$ into account, we get

$$\int_{\mathbb{R}^3} (w^- \cdot \nabla) w_3^+ \cdot |w_3^+|^2 \, dx = 0.$$

For the second integral term, applying the integration by parts and the incompressible conditions again yield

$$\int_{\mathbb{R}^3} \Delta w_3^+ |w_3^+|^2 \, dx = -\frac{3}{4}\int_{\mathbb{R}^3} \left|\nabla |w_3^+|^2\right|^2 dx.$$

We easily get

$$\frac{1}{4}\frac{d}{dt}\int_{\mathbb{R}^3} |w_3^+|^4 \, dx + \frac{3}{4}\int_{\mathbb{R}^3} \left|\nabla |w_3^+|^2\right|^2 dx = -\int_{\mathbb{R}^3} \frac{\partial \pi}{\partial x_3} w_3^+ |w_3^+|^2 \, dx. \tag{2.2}$$

Similarly, taking the inner product of the second equation of (2.1) with $(0, 0, w_3^- |w_3^-|^2)$, we obtain

$$\frac{1}{4}\frac{d}{dt}\int_{\mathbb{R}^3} |w_3^-|^4 \, dx + \frac{3}{4}\int_{\mathbb{R}^3} \left|\nabla |w_3^-|^2\right|^2 dx = -\int_{\mathbb{R}^3} \frac{\partial \pi}{\partial x_3} w_3^- |w_3^-|^2 \, dx. \tag{2.3}$$

Summing (2.2) and (2.3) together yields

$$\frac{1}{4}\frac{d}{dt}\int_{\mathbb{R}^3} \left(|w_3^+|^4 + |w_3^-|^4\right) dx + \frac{3}{4}\int_{\mathbb{R}^3} \left(\left|\nabla |w_3^+|^2\right|^2 + \left|\nabla |w_3^-|^2\right|^2\right) dx$$

$$
\begin{aligned}
&= -\int_{\mathbb{R}^3} \frac{\partial \pi}{\partial x_3} w_3^+ \left|w_3^+\right|^2 dx - \int_{\mathbb{R}^3} \frac{\partial \pi}{\partial x_3} w_3^- \left|w_3^-\right|^2 dx \\
&= J_1 + J_2. \qquad (2.4)
\end{aligned}
$$

In what follows, we will deal with each term on the right-hand side of (2.4) separately. We estimate $\left\|\frac{\partial \pi}{\partial x_3} \cdot \left|w_3^+\right|^2\right\|_{L^2}$ as follows :

$$
\begin{aligned}
\left\|\frac{\partial \pi}{\partial x_3} \cdot \left|w_3^+\right|^2\right\|_{L^2} &\le C \left\|\frac{\partial \pi}{\partial x_3}\right\|_{\dot{\mathcal{M}}_{2,\frac{3}{r}}} \left\|\left|w_3^+\right|^2\right\|_{\dot{B}_{2,1}^r} \\
&\le C \left\|\frac{\partial \pi}{\partial x_3}\right\|_{\dot{\mathcal{M}}_{2,\frac{3}{r}}} \left\|\nabla \left|w_3^+\right|^2\right\|_{L^2}^r \left\|\left|w_3^+\right|^2\right\|_{L^2}^{1-r}.
\end{aligned}
$$

Here we have used the following inequality due to Machihara and Ozawa [18]

$$
\|f\|_{\dot{B}_{2,1}^r} \le C \|f\|_{L^2}^{1-r} \|\nabla f\|_{L^2}^r \quad \text{for } 0 < r < 1.
$$

Hence, it follows from the Hölder inequality and Young's inequality that

$$
\begin{aligned}
|J_1| &\le \int_{\mathbb{R}^3} \left|\frac{\partial \pi}{\partial x_3}\right| |w_3^+|^3 dx \\
&\le C \left\|\frac{\partial \pi}{\partial x_3} \cdot \left|w_3^+\right|^2\right\|_{L^2} \left\|w_3^+\right\|_{L^2} \\
&\le C \left(\left\|\frac{\partial \pi}{\partial x_3}\right\|_{\dot{\mathcal{M}}_{2,\frac{3}{r}}}^{\frac{2}{1-r}} \left\|w_3^+\right\|_{L^4}^4\right)^{\frac{1-r}{2}} \left(\left\|\nabla \left|w_3^+\right|^2\right\|_{L^2}^2\right)^{\frac{r}{2}} \left(\left\|w^+\right\|_{L^2}^2\right)^{\frac{1}{2}} \\
&\le C \left\|\frac{\partial \pi}{\partial x_3}\right\|_{\dot{\mathcal{M}}_{2,\frac{3}{r}}}^{\frac{2}{1-r}} \left\|w_3^+\right\|_{L^4}^4 + \frac{1}{2} \left\|\nabla \left|w_3^+\right|^2\right\|_{L^2}^2 + C \left\|w^+\right\|_{L^2}^2, \qquad (2.5)
\end{aligned}
$$

Note that the weak solution $(u, b) \in L^\infty(0, T; L^2(\mathbb{R}^3))$, this leads to

$$
(w^+, w^-) \in L^\infty(0, T; L^2(\mathbb{R}^3)).
$$

Similarly, one can prove that

$$
|J_2| \le C \left\|\frac{\partial \pi}{\partial x_3}\right\|_{\dot{\mathcal{M}}_{2,\frac{3}{r}}}^{\frac{2}{1-r}} \left\|w_3^-\right\|_{L^4}^4 + \frac{1}{2} \left\|\nabla \left|w_3^-\right|^2\right\|_{L^2}^2 + C \left\|w^-\right\|_{L^2}^2. \qquad (2.6)
$$

Substituting (2.5) and (2.6) into (2.4), we obtain

$$
\begin{aligned}
&\frac{1}{4}\frac{d}{dt} \int_{\mathbb{R}^3} \left(\left|w_3^+\right|^4 + \left|w_3^-\right|^4\right) dx + \frac{1}{4} \int_{\mathbb{R}^3} \left(\left|\nabla \left|w_3^+\right|^2\right|^2 + \left|\nabla \left|w_3^-\right|^2\right|^2\right) dx \\
&\le C \left(\left\|\frac{\partial \pi}{\partial x_3}\right\|_{\dot{\mathcal{M}}_{2,\frac{3}{r}}}^{\frac{2}{1-r}} + 1\right) \left(\left\|w_3^+\right\|_{L^4}^4 + \left\|w_3^-\right\|_{L^4}^4\right), \qquad (2.7)
\end{aligned}
$$

for all $0 \leq t < T$. Setting

$$J = \left\| \frac{\partial \pi}{\partial x_3} \right\|_{\dot{\mathcal{M}}_{2,\frac{3}{r}}}^{\frac{2}{1-r}} \left( e + \|w_3^+\|_{L^4}^4 + \|w_3^-\|_{L^4}^4 \right).$$

On the other hand, we see that

$$\begin{aligned} 1 + \ln(1 + \|b\|_{L^4}) &\leq 1 + \ln\left(1 + \|b\|_{L^4}^4 + \frac{9}{8}\right) \\ &\leq 1 + \ln\left(e + \|b\|_{L^4}^4\right), \end{aligned}$$

where we have used the following inequality

$$x \leq x^4 + \frac{9}{8} \text{ for all } x \geq 0.$$

Consequently, $J$ can be estimated as follows:

$$\begin{aligned} J &= \frac{\left\| \frac{\partial \pi}{\partial x_3} \right\|_{\dot{\mathcal{M}}_{2,\frac{3}{r}}}^{\frac{2}{1-r}}}{1 + \ln(1 + \|b\|_{L^4})} (e + \|w_3^+\|_{L^4}^4 + \|w_3^-\|_{L^4}^4) \left[1 + \ln(1 + \|b\|_{L^4})\right] \\ &\leq \frac{\left\| \frac{\partial \pi}{\partial x_3} \right\|_{\dot{\mathcal{M}}_{2,\frac{3}{r}}}^{\frac{2}{1-r}}}{1 + \ln(1 + \|b\|_{L^4})} (e + \|w_3^+\|_{L^4}^4 + \|w_3^-\|_{L^4}^4) \left[1 + \ln(e + \|b\|_{L^4}^4)\right] \qquad (2.8) \\ &\leq \frac{\left\| \frac{\partial \pi}{\partial x_3} \right\|_{\dot{\mathcal{M}}_{2,\frac{3}{r}}}^{\frac{2}{1-r}}}{1 + \ln(1 + \|b\|_{L^4})} (e + \|w_3^+\|_{L^4}^4 + \|w_3^-\|_{L^4}^4) \left[1 + \ln(e + \|w_3^+\|_{L^4}^4 + \|w_3^-\|_{L^4}^4)\right]. \end{aligned}$$

Inserting (2.8) into (2.7) and setting

$$F(t) = e + \|w_3^+(t)\|_{L^4}^4 + \|w_3^-(t)\|_{L^4}^4,$$

we obtain

$$\frac{dF}{dt} \leq C \frac{\left\| \frac{\partial \pi}{\partial x_3} \right\|_{\dot{\mathcal{M}}_{2,\frac{3}{r}}}^{\frac{2}{1-r}}}{1 + \ln(1 + \|b\|_{L^4})} (1 + \ln F) F + CF.$$

for all $t \in [0, T]$. Thank's to Gronwall inequality, we get

$$F(t) \leq F(0) \exp\left( C \int_0^t \frac{\left\| \frac{\partial \pi}{\partial x_3}(s) \right\|_{\dot{\mathcal{M}}_{2,\frac{3}{r}}}^{\frac{2}{1-r}}}{1 + \ln(1 + \|b(s)\|_{L^4})} (1 + \ln F(s)) ds \right) \exp(CT),$$

which implies

$$1 + \ln F(t) \leq CT + \ln F(0) + C \int_0^t \frac{\left\| \frac{\partial \pi}{\partial x_3}(s) \right\|_{\dot{\mathcal{M}}_{2,\frac{3}{r}}}^{\frac{2}{1-r}}}{1 + \ln(1 + \|b(s)\|_{L^4})} (1 + \ln F(s)) ds.$$

Applying Gronwall's inequality again, one has

$$\ln F(t) \leq c(u_0, b_0, T)\exp\left( C\int_0^T \frac{\left\|\frac{\partial\pi}{\partial x_3}(s)\right\|_{\dot{\mathcal{M}}_{2,\frac{3}{r}}}^{\frac{2}{1-r}}}{1+\ln(1+\|b(s)\|_{L^4})}ds \right),$$

which implies that

$$\sup_{0\leq t\leq T}\left(\left\|w^+(.,t)\right\|_{L^4} + \left\|w^-(.,t)\right\|_{L^4}\right) < \infty \tag{2.9}$$

Hence, it follows from the triangle inequality and (2.9) that

$$\begin{aligned}
\sup_{0\leq t\leq T}\|u(.,t)\|_{L^4} &= \frac{1}{2}\sup_{0\leq t\leq T}\|(u+b)(.,t)+(u-b)(.,t)\|_{L^4}\\
&\leq \frac{1}{2}\sup_{0\leq t\leq T}\left(\|(u+b)(.,t)\|_{L^4}+\|(u-b)(.,t)\|_{L4}\right)\\
&\leq \frac{1}{2}\sup_{0\leq t\leq T}\left(\left\|w^+(.,t)\right\|_{L^4}+\left\|w^-(.,t)\right\|_{L^4}\right) < \infty
\end{aligned}$$

and

$$\begin{aligned}
\sup_{0\leq t\leq T}\|b(.,t)\|_{L^4} &= \frac{1}{2}\sup_{0\leq t\leq T}\|(u+b)(.,t)-(u-b)(.,t)\|_{L^4}\\
&\leq \frac{1}{2}\sup_{0\leq t\leq T}\left(\|(u+b)(.,t)\|_{L^4}+\|(u-b)(.,t)\|_{L^4}\right)\\
&\leq \frac{1}{2}\sup_{0\leq t\leq T}\left(\left\|w^+(.,t)\right\|_{L^4}+\left\|w^-(.,t)\right\|_{L^4}\right) < \infty.
\end{aligned}$$

Thus,

$$\sup_{0\leq t\leq T}\left(\|u(.,t)\|_{L^4}+\|b(.,t)\|_{L^4}\right) < \infty. \tag{2.10}$$

This completes the proof of Theorem 1.1.

## Acknowledgments

Part of the work was carried out while the first author was long term visitor at University of Catania. The hospitality and support of Catania University are graciously acknowledged.

## Conflict of Interest

All authors declare no conflicts of interest in this paper.

## References

1. L. Berselli and G. Galdi, *Regularity criteria involving the pressure for the weak solutions to the Navier-Stokes equations,* Proc. Amer. Math. Soc., **130** (2002), 3585-3595.

2. S. Benbernou, M Terbeche, and M.A. Ragusa, *A logarithmically improved regularity criterion for the MHD equations in terms of one directional derivative of the pressure,* Applicable Analysis, http://dx.doi.org/10.1080/00036811.2016.1207246.

3. C. Cao and J. Wu, *Two regularity criteria for the 3D MHD equations,* J. Differential Equations, **248** (2010), 2263-2274.

4. Q. Chen, C. Miao, and Z. Zhang, *On the regularity criterion of weak solution for the 3D viscous magneto-hydrodynamics equations,* Comm. Math. Phys., **284** (2008), 919-930.

5. H. Duan, *On regularity criteria in terms of pressure for the 3D viscous MHD equations,* Appl. Anal., **91** (2012), 947-952.

6. G. Duvaut and J.-L. Lions, *Inéquations en thermoélasticité et magnétohydrodynamique,* Arch. Ration. Mech. Anal., **46** (1972), 241-279.

7. S. Gala, *Extension criterion on regularity for weak solutions to the 3D MHD equations,* Math.. Meth. Appl. Sci., **33** (2010), 1496-1503.

8. C. He and Y. Wang, *Remark on the regularity for weak solutions to the magnetohydrodynamic equations,* Math. Methods Appl. Sci., **31** (2008), 1667-1684.

9. L. Ni, Z. Guo, and Y. Zhou, *Some new regularity criteria for the 3D MHD equations,* J. Math. Anal. Appl., **396** (2012), 108-118.

10. C. He and Z. Xin, *On the regularity of weak solutions to the magnetohydrodynamic equations,* J. Differential Equations, **213** (2005), 235-254.

11. E. Ji and J. Lee, *Some regularity criteria for the 3D incompressible magnetohydrodynamics,* J. Math. Anal. Appl., **369** (2010), 317-322.

12. X. Jia and Y. Zhou, *Regularity criteria for the 3D MHD equations via partial derivatives,* Kinet. Relat. Models, **5** (2012), 505-516.

13. X. Jia and Y. Zhou, *Regularity criteria for the 3D MHD equations via partial derivatives,* II. Kinet. Relat. Models, **7** (2014), no. 2, 291-304.

14. X. Jia and Y. Zhou, *Ladyzhenskaya-Prodi-Serrin type regularity criteria for the 3D incompressible MHD equations in terms of* $3 \times 3$ *mixture matrices,* Nonlinearity, **28** (2015), 3289-3307.

15. X. Jia and Y. Zhou, *A new regularity criterion for the 3D incompressible MHD equations in terms of one component of the gradient of pressure,* J. Math. Anal. Appl., **396** (2012), 345-350.

16. P.G. Lemarié-Rieusset, *The Navier-Stokes equations in the critical Morrey-Campanato space,* Rev. Mat. Iberoam., **23** (2007), no. 3, 897-930.

17. H. Lin and L. Du, *Regularity criteria for incompressible magnetohydrodynamics equations in three dimensions,* Nonlinearity, **26** (2013), 219-239.

18. S. Machihara and T. Ozawa, *Interpolation inequalities in Besov spaces,* Proc. Amer. Math. Soc., **131** (2003), 1553-1556.

19. M. Sermange and R. Temam, *Some mathematical questions related to the MHD equations,* Comm. Pure Appl. Math., **36** (1983), 635-664.

20. J. Wu, *Viscous and inviscid magnetohydrodynamics equations,* J. Anal. Math., **73** (1997), 251-265.

21. J. Wu, *Bounds and new approaches for the 3D MHD equations,* J. Nonlinear Sci., **12** (2002), 395-413.

22. J. Wu, *Regularity results for weak solutions of the 3D MHD equations,* Discrete Contin. Dyn. Syst., **10** (2004), 543-556.

23. K. Yamazaki, *Remarks on the regularity criteria of generalized MHD and Navier-Stokes systems,* J. Math. Phys., **54** (2013), 011502, 16pp.

24. Z. Zhang, P. Li, and G. Yu, *Regularity criteria for the 3D MHD equations via one directional derivative of the pressure,* J. Math. Anal. Appl., **401** (2013), 66-71.

25. Y. Zhou, *Remarks on regularities for the 3D MHD equations,* Discrete Contin. Dyn. Syst., **12** (2005), 881-886.

26. Y. Zhou, *Regularity criteria for the 3D MHD equations in terms of the pressure,* Int. J. Non-Linear Mech., **41** (2006), 1174-1180.

27. Y. Zhou and S. Gala, *Regularity criteria for the solutions to the 3D MHD equations in multiplier space,* Z. Angrew. Math. Phys., **61** (2010), 193-199.

28. Y. Zhou and S. Gala, *Regularity Criteria in Terms of the Pressure for the Navier-Stokes Equations in the Critical Morrey-Campanato Space,* Z. Anal. Anwend., **30** (2011), 83-93.

# 12

# Exact solutions of the generalized (2+1)-dimensional BKP equation by the $G'/G$-expansion method and the first integral method

**Huaji Cheng and Yanxia Hu***

School of Mathematics and Physics, North China Electric Power University, Beijing, 102206, China

* **Correspondence:** yxiahu@163.com

**Abstract:** In this paper, the $G'/G$-expansion method and the first integral method are performed to the generalized (2+1)-dimensional BKP equation. Rational function solutions, periodic function solutions and hyperbolic function solutions of the equation are obtained under some parametric conditions.

**Keywords:** Generalized (2+1)-dimensional BKP equation; $G'/G$-expansion method; first integral method
**Mathematics Subject Classification:** 34A05,34A34

## 1. Introduction

During the past decades, the exact solutions of nonlinear partial differential equations have been investigated by many authors. Meanwhile, many powerful methods have been proposed by them, such as Backlund transformation method [1], multiple exp-function method [2], homogeneous balance principle [3], tanh-sech method [4], $G'/G$-expansion method [5–7], the first integral method [8, 9] and so on.

The $G'/G$-expansion method was first presented by Wang [5] which can be used to deal with all types of nonlinear evolution equations. The first integral method was first proposed by Feng [8] for obtaining the exact solutions of Burgers-KdV equation which is based on the ring theory of commutative algebra. The basic idea of the first integral method is to construct a first integral with polynomial coefficients of an explicit form to an equivalent autonomous planer system by using the division theorem. Both the $G'/G$-expansion method and the first integral method are powerful methods for computing the exact solutions of nonlinear partial differential equations. They are direct, elementary and effective algebraic methods.

In this paper, we consider the following generalized (2+1)-dimensional BKP equation [10]

$$\begin{cases}(w^n)_t + (w^m)_{xxx} + (w^m)_{yyy} + \alpha(uw)_x + \beta(vw)_y = 0,\\ u_y = w_x,\\ v_x = w_y,\end{cases} \tag{1.1}$$

where $\alpha, \beta$ are arbitrary constants and $\alpha + \beta \neq 0, m, n$ are integers and $m, n \geq 2$. In [10], authors studied traveling wave solutions in the parameter space of this system by bifurcation theory of dynamical systems and they obtained some exact explicit parametric representations of periodic cusp wave solutions, solitary wave solutions and compacton solutions. In this paper, we continue to consider the problem of solving system (1.1) by using the $G'/G$-expansion method and the first integral method and we obtain the rational function solutions, periodic function solutions and the hyperbolic function solutions of (1.1) under some parametric conditions and the values of $m$, $n$ in several cases.

Specially, when $m = 1, n = 1, \alpha = \beta = 6$, (1.1) becomes

$$\begin{cases}w_t + w_{xxx} + w_{yyy} + 6(uw)_x + 6(vw)_y = 0,\\ u_y = w_x,\\ v_x = w_y.\end{cases}$$

It is the famous (2+1)-dimensional BKP equation which was introduced by Date et al. [11] and describes the processes of interaction of exponentially localized structures. It is one of a hierarchy of integrable systems emerging from a bilinear identity related to a Clifford algebra which is generated by two neutral fermion fields [12]. This equation has been studied by using many methods, such as the sine-cosine method [13], the $G'/G$-expansion method [6], the improved $G'/G$-expansion method [14] and so on.

The aim of this paper is to extract the exact solutions of the generalized (2+1)-dimensional BKP equation by using the $G'/G$-expansion method and the first integral method. The paper is arranged as follows: In section 2, we apply the $G'/G$-expansion method to this equation. In section 3, we apply the first integral method to solve this equation. In section 4, we give the conclusion of the paper.

## 2. Application of the $G'/G$-expansion method to the generalized (2+1)-dimensional BKP equation

We suppose the wave transformations

$$w(x, y, t) = w(\xi), u(x, y, t) = u(\xi), v(x, y, t) = v(\xi), \qquad \xi = k_1 x + l_1 y + \lambda_1 t \tag{2.1}$$

where $k_1, l_1, \lambda_1$ are constants. By using the wave transformations (2.1), (1.1) can be converted into ODEs

$$\begin{cases}\lambda_1(w^n)' + (k_1^3 + l_1^3)(w^m)''' + \alpha k_1(u'w + uw') + \beta l_1(v'w + vw') = 0,\\ l_1 u' = k_1 w',\\ k_1 v' = l_1 w',\end{cases} \tag{2.2}$$

where "′" is the derivative with respect to $\xi$. Integrating the second and third equation of (2.2) and neglecting integral constants, we obtain

$$\begin{cases} l_1 u = k_1 w, \\ k_1 v = l_1 w. \end{cases}$$

Substituting the above equations into the first equation of (2.2) and integrating it, then it becomes

$$\lambda_1 w^n + (k_1^3 + l_1^3)(w^m)'' + (\frac{\alpha k_1^2}{l_1} + \frac{\beta l_1^2}{k_1})w^2 = g, \tag{2.3}$$

where $g$ is an integral constant. We assume that (2.3) has the following formal solutions [7, 15]:

$$w(\xi) = D\left(\frac{G'}{G}\right)^N, \quad D \neq 0, \tag{2.4}$$

where $D$ is a constant to be determined later. $N$ is determined by balancing the linear term of the highest order derivatives with the highest order nonlinear term in (2.3) and $G$ satisfies a second order constant coefficient ODE which is

$$G''(\xi) + \lambda G'(\xi) + \mu G(\xi) = 0, \tag{2.5}$$

where $\lambda$, $\mu$ are constants and will be determined later. Next, we will obtain the exact solutions of (1.1) by considering the values of $m$ and $n$ in several cases.

### 2.1. $m \neq n, m > 2, n > 2$

Balancing $(w^m)''$ with $w^n$ of (2.3), we have $mN + 2 = nN$, i.e., $N = 2/(n - m)$. Thus, we assume

$$w(\xi) = D_1\left(\frac{G'}{G}\right)^{\frac{2}{n-m}}, \quad D_1 \neq 0 \tag{2.6}$$

where $D_1$ is a constant to be determined later. Then, we have

$$\begin{aligned} w^n &= D_1^n\left(\frac{G'}{G}\right)^{\frac{2n}{n-m}}, \qquad w^2 = D_1^2\left(\frac{G'}{G}\right)^{\frac{4}{n-m}}, \\ (w^m)'' &= \frac{2m}{n-m}D_1^m\Big[(\frac{2m}{n-m}+1)\left(\frac{G'}{G}\right)^{\frac{2m}{n-m}+2} + (\frac{4m}{n-m}+1)\lambda\left(\frac{G'}{G}\right)^{\frac{2m}{n-m}+1} \\ &+ \frac{2m}{n-m}(2\mu+\lambda^2)\left(\frac{G'}{G}\right)^{\frac{2m}{n-m}} + (\frac{4m}{n-m}-1)\lambda\mu\left(\frac{G'}{G}\right)^{\frac{2m}{n-m}-1} \\ &+ (\frac{2m}{n-m}-1)\mu^2\left(\frac{G'}{G}\right)^{\frac{2m}{n-m}-2}\Big]. \end{aligned}$$

Substituting the above formulas into (2.3) and collecting all terms with the same order of $G'/G$ together, we can convert the left-hand side of (2.3) into a polynomial in $G'/G$. Then, setting each coefficient of each polynomial to zero, we can derive a set of algebraic equation for $\lambda, \mu$ and $D_1$:

$\left(\frac{G'}{G}\right)^{\frac{2m}{n-m}+2}$ coeff:

$$(k_1^3 + l_1^3)(\frac{2m}{n-m}+1)\frac{2m}{n-m}D_1^m + \lambda_1 D_1^n = 0, \tag{2.7}$$

$\left(\frac{G'}{G}\right)^{\frac{2m}{n-m}+1}$ coeff:

$$(k_1^3 + l_1^3)(\frac{4m}{n-m} + 1)\frac{2m}{n-m}\lambda D_1^m = 0. \tag{2.8}$$

Here, we need to consider the value of $4/(n-m)$ in the following cases:

**Case 1.** $\frac{4}{n-m} = \frac{2m}{n-m} - 1$

$\left(\frac{G'}{G}\right)^{\frac{2m}{n-m}}$ coeff:

$$(k_1^3 + l_1^3)(\frac{2m}{n-m})^2(2\mu + \lambda^2)D_1^m = 0, \tag{2.9}$$

$\left(\frac{G'}{G}\right)^{\frac{2m}{n-m}-1}$ coeff:

$$(k_1^3 + l_1^3)(\frac{4m}{n-m} - 1)\frac{2m}{n-m}\lambda\mu D_1^m + (\frac{\alpha k_1^2}{l_1} + \frac{\beta l_1^2}{k_1})D_1^2 = 0, \tag{2.10}$$

$\left(\frac{G'}{G}\right)^{\frac{2m}{n-m}-2}$ coeff:

$$(k_1^3 + l_1^3)(\frac{2m}{n-m} - 1)\frac{2m}{n-m}\mu^2 D_1^m = 0. \tag{2.11}$$

Solving the set of (2.7)-(2.11), we obtain

$$\lambda = \mu = 0, \quad g = 0, \quad \frac{\alpha k_1^2}{l_1} + \frac{\beta l_1^2}{k_1} = 0, \quad D_1 = \left( - \frac{(k_1^3 + l_1^3)(\frac{2m}{n-m} + 1)\frac{2m}{n-m}}{\lambda_1} \right)^{1/(n-m)}. \tag{2.12}$$

**Case 2.** $\frac{4}{n-m} = \frac{2m}{n-m} - 2$

$\left(\frac{G'}{G}\right)^{\frac{2m}{n-m}}$ coeff:

$$(k_1^3 + l_1^3)(\frac{2m}{n-m})^2(2\mu + \lambda^2)D_1^m = 0, \tag{2.13}$$

$\left(\frac{G'}{G}\right)^{\frac{2m}{n-m}-1}$ coeff:

$$(k_1^3 + l_1^3)(\frac{4m}{n-m} - 1)\frac{2m}{n-m}\lambda\mu D_1^m = 0, \tag{2.14}$$

$\left(\frac{G'}{G}\right)^{\frac{2m}{n-m}-2}$ coeff:

$$(k_1^3 + l_1^3)(\frac{2m}{n-m} - 1)\frac{2m}{n-m}\mu^2 D_1^m + (\frac{\alpha k_1^2}{l_1} + \frac{\beta l_1^2}{k_1})D_1^2 = 0. \tag{2.15}$$

Solving the set of (2.7)-(2.8) and (2.13)-(2.15), we get the same results as those of Case 1.

**Case 3.** $\frac{4}{n-m} \neq \frac{2m}{n-m} - 1$ and $\frac{4}{n-m} \neq \frac{2m}{n-m} - 2$

$\left(\frac{G'}{G}\right)^{\frac{2m}{n-m}}$ coeff:

$$(k_1^3 + l_1^3)(\frac{2m}{n-m})^2(2\mu + \lambda^2)D_1^m = 0, \tag{2.16}$$

$\left(\frac{G'}{G}\right)^{\frac{2m}{n-m}-1}$ coeff:

$$(k_1^3 + l_1^3)(\frac{4m}{n-m} - 1)\frac{2m}{n-m}\lambda\mu D_1^m = 0, \tag{2.17}$$

$\left(\frac{G'}{G}\right)^{\frac{2m}{n-m}-2}$ coeff:

$$(k_1^3 + l_1^3)(\frac{2m}{n-m} - 1)\frac{2m}{n-m}\mu^2 D_1^m = 0, \tag{2.18}$$

$\left(\frac{G'}{G}\right)^{\frac{4}{n-m}}$ coeff:

$$(\frac{\alpha k_1^2}{l_1} + \frac{\beta l_1^2}{k_1})D_1^2 = 0. \tag{2.19}$$

Solving the set of (2.7)-(2.8) and (2.16)-(2.19), we obtain the same results as those of former cases. Substituting (2.12) into (2.5) and (2.6), then, we can get the rational function solutions

$$\begin{aligned}
w(x,y,t) &= \left(-\frac{(k_1^3 + l_1^3)(\frac{2m}{n-m} + 1)\frac{2m}{n-m}}{\lambda_1}\right)^{\frac{1}{n-m}}\left(\frac{C_1}{C_1(k_1x + l_1y + \lambda_1 t) + C_2}\right)^{\frac{2}{n-m}}, \\
v(x,y,t) &= \frac{l_1}{k_1}\left(-\frac{(k_1^3 + l_1^3)(\frac{2m}{n-m} + 1)\frac{2m}{n-m}}{\lambda_1}\right)^{\frac{1}{n-m}}\left(\frac{C_1}{C_1(k_1x + l_1y + \lambda_1 t) + C_2}\right)^{\frac{2}{n-m}}, \\
u(x,y,t) &= \frac{k_1}{l_1}\left(-\frac{(k_1^3 + l_1^3)(\frac{2m}{n-m} + 1)\frac{2m}{n-m}}{\lambda_1}\right)^{\frac{1}{n-m}}\left(\frac{C_1}{C_1(k_1x + l_1y + \lambda_1 t) + C_2}\right)^{\frac{2}{n-m}},
\end{aligned}$$

where $C_1, C_2$ are arbitrary constants and $\alpha k_1^3 + \beta l_1^3 = 0$.

*2.2.* $m = 2, n > 2$

(2.3) becomes

$$\lambda_1 w^n + (k_1^3 + l_1^3)(w^2)'' + (\frac{\alpha k_1^2}{l_1} + \frac{\beta l_1^2}{k_1})w^2 = g. \tag{2.20}$$

Balancing $(w^2)''$ with $w^n$, we have $N = 2/(n-2)$. Thus, (2.20) has the following formal solutions

$$w(\xi) = D_2\left(\frac{G'}{G}\right)^{\frac{2}{n-2}}, \quad D_2 \neq 0, \tag{2.21}$$

where $D_2$ is a constant to be determined later and $G$ satisfies (2.5). Similarly, we can get a set of algebraic equations:

$\left(\frac{G'}{G}\right)^{\frac{4}{n-2}+2}$ coeff:

$$(k_1^3+l_1^3)(\frac{4}{n-2}+1)\frac{4}{n-2}D_2^2+\lambda_1 D_2^n=0, \tag{2.22}$$

$\left(\frac{G'}{G}\right)^{\frac{4}{n-2}+1}$ coeff:

$$(k_1^3+l_1^3)(\frac{8}{n-2}+1)\frac{4}{n-2}\lambda D_2^2=0, \tag{2.23}$$

$\left(\frac{G'}{G}\right)^{\frac{4}{n-2}}$ coeff:

$$(k_1^3+l_1^3)(\frac{4}{n-2})^2(2\mu+\lambda^2)D_2^2+(\frac{\alpha k_1^2}{l_1}+\frac{\beta l_1^2}{k_1})D_2^2=0, \tag{2.24}$$

**I. The case** $g=0$

$\left(\frac{G'}{G}\right)^{\frac{4}{n-2}-1}$ coeff:

$$(k_1^3+l_1^3)(\frac{8}{n-2}-1)\frac{4}{n-2}\lambda\mu D_2^2=0, \tag{2.25}$$

$\left(\frac{G'}{G}\right)^{\frac{4}{n-2}-2}$ coeff:

$$(k_1^3+l_1^3)(\frac{4}{n-2}-1)\frac{4}{n-2}\mu^2 D_2^2=0. \tag{2.26}$$

Solving that set of (2.22)-(2.26), we obtain

$$\lambda=\mu=0,\quad \frac{\alpha k_1^2}{l_1}+\frac{\beta l_1^2}{k_1}=0,\quad D_2=\left(-\frac{(k_1^3+l_1^3)(\frac{4}{n-2}+1)\frac{4}{n-2}}{\lambda_1}\right)^{1/(n-2)}. \tag{2.27}$$

Specially, when $\frac{4}{n-2}-1=0$, i.e., $n=6$, we obtain

$$\lambda=0,\quad \mu=-\frac{\frac{\alpha k_1^2}{l_1}+\frac{\beta l_1^2}{k_1}}{2(k_1^3+l_1^3)},\quad D_2=\left(\frac{-2(k_1^3+l_1^3)}{\lambda_1}\right)^{1/4}. \tag{2.28}$$

Substituting (2.27) into (2.5) and (2.21), then, we can get the rational function solutions

$$w(x,y,t) = \left(-\frac{(k_1^3+l_1^3)(\frac{4}{n-2}+1)\frac{4}{n-2}}{\lambda_1}\right)^{\frac{1}{n-2}}\left(\frac{C_3}{C_3(k_1x+l_1y+\lambda_1t)+C_4}\right)^{\frac{2}{n-2}},$$

$$
\begin{aligned}
v(x,y,t) &= \frac{l_1}{k_1}\Big(-\frac{(k_1^3+l_1^3)(\frac{4}{n-2}+1)\frac{4}{n-2}}{\lambda_1}\Big)^{\frac{1}{n-2}}\Big(\frac{C_3}{C_3(k_1x+l_1y+\lambda_1t)+C_4}\Big)^{\frac{2}{n-2}},\\
u(x,y,t) &= \frac{k_1}{l_1}\Big(-\frac{(k_1^3+l_1^3)(\frac{4}{n-2}+1)\frac{4}{n-2}}{\lambda_1}\Big)^{\frac{1}{n-2}}\Big(\frac{C_3}{C_3(k_1x+l_1y+\lambda_1t)+C_4}\Big)^{\frac{2}{n-2}},
\end{aligned}
$$

where $C_3, C_4$ are arbitrary constants and $\alpha k_1^3+\beta l_1^3=0$. Substituting (2.28) into (2.5) and (2.21), then, we have

$$G''+\Big(-\frac{\frac{\alpha k_1^2}{l_1}+\frac{\beta l_1^2}{k_1}}{2(k_1^3+l_1^3)}\Big)G=0.$$

**Case 1.** $\frac{\frac{\alpha k_1^2}{l_1}+\frac{\beta l_1^2}{k_1}}{2(k_1^3+l_1^3)}>0$

We obtain the hyperbolic function solutions

$$
\begin{aligned}
w(x,y,t) = {} & \Big(\frac{-(\frac{\alpha k_1^2}{l_1}+\frac{\beta l_1^2}{k_1})^2}{2\lambda_1(k_1^3+l_1^3)}\Big)^{1/4}\\
& \left\{\frac{C_5\sinh\Big(\frac{\frac{\alpha k_1^2}{l_1}+\frac{\beta l_1^2}{k_1}}{2(k_1^3+l_1^3)}\Big)^{1/2}(k_1x+l_1y+\lambda_1t)+C_6\cosh\Big(\frac{\frac{\alpha k_1^2}{l_1}+\frac{\beta l_1^2}{k_1}}{2(k_1^3+l_1^3)}\Big)^{1/2}(k_1x+l_1y+\lambda_1t)}{C_5\cosh\Big(\frac{\frac{\alpha k_1^2}{l_1}+\frac{\beta l_1^2}{k_1}}{2(k_1^3+l_1^3)}\Big)^{1/2}(k_1x+l_1y+\lambda_1t)+C_6\sinh\Big(\frac{\frac{\alpha k_1^2}{l_1}+\frac{\beta l_1^2}{k_1}}{2(k_1^3+l_1^3)}\Big)^{1/2}(k_1x+l_1y+\lambda_1t)}\right\}^{1/2},\\
v(x,y,t) = {} & \frac{l_1}{k_1}\Big(\frac{-(\frac{\alpha k_1^2}{l_1}+\frac{\beta l_1^2}{k_1})^2}{2\lambda_1(k_1^3+l_1^3)}\Big)^{1/4}\\
& \left\{\frac{C_5\sinh\Big(\frac{\frac{\alpha k_1^2}{l_1}+\frac{\beta l_1^2}{k_1}}{2(k_1^3+l_1^3)}\Big)^{1/2}(k_1x+l_1y+\lambda_1t)+C_6\cosh\Big(\frac{\frac{\alpha k_1^2}{l_1}+\frac{\beta l_1^2}{k_1}}{2(k_1^3+l_1^3)}\Big)^{1/2}(k_1x+l_1y+\lambda_1t)}{C_5\cosh\Big(\frac{\frac{\alpha k_1^2}{l_1}+\frac{\beta l_1^2}{k_1}}{2(k_1^3+l_1^3)}\Big)^{1/2}(k_1x+l_1y+\lambda_1t)+C_6\sinh\Big(\frac{\frac{\alpha k_1^2}{l_1}+\frac{\beta l_1^2}{k_1}}{2(k_1^3+l_1^3)}\Big)^{1/2}(k_1x+l_1y+\lambda_1t)}\right\}^{1/2},\\
u(x,y,t) = {} & \frac{k_1}{l_1}\Big(\frac{-(\frac{\alpha k_1^2}{l_1}+\frac{\beta l_1^2}{k_1})^2}{2\lambda_1(k_1^3+l_1^3)}\Big)^{1/4}\\
& \left\{\frac{C_5\sinh\Big(\frac{\frac{\alpha k_1^2}{l_1}+\frac{\beta l_1^2}{k_1}}{2(k_1^3+l_1^3)}\Big)^{1/2}(k_1x+l_1y+\lambda_1t)+C_6\cosh\Big(\frac{\frac{\alpha k_1^2}{l_1}+\frac{\beta l_1^2}{k_1}}{2(k_1^3+l_1^3)}\Big)^{1/2}(k_1x+l_1y+\lambda_1t)}{C_5\cosh\Big(\frac{\frac{\alpha k_1^2}{l_1}+\frac{\beta l_1^2}{k_1}}{2(k_1^3+l_1^3)}\Big)^{1/2}(k_1x+l_1y+\lambda_1t)+C_6\sinh\Big(\frac{\frac{\alpha k_1^2}{l_1}+\frac{\beta l_1^2}{k_1}}{2(k_1^3+l_1^3)}\Big)^{1/2}(k_1x+l_1y+\lambda_1t)}\right\}^{1/2},
\end{aligned}
$$

where $C_5, C_6$ are arbitrary constants.

**Case 2.** $\frac{\frac{\alpha k_1^2}{l_1}+\frac{\beta l_1^2}{k_1}}{2(k_1^3+l_1^3)}<0$

We obtain the hyperbolic function solutions

$$w(x,y,t) = \Big(\frac{-(\frac{\alpha k_1^2}{l_1}+\frac{\beta l_1^2}{k_1})^2}{2\lambda_1(k_1^3+l_1^3)}\Big)^{1/4}$$

$$\left\{\frac{-C_7\sin\left(\frac{-(\frac{\alpha k_1^2}{l_1}+\frac{\beta l_1^2}{k_1})}{2(k_1^3+l_1^3)}\right)^{1/2}(k_1x+l_1y+\lambda_1 t)+C_8\cos\left(\frac{-(\frac{\alpha k_1^2}{l_1}+\frac{\beta l_1^2}{k_1})}{2(k_1^3+l_1^3)}\right)^{1/2}(k_1x+l_1y+\lambda_1 t)}{C_7\cos\left(\frac{-(\frac{\alpha k_1^2}{l_1}+\frac{\beta l_1^2}{k_1})}{2(k_1^3+l_1^3)}\right)^{1/2}(k_1x+l_1y+\lambda_1 t)+C_8\sin\left(\frac{-(\frac{\alpha k_1^2}{l_1}+\frac{\beta l_1^2}{k_1})}{2(k_1^3+l_1^3)}\right)^{1/2}(k_1x+l_1y+\lambda_1 t)}\right\}^{1/2},$$

$$v(x,y,t)=\frac{l_1}{k_1}\left(\frac{-(\frac{\alpha k_1^2}{l_1}+\frac{\beta l_1^2}{k_1})^2}{2\lambda_1(k_1^3+l_1^3)}\right)^{1/4}$$

$$\left\{\frac{-C_7\sin\left(\frac{-(\frac{\alpha k_1^2}{l_1}+\frac{\beta l_1^2}{k_1})}{2(k_1^3+l_1^3)}\right)^{1/2}(k_1x+l_1y+\lambda_1 t)+C_8\cos\left(\frac{-(\frac{\alpha k_1^2}{l_1}+\frac{\beta l_1^2}{k_1})}{2(k_1^3+l_1^3)}\right)^{1/2}(k_1x+l_1y+\lambda_1 t)}{C_7\cos\left(\frac{-(\frac{\alpha k_1^2}{l_1}+\frac{\beta l_1^2}{k_1})}{2(k_1^3+l_1^3)}\right)^{1/2}(k_1x+l_1y+\lambda_1 t)+C_8\sin\left(\frac{-(\frac{\alpha k_1^2}{l_1}+\frac{\beta l_1^2}{k_1})}{2(k_1^3+l_1^3)}\right)^{1/2}(k_1x+l_1y+\lambda_1 t)}\right\}^{1/2},$$

$$u(x,y,t)=\frac{k_1}{l_1}\left(\frac{-(\frac{\alpha k_1^2}{l_1}+\frac{\beta l_1^2}{k_1})^2}{2\lambda_1(k_1^3+l_1^3)}\right)^{1/4}$$

$$\left\{\frac{-C_7\sin\left(\frac{-(\frac{\alpha k_1^2}{l_1}+\frac{\beta l_1^2}{k_1})}{2(k_1^3+l_1^3)}\right)^{1/2}(k_1x+l_1y+\lambda_1 t)+C_8\cos\left(\frac{-(\frac{\alpha k_1^2}{l_1}+\frac{\beta l_1^2}{k_1})}{2(k_1^3+l_1^3)}\right)^{1/2}(k_1x+l_1y+\lambda_1 t)}{C_7\cos\left(\frac{-(\frac{\alpha k_1^2}{l_1}+\frac{\beta l_1^2}{k_1})}{2(k_1^3+l_1^3)}\right)^{1/2}(k_1x+l_1y+\lambda_1 t)+C_8\sin\left(\frac{-(\frac{\alpha k_1^2}{l_1}+\frac{\beta l_1^2}{k_1})}{2(k_1^3+l_1^3)}\right)^{1/2}(k_1x+l_1y+\lambda_1 t)}\right\}^{1/2},$$

where $C_7, C_8$ are arbitrary constants.

**Case 3.** $\frac{\frac{\alpha k_1^2}{l_1}+\frac{\beta l_1^2}{k_1}}{2(k_1^3+l_1^3)}=0$

We obtain the rational function solutions

$$w(x,y,t)=\left(\frac{-2(k_1^3+l_1^3)}{\lambda_1}\right)^{1/4}\left\{\frac{C_9}{C_9(k_1x+l_1y+\lambda_1 t)+C_{10}}\right\}^{1/2},$$

$$v(x,y,t)=\frac{l_1}{k_1}\left(\frac{-2(k_1^3+l_1^3)}{\lambda_1}\right)^{1/4}\left\{\frac{C_9}{C_9(k_1x+l_1y+\lambda_1 t)+C_{10}}\right\}^{1/2},$$

$$u(x,y,t)=\frac{k_1}{l_1}\left(\frac{-2(k_1^3+l_1^3)}{\lambda_1}\right)^{1/4}\left\{\frac{C_9}{C_9(k_1x+l_1y+\lambda_1 t)+C_{10}}\right\}^{1/2},$$

where $C_9, C_{10}$ are arbitrary constants.

**II. The case** $g\neq 0$

When $\frac{4}{n-2}-2=0$, i.e, $n=4$.

$\left(\frac{G'}{G}\right)$ coeff:

$$6(k_1^3+l_1^3)\lambda\mu D_2^2=0, \tag{2.29}$$

$\left(\frac{G'}{G}\right)^0$ coeff:

$$2(k_1^3+l_1^3)\mu^2 D_2^2=g. \tag{2.30}$$

Solving the set of (2.22)-(2.24) and (2.29)-(2.30), we obtain

$$\lambda = 0, \quad \mu = -\frac{\frac{\alpha k_1^2}{l_1} + \frac{\beta l_1^2}{k_1}}{2(k_1^3 + l_1^3)}, \quad D_2 = \left(\frac{-6(k_1^3 + l_1^3)}{\lambda_1}\right)^{1/2}, \quad \frac{\alpha k_1^2}{l_1} + \frac{\beta l_1^2}{k_1} \neq 0, \quad \frac{-3(\frac{\alpha k_1^2}{l_1} + \frac{\beta l_1^2}{k_1})^2}{\lambda_1} = g. \tag{2.31}$$

Similarly, we can obtain the hyperbolic function solutions and trigonometric function solutions

**Case 1.** $\frac{\frac{\alpha k_1^2}{l_1} + \frac{\beta l_1^2}{k_1}}{2(k_1^3+l_1^3)} > 0$

$$w(x,y,t) = \left(\frac{-3(\frac{\alpha k_1^2}{l_1} + \frac{\beta l_1^2}{k_1})}{\lambda_1}\right)^{1/2} \left\{\frac{C_{11}\sinh\left(\frac{\frac{\alpha k_1^2}{l_1} + \frac{\beta l_1^2}{k_1}}{2(k_1^3+l_1^3)}\right)^{1/2}(k_1x + l_1y + \lambda_1 t) + C_{12}\cosh\left(\frac{\frac{\alpha k_1^2}{l_1} + \frac{\beta l_1^2}{k_1}}{2(k_1^3+l_1^3)}\right)^{1/2}(k_1x + l_1y + \lambda_1 t)}{C_{11}\cosh\left(\frac{\frac{\alpha k_1^2}{l_1} + \frac{\beta l_1^2}{k_1}}{2(k_1^3+l_1^3)}\right)^{1/2}(k_1x + l_1y + \lambda_1 t) + C_{12}\sinh\left(\frac{\frac{\alpha k_1^2}{l_1} + \frac{\beta l_1^2}{k_1}}{2(k_1^3+l_1^3)}\right)^{1/2}(k_1x + l_1y + \lambda_1 t)}\right\} \tag{2.32}$$

$$v(x,y,t) = \frac{l_1}{k_1}\left(\frac{-3(\frac{\alpha k_1^2}{l_1} + \frac{\beta l_1^2}{k_1})}{\lambda_1}\right)^{1/2} \left\{\frac{C_{11}\sinh\left(\frac{\frac{\alpha k_1^2}{l_1} + \frac{\beta l_1^2}{k_1}}{2(k_1^3+l_1^3)}\right)^{1/2}(k_1x + l_1y + \lambda_1 t) + C_{12}\cosh\left(\frac{\frac{\alpha k_1^2}{l_1} + \frac{\beta l_1^2}{k_1}}{2(k_1^3+l_1^3)}\right)^{1/2}(k_1x + l_1y + \lambda_1 t)}{C_{11}\cosh\left(\frac{\frac{\alpha k_1^2}{l_1} + \frac{\beta l_1^2}{k_1}}{2(k_1^3+l_1^3)}\right)^{1/2}(k_1x + l_1y + \lambda_1 t) + C_{12}\sinh\left(\frac{\frac{\alpha k_1^2}{l_1} + \frac{\beta l_1^2}{k_1}}{2(k_1^3+l_1^3)}\right)^{1/2}(k_1x + l_1y + \lambda_1 t)}\right\},$$

$$u(x,y,t) = \frac{k_1}{l_1}\left(\frac{-3(\frac{\alpha k_1^2}{l_1} + \frac{\beta l_1^2}{k_1})}{\lambda_1}\right)^{1/2} \left\{\frac{C_{11}\sinh\left(\frac{\frac{\alpha k_1^2}{l_1} + \frac{\beta l_1^2}{k_1}}{2(k_1^3+l_1^3)}\right)^{1/2}(k_1x + l_1y + \lambda_1 t) + C_{12}\cosh\left(\frac{\frac{\alpha k_1^2}{l_1} + \frac{\beta l_1^2}{k_1}}{2(k_1^3+l_1^3)}\right)^{1/2}(k_1x + l_1y + \lambda_1 t)}{C_{11}\cosh\left(\frac{\frac{\alpha k_1^2}{l_1} + \frac{\beta l_1^2}{k_1}}{2(k_1^3+l_1^3)}\right)^{1/2}(k_1x + l_1y + \lambda_1 t) + C_{12}\sinh\left(\frac{\frac{\alpha k_1^2}{l_1} + \frac{\beta l_1^2}{k_1}}{2(k_1^3+l_1^3)}\right)^{1/2}(k_1x + l_1y + \lambda_1 t)}\right\},$$

where $C_{11}, C_{12}$ are arbitrary constants and $\lambda_1(k_1^3 + l_1^3) < 0$.

**Case 2.** $\frac{\frac{\alpha k_1^2}{l_1} + \frac{\beta l_1^2}{k_1}}{2(k_1^3+l_1^3)} < 0$

$$w(x,y,t) = \left(\frac{-3(\frac{\alpha k_1^2}{l_1} + \frac{\beta l_1^2}{k_1})}{\lambda_1}\right)^{1/2} \left\{\frac{-C_{13}\sin\left(\frac{-(\frac{\alpha k_1^2}{l_1} + \frac{\beta l_1^2}{k_1})}{2(k_1^3+l_1^3)}\right)^{1/2}(k_1x + l_1y + \lambda_1 t) + C_{14}\cos\left(\frac{-(\frac{\alpha k_1^2}{l_1} + \frac{\beta l_1^2}{k_1})}{2(k_1^3+l_1^3)}\right)^{1/2}(k_1x + l_1y + \lambda_1 t)}{C_{13}\cos\left(\frac{-(\frac{\alpha k_1^2}{l_1} + \frac{\beta l_1^2}{k_1})}{2(k_1^3+l_1^3)}\right)^{1/2}(k_1x + l_1y + \lambda_1 t) + C_{14}\sin\left(\frac{-(\frac{\alpha k_1^2}{l_1} + \frac{\beta l_1^2}{k_1})}{2(k_1^3+l_1^3)}\right)^{1/2}(k_1x + l_1y + \lambda_1 t)}\right\} \tag{2.33}$$

$$v(x,y,t) = \frac{l_1}{k_1}\left(\frac{-3(\frac{\alpha k_1^2}{l_1} + \frac{\beta l_1^2}{k_1})}{\lambda_1}\right)^{1/2}$$

$$\left\{\frac{-C_{13}\sin\left(\frac{-(\frac{\alpha k_1^2}{l_1}+\frac{\beta l_1^2}{k_1})}{2(k_1^3+l_1^3)}\right)^{1/2}(k_1x+l_1y+\lambda_1t)+C_{14}\cos\left(\frac{-(\frac{\alpha k_1^2}{l_1}+\frac{\beta l_1^2}{k_1})}{2(k_1^3+l_1^3)}\right)^{1/2}(k_1x+l_1y+\lambda_1t)}{C_{13}\cos\left(\frac{-(\frac{\alpha k_1^2}{l_1}+\frac{\beta l_1^2}{k_1})}{2(k_1^3+l_1^3)}\right)^{1/2}(k_1x+l_1y+\lambda_1t)+C_{14}\sin\left(\frac{-(\frac{\alpha k_1^2}{l_1}+\frac{\beta l_1^2}{k_1})}{2(k_1^3+l_1^3)}\right)^{1/2}(k_1x+l_1y+\lambda_1t)}\right\},$$

$$u(x,y,t) = \frac{k_1}{l_1}\left(\frac{-3(\frac{\alpha k_1^2}{l_1}+\frac{\beta l_1^2}{k_1})}{\lambda_1}\right)^{1/2}$$

$$\left\{\frac{-C_{13}\sin\left(\frac{-(\frac{\alpha k_1^2}{l_1}+\frac{\beta l_1^2}{k_1})}{2(k_1^3+l_1^3)}\right)^{1/2}(k_1x+l_1y+\lambda_1t)+C_{14}\cos\left(\frac{-(\frac{\alpha k_1^2}{l_1}+\frac{\beta l_1^2}{k_1})}{2(k_1^3+l_1^3)}\right)^{1/2}(k_1x+l_1y+\lambda_1t)}{C_{13}\cos\left(\frac{-(\frac{\alpha k_1^2}{l_1}+\frac{\beta l_1^2}{k_1})}{2(k_1^3+l_1^3)}\right)^{1/2}(k_1x+l_1y+\lambda_1t)+C_{14}\sin\left(\frac{-(\frac{\alpha k_1^2}{l_1}+\frac{\beta l_1^2}{k_1})}{2(k_1^3+l_1^3)}\right)^{1/2}(k_1x+l_1y+\lambda_1t)}\right\},$$

where $C_{13}, C_{14}$ are arbitrary constants and $\lambda_1(k_1^3 + l_1^3) < 0$.

*2.3.* $m > 2, n = 2$

(2.3) becomes

$$\left(\lambda_1 + \frac{\alpha k_1^2}{l_1} + \frac{\beta l_1^2}{k_1}\right)w^2 + (k_1^3 + l_1^3)(w^m)'' = g. \tag{2.34}$$

Balancing $(w^m)''$ with $w^2$, we have $N = 2/(2-m)$. Thus, (2.34) has the following formal solution

$$w(\xi) = D_3\left(\frac{G'}{G}\right)^{\frac{2}{2-m}}, \quad D_3 \neq 0, \tag{2.35}$$

where $D_3$ is a constant to be determined later and $G$ satisfies (2.5). Similarly, we can get a set of algebraic equations:

$\left(\frac{G'}{G}\right)^{\frac{2m}{2-m}+2}$ coeff:

$$(k_1^3 + l_1^3)(\frac{2m}{2-m} + 1)(\frac{2m}{2-m})D_3^m + \left(\lambda_1 + \frac{\alpha k_1^2}{l_1} + \frac{\beta l_1^2}{k_1}\right)D_3^2 = 0,$$

$\left(\frac{G'}{G}\right)^{\frac{2m}{2-m}+1}$ coeff:

$$(k_1^3 + l_1^3)(\frac{4m}{2-m} + 1)\frac{2m}{2-m}\lambda D_3^m = 0,$$

$\left(\frac{G'}{G}\right)^{\frac{2m}{2-m}}$ coeff:

$$(k_1^3 + l_1^3)(\frac{2m}{2-m})^2(2\mu + \lambda^2)D_3^m = 0,$$

$\left(\frac{G'}{G}\right)^{\frac{2m}{2-m}-1}$ coeff:

$$(k_1^3 + l_1^3)(\frac{4m}{2-m} - 1)\frac{2m}{2-m}\lambda\mu D_3^m = 0,$$

$\left(\frac{G'}{G}\right)^{\frac{2m}{2-m}-2}$ coeff:

$$(k_1^3+l_1^3)(\frac{2m}{2-m}-1)\frac{2m}{2-m}\mu^2 D_3^m=0.$$

Solving the above algebraic equations, we obtain

$$\lambda=\mu=0,\quad g=0,\quad D_3=\left(-\frac{(k_1^3+l_1^3)(\frac{2m}{2-m}+1)\frac{2m}{2-m}}{\lambda_1+\frac{\alpha k_1^2}{l_1}+\frac{\beta l_1^2}{k_1}}\right)^{1/(2-m)}. \tag{2.36}$$

Substituting (2.36) into (2.5) and (2.35), then, when $m \neq n = 2$, we have the rational function solutions

$$w(x,y,t) = \left(-\frac{(k_1^3+l_1^3)(\frac{2m}{2-m}+1)\frac{2m}{2-m}}{\lambda_1+\frac{\alpha k_1^2}{l_1}+\frac{\beta l_1^2}{k_1}}\right)^{\frac{1}{2-m}}\left(\frac{C_{15}}{C_{15}(k_1x+l_1y+\lambda_1 t)+C_{16}}\right)^{\frac{2}{2-m}},$$

$$v(x,y,t) = \frac{l_1}{k_1}\left(-\frac{(k_1^3+l_1^3)(\frac{2m}{2-m}+1)\frac{2m}{2-m}}{\lambda_1+\frac{\alpha k_1^2}{l_1}+\frac{\beta l_1^2}{k_1}}\right)^{\frac{1}{2-m}}\left(\frac{C_{15}}{C_{15}(k_1x+l_1y+\lambda_1 t)+C_{16}}\right)^{\frac{2}{2-m}},$$

$$u(x,y,t) = \frac{k_1}{l_1}\left(-\frac{(k_1^3+l_1^3)(\frac{2m}{2-m}+1)\frac{2m}{2-m}}{\lambda_1+\frac{\alpha k_1^2}{l_1}+\frac{\beta l_1^2}{k_1}}\right)^{\frac{1}{2-m}}\left(\frac{C_{15}}{C_{15}(k_1x+l_1y+\lambda_1 t)+C_{16}}\right)^{\frac{2}{2-m}},$$

where $C_{15}, C_{16}$ are arbitrary constants and $\lambda_1+\frac{\alpha k_1^2}{l_1}+\frac{\beta l_1^2}{k_1}\neq 0$.

### 2.4. $m = n = 2$

Now, (2.3) can be converted into a second order ODE

$$\left(\lambda_1+\frac{\alpha k_1^2}{l_1}+\frac{\beta l_1^2}{k_1}\right)w^2+(k_1^3+l_1^3)(w^2)''=g. \tag{2.37}$$

Obviously, the characteristic equation of (2.37) is $r^2+\left(\frac{\lambda_1+\frac{\alpha k_1^2}{l_1}+\frac{\beta l_1^2}{k_1}}{k_1^3+l_1^3}\right)=0$, $r$ is the characteristic value.

**Case 1.** $\frac{\lambda_1+\frac{\alpha k_1^2}{l_1}+\frac{\beta l_1^2}{k_1}}{k_1^3+l_1^3}<0$

We can obtain the exact solution

$$w(x,y,t) = \left(C_{17}e^{\left(-\frac{\lambda_1+\frac{\alpha k_1^2}{l_1}+\frac{\beta l_1^2}{k_1}}{k_1^3+l_1^3}\right)^{1/2}(k_1x+l_1y+\lambda_1 t)}+C_{18}e^{-\left(-\frac{\lambda_1+\frac{\alpha k_1^2}{l_1}+\frac{\beta l_1^2}{k_1}}{k_1^3+l_1^3}\right)^{1/2}(k_1x+l_1y+\lambda_1 t)}+\frac{g}{\lambda_1+\frac{\alpha k_1^2}{l_1}+\frac{\beta l_1^2}{k_1}}\right)^{1/2} \tag{2.38}$$

$$v(x,y,t) = \frac{l_1}{k_1}\left(C_{17}e^{\left(-\frac{\lambda_1+\frac{\alpha k_1^2}{l_1}+\frac{\beta l_1^2}{k_1}}{k_1^3+l_1^3}\right)^{1/2}(k_1x+l_1y+\lambda_1 t)}+C_{18}e^{-\left(-\frac{\lambda_1+\frac{\alpha k_1^2}{l_1}+\frac{\beta l_1^2}{k_1}}{k_1^3+l_1^3}\right)^{1/2}(k_1x+l_1y+\lambda_1 t)}+\frac{g}{\lambda_1+\frac{\alpha k_1^2}{l_1}+\frac{\beta l_1^2}{k_1}}\right)^{1/2},$$

$$u(x,y,t) = \frac{k_1}{l_1}\left(C_{17}e^{\left(-\frac{\lambda_1+\frac{\alpha k_1^2}{l_1}+\frac{\beta l_1^2}{k_1}}{k_1^3+l_1^3}\right)^{1/2}(k_1x+l_1y+\lambda_1 t)}+C_{18}e^{-\left(-\frac{\lambda_1+\frac{\alpha k_1^2}{l_1}+\frac{\beta l_1^2}{k_1}}{k_1^3+l_1^3}\right)^{1/2}(k_1x+l_1y+\lambda_1 t)}+\frac{g}{\lambda_1+\frac{\alpha k_1^2}{l_1}+\frac{\beta l_1^2}{k_1}}\right)^{1/2},$$

where $C_{17}, C_{18}$ are arbitrary constants.

**Case 2.** $\frac{\lambda_1+\frac{\alpha k_1^2}{l_1}+\frac{\beta l_1^2}{k_1}}{k_1^3+l_1^3} > 0$

We can obtain the periodic function solutions

$$\begin{aligned}
w(x,y,t) &= \Bigg(C_{19}\cos\left[\left(\frac{\lambda_1+\frac{\alpha k_1^2}{l_1}+\frac{\beta l_1^2}{k_1}}{k_1^3+l_1^3}\right)^{1/2}(k_1x+l_1y+\lambda_1 t)\right] \\
&\quad +C_{20}\sin\left[\left(\frac{\lambda_1+\frac{\alpha k_1^2}{l_1}+\frac{\beta l_1^2}{k_1}}{k_1^3+l_1^3}\right)^{1/2}(k_1x+l_1y+\lambda_1 t)\right]+\frac{g}{\lambda_1+\frac{\alpha k_1^2}{l_1}+\frac{\beta l_1^2}{k_1}}\Bigg)^{1/2}, \qquad (2.39)\\
v(x,y,t) &= \frac{l_1}{k_1}\Bigg(C_{19}\cos\left[\left(\frac{\lambda_1+\frac{\alpha k_1^2}{l_1}+\frac{\beta l_1^2}{k_1}}{k_1^3+l_1^3}\right)^{1/2}(k_1x+l_1y+\lambda_1 t)\right] \\
&\quad +C_{20}\sin\left[\left(\frac{\lambda_1+\frac{\alpha k_1^2}{l_1}+\frac{\beta l_1^2}{k_1}}{k_1^3+l_1^3}\right)^{1/2}(k_1x+l_1y+\lambda_1 t)\right]+\frac{g}{\lambda_1+\frac{\alpha k_1^2}{l_1}+\frac{\beta l_1^2}{k_1}}\Bigg)^{1/2},\\
u(x,y,t) &= \frac{k_1}{l_1}\Bigg(C_{19}\cos\left[\left(\frac{\lambda_1+\frac{\alpha k_1^2}{l_1}+\frac{\beta l_1^2}{k_1}}{k_1^3+l_1^3}\right)^{1/2}(k_1x+l_1y+\lambda_1 t)\right] \\
&\quad +C_{20}\sin\left[\left(\frac{\lambda_1+\frac{\alpha k_1^2}{l_1}+\frac{\beta l_1^2}{k_1}}{k_1^3+l_1^3}\right)^{1/2}(k_1x+l_1y+\lambda_1 t)\right]+\frac{g}{\lambda_1+\frac{\alpha k_1^2}{l_1}+\frac{\beta l_1^2}{k_1}}\Bigg)^{1/2},
\end{aligned}$$

where $C_{19}, C_{20}$ are arbitrary constants.

**Case 3.** $\frac{\lambda_1+\frac{\alpha k_1^2}{l_1}+\frac{\beta l_1^2}{k_1}}{k_1^3+l_1^3} = 0$

We can obtain the rational function solutions

$$\begin{aligned}
w(x,y,t) &= \left(\frac{g}{2(k_1^3+l_1^3)}(k_1x+l_1y+\lambda_1 t)^2+C_{21}(k_1x+l_1y+\lambda_1 t)+C_{22}\right)^{1/2}, \qquad (2.40)\\
v(x,y,t) &= \frac{l_1}{k_1}\left(\frac{g}{2(k_1^3+l_1^3)}(k_1x+l_1y+\lambda_1 t)^2+C_{21}(k_1x+l_1y+\lambda_1 t)+C_{22}\right)^{1/2},\\
u(x,y,t) &= \frac{k_1}{l_1}\left(\frac{g}{2(k_1^3+l_1^3)}(k_1x+l_1y+\lambda_1 t)^2+C_{21}(k_1x+l_1y+\lambda_1 t)+C_{22}\right)^{1/2},
\end{aligned}$$

where $C_{21}, C_{22}$ are arbitrary constants.

## 3. Application of the first integral method to the generalized (2+1)-dimensional BKP equation

### 3.1. $m \neq n$

For simplicity, we let $g = 0$ and propose a transformation $w = \varphi^{\frac{2}{n-m}}$. Then, (2.3) is converted to

$$\lambda_1\varphi^4+(k_1^3+l_1^3)\left(\frac{2m}{n-m}-1\right)\frac{2m}{n-m}(\varphi')^2+(k_1^3+l_1^3)\frac{2m}{n-m}\varphi\varphi''+\left(\frac{\alpha k_1^2}{l_1}+\frac{\beta l_1^2}{k_1}\right)\varphi^{2-\frac{2m-4}{n-m}}=0. \qquad (3.1)$$

Let $x = \varphi, y = \frac{d\varphi}{d\xi}$, thus (3.1) is equivalent to the two dimensional autonomous system

$$\begin{cases} x' = y, \\ y' = -\left(\dfrac{\lambda_1 x^4 + (\frac{\alpha k_1^2}{l_1} + \frac{\beta l_1^2}{k_1})x^{2-\frac{2m-4}{n-m}} + (k_1^3 + l_1^3)(\frac{2m}{n-m} - 1)\frac{2m}{n-m}y^2}{(k_1^3 + l_1^3)\frac{2m}{n-m}x}\right). \end{cases} \tag{3.2}$$

Making the transformation $d\eta = \frac{d\xi}{(k_1^3+l_1^3)\frac{2m}{n-m}x}$, then, (3.2) becomes

$$\begin{cases} \frac{dx}{d\eta} = (k_1^3 + l_1^3)\frac{2m}{n-m}xy, \\ \frac{dy}{d\eta} = -\left(\lambda_1 x^4 + (\frac{\alpha k_1^2}{l_1} + \frac{\beta l_1^2}{k_1})x^{2-\frac{2m-4}{n-m}} + (k_1^3 + l_1^3)(\frac{2m}{n-m} - 1)\frac{2m}{n-m}y^2\right). \end{cases} \tag{3.3}$$

Then, we will apply the Division Theorem to seek the first integral of system (3.3). Suppose that $x = x(\eta), y = y(\eta)$ are the nontrivial solutions to (3.3), and $p(x, y) = \sum_{i=0}^{M} a_i(x)y^i$ is an irreducible polynomial in $C[x, y]$, where $a_i(x)(i = 0, 1..., M)$ are polynomials of $x$ and $a_i(x) \neq 0$. Let $p(x(\eta), y(\eta)) = 0$ be the first integral to system (3.3). $\frac{dp}{d\eta}$ is a polynomial in $x$, $y$ and $\frac{dp}{d\eta}\big|_{(3.3)} = 0$. According to the Division Theorem, there exists a polynomial $g(x) + h(x)y$ in $C[x, y]$, such that

$$\begin{aligned} \frac{dp}{d\eta}\Big|_{(3.3)} &= \left(\frac{\partial p}{\partial x}\frac{dx}{d\eta} + \frac{\partial p}{\partial y}\frac{dy}{d\eta}\right)\Big|_{(3.3)} \\ &= \sum_{i=0}^{M}[a_i'(x)y^i \cdot (k_1^3 + l_1^3)\frac{2m}{n-m}xy] \\ &\quad - \sum_{i=0}^{2}\left[ia_i(x)y^{i-1}(\lambda_1 x^4 + (\frac{\alpha k_1^2}{l_1} + \frac{\beta l_1^2}{k_1})x^{2-\frac{2m-4}{n-m}} + (k_1^3 + l_1^3)(\frac{2m}{n-m} - 1)\frac{2m}{n-m}y^2)\right] \\ &= [g(x) + h(x)y]\left[\sum_{i=0}^{M} a_i(x)y^i\right]. \end{aligned} \tag{3.4}$$

Here, let $M = 1$, thus, $p(x, y) = a_0(x) + a_1(x)y$. By comparing with the coefficients of $y^i$ of both sides of (3.4), we have

$$(k_1^3 + l_1^3)\frac{2m}{n-m}xa_1'(x) = h(x)a_1(x) + (k_1^3 + l_1^3)(\frac{2m}{n-m} - 1)\frac{2m}{n-m}a_1(x), \tag{3.5}$$

$$(k_1^3 + l_1^3)\frac{2m}{n-m}xa_0'(x) = g(x)a_1(x) + h(x)a_0(x), \tag{3.6}$$

$$g(x)a_0(x) = -\left(\lambda_1 x^4 + (\frac{\alpha k_1^2}{l_1} + \frac{\beta l_1^2}{k_1})x^{2-\frac{2m-4}{n-m}}\right)a_1(x). \tag{3.7}$$

Since $a_i(x)(i = 0, 1)$ are polynomials, then from (3.5), we deduce that $h(x) = -(k_1^3 + l_1^3)(\frac{2m}{n-m} - 1)\frac{2m}{n-m}$ and $a_1(x)$ is a constant. For simplicity, take $a_1(x) = 1$. Balancing the degrees of $g(x)$ and $a_0(x)$, we conclude that $\deg(g(x)) = \deg(a_0(x))$. Then, we derive $\deg(g(x)) = \deg(a_0(x)) = j, (j \in Z^+, j \geq 2)$.

When $2 - \frac{2m-4}{n-m} = 4$ $(n = 2)$ and $\deg(g(x)) = \deg(a_0(x)) = 2$, we suppose that

$$g(x) = A_0 + A_1 x + A_2 x^2,$$

$$a_0(x) = B_0 + B_1 x + B_2 x^2, \ (A_2 \neq 0, B_2 \neq 0), \tag{3.8}$$

where $A_i, B_i, (i = 0, 1, 2)$ are all constants to be determined. Substituting (3.8) into (3.6), we obtain

$$g(x) = (k_1^3 + l_1^3)\frac{2m}{2-m}\left[(\frac{2m}{2-m} - 1)B_0 + \frac{2m}{2-m}B_1 x + (\frac{2m}{2-m} + 1)B_2 x^2\right].$$

Substituting $a_0(x), a_1(x)$ and $g(x)$ into (3.7), and setting all the coefficients of powers $x$ to be zero, we can get a system of nonlinear algebraic equations. After solving it, we can get the following solutions

$$B_0 = B_1 = 0, \qquad B_2 = \pm\left(-\frac{\lambda_1 + \frac{\alpha k_1^2}{l_1} + \frac{\beta l_1^2}{k_1}}{(k_1^3 + l_1^3)(\frac{2m}{2-m} + 1)\frac{2m}{2-m}}\right)^{1/2}. \tag{3.9}$$

Using the conditions (3.9) in $p(x, y) = a_0(x) + a_1(x)y = 0$, we obtain

$$y \pm \left(-\frac{\lambda_1 + \frac{\alpha k_1^2}{l_1} + \frac{\beta l_1^2}{k_1}}{(k_1^3 + l_1^3)(\frac{2m}{2-m} + 1)\frac{2m}{2-m}}\right)^{1/2} x^2 = 0. \tag{3.10}$$

Combining (3.3) with (3.10), we find

$$\frac{dx}{d\eta} = \pm(k_1^3 + l_1^3)\frac{2m}{n-m}\left(-\frac{\lambda_1 + \frac{\alpha k_1^2}{l_1} + \frac{\beta l_1^2}{k_1}}{(k_1^3 + l_1^3)(\frac{2m}{2-m} + 1)\frac{2m}{2-m}}\right)^{1/2} x^3.$$

Thus, (3.10) can be reduced to

$$\frac{d\varphi}{d\xi} = \pm\left(-\frac{\lambda_1 + \frac{\alpha k_1^2}{l_1} + \frac{\beta l_1^2}{k_1}}{(k_1^3 + l_1^3)(\frac{2m}{2-m} + 1)\frac{2m}{2-m}}\right)^{1/2} \varphi^2.$$

Then, we have

$$\varphi(\xi) = \left[\pm\left(-\frac{\lambda_1 + \frac{\alpha k_1^2}{l_1} + \frac{\beta l_1^2}{k_1}}{(k_1^3 + l_1^3)(\frac{2m}{2-m} + 1)\frac{2m}{2-m}}\right)^{1/2} \xi + C_{23}\right]^{-1}.$$

Thus, we can have the rational function solutions

$$\begin{aligned} w(x,y,t) &= \left[\pm\left(-\frac{\lambda_1 + \frac{\alpha k_1^2}{l_1} + \frac{\beta l_1^2}{k_1}}{(k_1^3 + l_1^3)(\frac{2m}{2-m} + 1)\frac{2m}{2-m}}\right)^{1/2} (k_1 x + l_1 y + \lambda_1 t) + C_{23}\right]^{2/(m-2)}, \\ v(x,y,t) &= \frac{l_1}{k_1}\left[\pm\left(-\frac{\lambda_1 + \frac{\alpha k_1^2}{l_1} + \frac{\beta l_1^2}{k_1}}{(k_1^3 + l_1^3)(\frac{2m}{2-m} + 1)\frac{2m}{2-m}}\right)^{1/2} (k_1 x + l_1 y + \lambda_1 t) + C_{23}\right]^{2/(m-2)}, \\ u(x,y,t) &= \frac{k_1}{l_1}\left[\pm\left(-\frac{\lambda_1 + \frac{\alpha k_1^2}{l_1} + \frac{\beta l_1^2}{k_1}}{(k_1^3 + l_1^3)(\frac{2m}{2-m} + 1)\frac{2m}{2-m}}\right)^{1/2} (k_1 x + l_1 y + \lambda_1 t) + C_{23}\right]^{2/(m-2)}, \end{aligned}$$

where $C_{23}$ is an arbitrary constant and $k_1^3 + l_1^3 \neq 0$.

*Remark* 1: When $\deg(g(x)) = \deg(a_0(x)) = 2$ and $2 - \frac{2m-4}{n-m} = i, (i \in Z, i < 4)$, there is no solution for them by using the method as that of $2 - \frac{2m-4}{n-m} = 4$.

*Remark* 2: When $\deg(g(x)) = \deg(a_0(x)) = j, (j \in Z, j > 2)$, there is no exact solution of (1.1) by using the method as that of $\deg(g(x)) = \deg(a_0(x)) = 2$.

*3.2.* $m = n$

Similarly, we propose a transformation denoted by $w = \phi^{\frac{2}{2-m}}$. Then, (2.3) can be converted to

$$\lambda_1\phi^2 + \left(\frac{\alpha k_1^2}{l_1} + \frac{\beta l_1^2}{k_1}\right)\phi^4 + (k_1^3 + l_1^3)(\frac{2m}{2-m} - 1)\frac{2m}{2-m}(\phi')^2 + (k_1^3 + l_1^3)\frac{2m}{2-m}\phi\phi'' - g\phi^{2-\frac{2m}{2-m}} = 0. \quad (3.11)$$

Let $x = \phi, y = \frac{d\phi}{d\xi}$, thus (3.11) is equivalent to the two dimensional autonomous system

$$\begin{cases} x' = y, \\ y' = -\left(\frac{\lambda_1 x^2 + (\frac{\alpha k_1^2}{l_1} + \frac{\beta l_1^2}{k_1})x^4 - gx^{2-\frac{2m}{2-m}} + (k_1^3+l_1^3)(\frac{2m}{2-m} - 1)\frac{2m}{2-m}y^2}{(k_1^3+l_1^3)\frac{2m}{2-m}x}\right). \end{cases} \quad (3.12)$$

Making the transformation $d\eta = \frac{d\xi}{(k_1^3+l_1^3)\frac{2m}{n-m}x}$, then, (3.12) becomes

$$\begin{cases} \frac{dx}{d\eta} = (k_1^3 + l_1^3)\frac{2m}{2-m}xy, \\ \frac{dy}{d\eta} = -\left(\lambda_1 x^2 + (\frac{\alpha k_1^2}{l_1} + \frac{\beta l_1^2}{k_1})x^4 - gx^{2-\frac{2m}{2-m}} + (k_1^3 + l_1^3)(\frac{2m}{2-m} - 1)\frac{2m}{2-m}y^2\right). \end{cases} \quad (3.13)$$

Similarly, let $M = 1$, we have

$$(k_1^3 + l_1^3)\frac{2m}{2-m}xa_1'(x) = h(x)a_1(x) + (k_1^3 + l_1^3)(\frac{2m}{2-m} - 1)\frac{2m}{2-m}a_1(x), \quad (3.14)$$

$$(k_1^3 + l_1^3)\frac{2m}{2-m}xa_0'(x) = g(x)a_1(x) + h(x)a_0(x), \quad (3.15)$$

$$g(x)a_0(x) = -\left(\lambda_1 x^2 + (\frac{\alpha k_1^2}{l_1} + \frac{\beta l_1^2}{k_1})x^4 - gx^{2-\frac{2m}{2-m}}\right)a_1(x). \quad (3.16)$$

According to $m \neq n$, we have $h(x) = -(k_1^3 + l_1^3)(\frac{2m}{2-m} - 1)\frac{2m}{2-m}$, $a_1(x) = 1$ and $\deg(g(x)) = \deg(a_0(x)) = j, (j \in Z^+, j \geq 2)$. Considering all cases, only when $\deg(g(x)) = \deg(a_0(x)) = 3$, i.e., $2 - \frac{2m}{2-m} = 6$ $(m = n = 4)$, there exist solutions of (1). We suppose that

$$\begin{aligned} g(x) &= a_0 + a_1x + a_2x^2 + a_3x^3, \\ a_0(x) &= b_0 + b_1x + b_2x^2 + b_3x^3, \quad (a_3 \neq 0, b_3 \neq 0), \end{aligned} \quad (3.17)$$

where $a_i, b_i, (i = 0, 1, 2, 3)$ are all constants to be determined. Substituting (3.17) into (3.15), we obtain

$$g(x) = 4(k_1^3 + l_1^3)\left(5b_0 + 4b_1x + 3b_2x^2 + 2b_3x^3\right).$$

Substituting $a_0(x), a_1(x)$ and $g(x)$ into (3.16), and setting all the coefficients of powers $x$ to be zero, we have

$$b_0 = b_2 = 0, \quad 16(k_1^3 + l_1^3)b_1^2 = -\lambda_1, \quad 8(k_1^3 + l_1^3)b_3^2 = g, \quad 24(k_1^3 + l_1^3)b_1b_3 = -(\frac{\alpha k_1^2}{l_1} + \frac{\beta l_1^2}{k_1}).$$

Solving it, we find

$$b_0 = b_2 = 0, \quad b_1 = \pm\sqrt{\frac{-\lambda_1}{16(k_1^3 + l_1^3)}}, \quad b_3 = \pm\sqrt{\frac{g}{8(k_1^3 + l_1^3)}}, \quad -9\lambda_1 g = 2(\frac{\alpha k_1^2}{l_1} + \frac{\beta l_1^2}{k_1})^2. \quad (3.18)$$

Using the conditions (3.18) in $p(x, y) = a_0(x) + a_1(x)y = 0$, we obtain

$$y = \pm\sqrt{\frac{-\lambda_1}{16(k_1^3 + l_1^3)}}x \pm \sqrt{\frac{g}{8(k_1^3 + l_1^3)}}x^3. \tag{3.19}$$

Then, (3.19) can be reduced to

$$\frac{d\phi}{d\xi} = \pm\sqrt{\frac{-\lambda_1}{16(k_1^3 + l_1^3)}}\xi \pm \sqrt{\frac{g}{8(k_1^3 + l_1^3)}}\xi^3. \tag{3.20}$$

Solving (3.20), we obtain

$$\phi(\xi) = \pm\left(\pm\sqrt{\frac{-2g}{\lambda_1}} + C_{24}e^{\pm\frac{1}{2}\sqrt{\frac{-\lambda_1}{k_1^3+l_1^3}}\xi}\right)^{-1/2}.$$

Thus, we can have the exact solution

$$\begin{aligned} w(x, y, t) = & \pm\left(\pm\sqrt{\frac{-2g}{\lambda_1}} + C_{24}e^{\pm\frac{1}{2}\sqrt{\frac{-\lambda_1}{k_1^3+l_1^3}}(k_1x+l_1y+\lambda_1t)}\right)^{1/2}, \\ v(x, y, t) = & \pm\frac{l_1}{k_1}\left(\pm\sqrt{\frac{-2g}{\lambda_1}} + C_{24}e^{\pm\frac{1}{2}\sqrt{\frac{-\lambda_1}{k_1^3+l_1^3}}(k_1x+l_1y+\lambda_1t)}\right)^{1/2}, \\ u(x, y, t) = & \pm\frac{k_1}{l_1}\left(\pm\sqrt{\frac{-2g}{\lambda_1}} + C_{24}e^{\pm\frac{1}{2}\sqrt{\frac{-\lambda_1}{k_1^3+l_1^3}}(k_1x+l_1y+\lambda_1t)}\right)^{1/2}, \end{aligned}$$

where $C_{24}$ is an arbitrary constant and $\alpha k_1^3 + \beta l_1^3 \neq 0$, $\lambda_1(k_1^3 + l_1^3) < 0$.

## 4. Conclusion

This paper considered the generalized (2+1)-dimensional BKP equation, by the aid of the $G'/G$-expansion method and the first integral method. Rational function solutions, periodic function solutions and hyperbolic function solutions are obtained under some parametric conditions and the values of $m$ and $n$ in several cases. In [10], authors gave some exact solutions of system (1.1) under some parametric conditions by using the bifurcation theory of dynamical systems. Here, we make a simple comparison:

1. When $m = 2, n = 3, g = 0$, in [10], authors gave the exact solution (3.20) in P2443 under the parametric conditions $\alpha + \beta < 0, c < 0$ and in this paper, we get $w(x, y, t) = \frac{-20(k_1^3+l_1^3)}{\lambda_1}\left(\frac{C_1}{C_1(k_1x+l_1y+\lambda_1t)+C_2}\right)^2$ under the parametric condition $\alpha k_1^3 + \beta l_1^3 = 0$.

2. When $m = 2, n = 2(k + 1), (k \in Z^+), g = 0$, in [10], authors gave the solitary wave solutions (3.9) in P2441 under the parametric conditions $\alpha + \beta < 0, c < 0$ and in this paper, we get $w(x, y, t) = \left(\sqrt{\frac{-2(k+2)(k_1^3+l_1^3)}{\lambda_1 k^2}}\frac{C_1}{C_1(k_1x+l_1y+\lambda_1t)+C_2}\right)^{1/k}$ under the parametric condition $\alpha k_1^3 + \beta l_1^3 = 0$.

3. When $m = 3, n = 4, g = 0$, in [10], authors gave the compacton solution (3.23) in P2443 under the parametric conditions $\alpha + \beta < 0, c < 0$ and in this paper, we get $w(x, y, t) = \frac{-42(k_1^3+l_1^3)}{\lambda_1}\left(\frac{C_1}{C_1(k_1x+l_1y+\lambda_1t)+C_2}\right)^2$ under the parametric condition $\alpha k_1^3 + \beta l_1^3 = 0$.

4. When $m = 3, n = 5, g = 0$, in [10], authors gave the exact solution (3.27) in P2443 under the parametric conditions $c < 0$ and in this paper, we get $w(x, y, t) = \sqrt{\frac{-12(k_1^3+l_1^3)}{\lambda_1} \frac{C_1}{C_1(k_1x+l_1y+\lambda_1t)+C_2}}$ under the parametric condition $\alpha k_1^3 + \beta l_1^3 = 0$.

5. When $m = 4, n = 6, g = 0$, in [10], authors gave the periodic cusp wave solutions (3.6) in P2441 under the parametric conditions $\alpha + \beta < 0, c < 0$ and in this paper, we get $w(x, y, t) = \sqrt{\frac{-20(k_1^3+l_1^3)}{\lambda_1} \frac{C_1}{C_1(k_1x+l_1y+\lambda_1t)+C_2}}$ under the parametric condition $\alpha k_1^3 + \beta l_1^3 = 0$.

6. When $m = 4, n = 2k + 1, (k \in Z^+), g = 0$, in [10], authors gave the exact solutions (3.17) and (3.18) in P2442 under the parametric conditions $\alpha + \beta < 0, c < 0$ and in this paper, we get $w(x, y, t) = \left(\frac{-8(2k+5)(k_1^3+l_1^3)}{\lambda_1(2k-3)^2}\right)^{\frac{1}{2k-3}} \left(\frac{C_1}{C_1(k_1x+l_1y+\lambda_1t)+C_2}\right)^{\frac{2}{2k-3}}$ under the parametric condition $\alpha k_1^3 + \beta l_1^3 = 0$.

7. When $m = 2, n = 4, g \neq 0$, in [10], authors gave the exact solutions (3.30) in P2443 under the parametric conditions $g < \frac{(\alpha+\beta)^2}{4c}, c > 0, g > 0, \alpha + \beta > 0$, (3.33) in P2444 under the parametric conditions $g < \frac{(\alpha+\beta)^2}{4c}, c < 0, g > 0, \alpha + \beta > 0$ or $g > \frac{(\alpha+\beta)^2}{4c}, c < 0, g > 0, \alpha + \beta < 0$ and (3.41), (3.43) in P2445 under the parametric conditions $g > \frac{(\alpha+\beta)^2}{4c}, c < 0, g < 0, \alpha + \beta < 0$ and in this paper, we get (2.32) under the parametric conditions $\lambda_1(k_1^3 + l_1^3) < 0$, $\frac{\frac{\alpha k_1^2}{l_1}+\frac{\beta l_1^2}{k_1}}{2(k_1^3+l_1^3)} > 0$ and (2.33) under the parametric conditions $\lambda_1(k_1^3 + l_1^3) < 0$, $\frac{\frac{\alpha k_1^2}{l_1}+\frac{\beta l_1^2}{k_1}}{2(k_1^3+l_1^3)} < 0$.

8. When $m = 2, n = 2, g \neq 0$, in [10], authors gave the exact solutions (3.36) and (3.38) in P2444 under the parametric conditions $\alpha + \beta - c > 0, g > 0$ and in this paper, we get the exact solutions (2.38) under the parametric conditions $\frac{\lambda_1+\frac{\alpha k_1^2}{l_1}+\frac{\beta l_1^2}{k_1}}{k_1^3+l_1^3} < 0$, (2.39) under the parametric conditions $\frac{\lambda_1+\frac{\alpha k_1^2}{l_1}+\frac{\beta l_1^2}{k_1}}{k_1^3+l_1^3} > 0$ and (2.40) under the parametric conditions $\frac{\lambda_1+\frac{\alpha k_1^2}{l_1}+\frac{\beta l_1^2}{k_1}}{k_1^3+l_1^3} = 0$.

In addition, when let $m, n$ be other values, we have got other exact solutions of (1.1) under some parametric conditions that haven't been given in [10]. Certainly, system (1.1) should be studied further, which will be left to a further discussion.

## Conflict of Interest

All authors declare no conflicts of interest in this paper.

## References

1. M. R. Miura, *Backlund Transformation,* New York: Springer-Verlag, Berlin, 1978.

2. W. Ma, T. Huang, and Y. Zhang, *A multiple exp-function method for nonlinear differential equations and its application,* Phys. Scr., **82** (2010), 065003.

3. W. Ma and J. H. Lee, *A transfortiom rational function method and exact solutions to (3+1)-dimensional Jimbo-Miwa equation,* Choas Solitons Fractals, **42** (2009), 1356-1363.

4. A. M. Wazwaz, *Two reliable methods for solving variants of the KdV equation with compact and noncompact structures,* Choas Solitons Fractals, **28** (2006), 454-462.

5. M. L. Wang, X. Z. Li, and J. L. Zhang, *The $G'/G$-expansion method and traveling wave solutions of nonlinear evolutions in mathematical physics,* Phys. Lett., **372** (2008), 417-423.
6. E. M. E. Zayed and Khaled A. Gepreel, *Some applications of the* $(\frac{G'}{G})$*-expansion method to non-linear partial differential equations,* Appl. Math. Comput., **212** (2009), 1-13.
7. H. Q. Zhang, *New application of the $G'/G$-expansion method,* Commun Nolinear Sci. Numer. Simul., **14** (2009), 3220-3225.
8. Z. S. Feng, *The first integral method to study the Burgers-Korteweg-de Vries equation*, J. Phys. A., **35** (2002), 343-349.
9. N. Taghizadeh and M. Mirzazadeh, *Exact solutions of some nonlinear evolution equations via the first integral method,* Ain Shams Engineering Journal, **4** (2013), 493-499.
10. Y. G. Xie, B. W. Zhou, and S. Q. Tang, *Bifurcations of traveling wave soluions for the generalized (2+1)-dimensional Boussinesq-Kadomtesv-Petviashvili equation,* Appl. Math. Comput., **217** (2010), 2433-2447.
11. E. Date, M. Jimbo, M. Kashiwara, and T. Miwa, *Transformation groups for soliton equations. IV. A new hierarchy of soliton equations of KP type,* Physica D., **4** (1982), 343-365.
12. H. C. Ma, Y. Wang, and Z. Y. Qin, *New exact complex traveling wave solutions for (2+1)-dimensional BKP equation,* Appl. Math. Comput., **208** (2009), 564-568.
13. F. Tascan and A. Bekir, *Analytic solutions of the (2+1)-dimensional nonlinear evolution equations using the sine-cosine method,* Appl. Math. Comput., **215** (2009), 3134-3139.
14. H. Q. Zhang, *A note on exact complex travelling wave solutions for (2+1)-dimensional B-type Kadomtsev-Petviashvili equation,* Appl. Math. Comput., **216** (2010), 2771-2777.
15. M. Mirzazadeh and M. Eslami, *Topological solitons of resonant nonlinear Schrödinger's equation with dual-power law nonlinearity by $G'/G$-expansion technique,* Optik, **125** (2014), 5480-5489.

# 13

# Modeling electromagnetism in and near composite material using two-scale behavior of the time-harmonic Maxwell equations

**Canot Hélène*and Frénod Emmanuel**

Université de Bretagne-Sud, UMR 6205, LMBA, F-56000 Vannes, France

* **Correspondence:** helene.canot@univ-ubs.fr

**Abstract:** The main purpose of this article is to study the two-scale behavior of the electromagnetic field in 3D in and near composite material. For this, time-harmonic Maxwell equations, for a conducting two-phase composite and the air above, are considered. Technique of two-scale convergence is used to obtain the homogenized problem.

**Keywords:** Harmonic Maxwell Equations; Two-scale Convergence; Asymptotic Expansion; Asymptotic Analysis; Electromagnetism; Homogenization; Effective Behavior; Frequencies; Composite Material; Electromagnetic Pulses

## 1. Introduction

We are interested in the time-harmonic Maxwell equations in and near a composite material with boundary conditions modeling electromagnetic field radiated by an electromagnetic pulse (EMP). An electromagnetic pulse is a short burst of electromagnetic energy. It may be generated by a natural occurrence such like a lightning strike, meteoric EMP, EMP caused by geomagnetic Storm or nuclear EMP. This focuses on what happens over a period of time of a millisecond during the peak of the first return stroke. We study the electromagnetic pulse caused by this lightning strike. This is the first step of a larger study which goal is to understand the behavior of the electromagnetic field and its interaction with a composite material.

EMP interference is generally damaging to electronic equipment. A lightning strike can damage physical objects such as aircraft structures, either through heating effects or disruptive effects of the very large magnetic field generated by the current. Structures and systems require some form of protection against lightning. Every commercial aircraft is struck by lightning at least once a year on average. Aircraft lightning protection is a major concern for aircraft manufacturers. Increasing its use of composite materials, up to 53% for the latest Airbus A350, and 50% for the Boeing B787, aircrafts offer increased vulnerability facing lightning. Earlier generation aircrafts, whose fuselages were predominantly composed of aluminum, behave like a Faraday cage and offer maximum protection for the internal equipment. Currently, in aircrafts, composite materials consisting of a resin enclosing carbon fibers have significant advantages in terms of weight gain and therefore fuel saving. Yet,because aluminium conducts 100 to 1000 times more than composite, we lose the Faraday effect. Modern aircrafts have seen also the increasing reliance on electronic avionics systems instead of mechanical controls and electromechanical instrumentation. For these reasons, aircraft manufacturers are very sensitive to lightning protection and pay special attention to aircraft certification through testing and analysis.

There are two types of lightning strikes to aircraft: the first one is the interception by the aircraft of a lightning leader. The second one, which makes about 90% of the cases, is when the aircraft initiates the lightning discharge by emitting two leaders when it is found in the intense electric field region produced by a thundercloud, our approach applies in this case. When the aircraft flies through a cloud region where the atmospheric electric field is large enough, an ionized channel, called a positive leader, merges from the aircraft in the direction of the ambient electric field. Laroche et al. [15], at an altitude of 6000 m, observed an atmospheric electric field close to 50 kV/m inside the storm clouds, 100 kV/m to the ground. When upward leader connects with the downward negative leader of the cloud, a return stroke is produced and a bright return stroked wave travels from aircraft to cloud. The lightning return strokes radiate powerful electromagnetic fields which may cause damage to aircraft electronic equipment. Our work is devoted to the study of the electromagnetic waves propagation in the air and in the composite material. In this artificial periodic material, the electromagnetic field satisfies the Maxwell equations.

We evaluate the electromagnetic field within and near a periodic structure when the period of this microstructure is small compared to the wavelength of the electromagnetic wave. Our model is composed by air above the composite fuselage and we study the behavior of the electromagnetic wave in the domain filled by the composite material, representing the skin aircraft, and the air. We build the 3D model, under simplifying assumptions, using linear time-harmonic Maxwell equations and constitutive relations for electric and magnetic fields. Composite materials consist of conducting carbon fibers, distributed as periodic inclusions in a matrix (epoxy resin). We impose a magnetic permeability $\mu_0$ uniform and an electrical permittivity $\epsilon = \epsilon_0 \epsilon^\star$, where $\epsilon^\star$ is the relative permittivity depending of the medium. In the future, we will enrich this model by adding complexity and we will consider non uniform magnetic permeability and electrical permittivity.

Now, we account for some characteristic values. In the first place we focus on the boundary conditions as we consider them as the source. Then, we use on the upper frontier, the magnetic field induced by the peak of the current of the first return stroke

$$\overline{H_d} = \frac{I}{2\pi r}, \tag{1}$$

with current intensity $I = 200$ kA and $r$ the radius of the lightning strike, this is the worst aggression that can suffer an aircraft, and we deduce a characteristic electric field $\overline{E} = 20$ kV/m. In our model we consider that we have very conductive - but not perfect conductors - carbon fibers and an epoxy resin whose conduction depends on its doping rate. The conductivity of the air is non-linear. Air is a strong insulator [29] with conductivity of the order of $10^{-14}\ S.m^{-1}$ but beyond some electric solicitation, the air loses its insulating nature and locally becomes suddenly conductive. The ionization phenomenon is the only cause that can make the air conductor of electricity. The ionized channel becomes very conductive.

Our mathematical context is periodic homogenization. We consider a microscopic scale $\varepsilon$, which represents the ratio between the diameter of the fiber and thickness of the composite material. So, we are trying to understand how the microscopic structure affects the macroscopic electromagnetic field behavior. Homogenization of Maxwell equations with periodically oscillating coefficients was studied in many papers. N. Wellander homogenized linear and non-linear Maxwell equations with perfect conducting boundary conditions using two-scale convergence in [26] and [27]. N. Wellander and B. Kristensson homogenized the full time-harmonic Maxwell equation with penetrable boundary conditions and at fixed frequency in [28]. The homogenized time-harmonic Maxwell equation for the scattering problem was done in F. Guenneau, S. Zolla and A. Nicolet [12]. Y. Amirat and V. Shelukhin perform two-scale homogenization time-harmonic Maxwell equations for a periodical structure in [5]. They calculate the effective dielectric $\varepsilon$ and effective electric conductivity $\sigma$. They proved that homogenized Maxwell equations are different in low and high frequencies. The result obtained by two-scale convergence approach takes into account the characteristic sizes of skin thickness and wavelength around the material.

On of the parameter we account for in our model: $\delta = \frac{1}{\sqrt{\overline{\omega}\,\overline{\sigma}\mu_0}}$, where $\overline{\sigma}$ is the characteristic conductivity and $\overline{\omega}$ the order of the magnitude of the pulsation shares much with the definition of theoretical thickness skin $\delta = \sqrt{\frac{2}{\overline{\omega}\overline{\sigma}\mu_0}}$. The thickness skin is the depth at which the surface current moves to a factor of $e^{-1}$. Indeed, at high frequency, the skin effect phenomenon appears because the current tends to concentrate at the periphery of the conductor. On the other side, at low frequencies the penetration depth is much greater than the thickness of the plate which means that a part of the electric field penetrates the composite plate. We use the theory of two-scale convergence introduced by G. Nguetseng [21] and developed by G. Allaire [3].

The paper is organized as follows : in Section 2 we specify the geometry of the model and the dimensionless equations converting the problem into an equivalent one with which we work in the following sections. In Section 3 we perform the mathematical analysis of the model. In particular, we introduce the weak formulation of the problem for the electric field and we regularize it using divergence term. We establish the existence and uniqueness result for the regularized Maxwell equations thanks to Lax-Milgram Theorem. We conclude this section by estimate of the electric field. The last section is devoted to the homogenization of the problems for electric field using the two-scale convergence concept.

## 2. Modeling

This section is dedicated to the complete mathematical model we will study in this paper. First, we consider a problem that seems relevant with the perspective of propagation of the electromagnetic field in the air and in the skin of aircraft fuselage made of composite material. Secondly, we make a scaling of this model and finally we operate simplifications. If desired, the reader can go directly to the mathematical analysis knowing that the problem to be studied is given by (63), (68) equipped with boundary conditions (66), (67).

*2.1. Notations and setting of the problem*

We consider set $\widetilde{\Omega} = \{(\widetilde{x}, \widetilde{y}, \widetilde{z}) \in \mathbb{R}^3,\ \widetilde{y} \in (-\overline{L}, d)\}$ for $\overline{L}$ and $d$ two positive constants, with two open subsets $\widetilde{\Omega}_a$ and $\widetilde{P}$ (see Figure 1). The air fills $\widetilde{\Omega}_a$ and we consider that the composite material, with two materials periodically distributed, stands in domain $\widetilde{P}$.

We assume that the thickness $\overline{L}$ of the composite material is much smaller than its horizontal size. We denote by $e$ the lateral size of the basic cell $\widetilde{Y}^e$ of the periodic microstructure of the material. The cell is composed of a carbon fiber in the resin. We define now more precisely the material, introducing:

$$\widetilde{P} = \{(\widetilde{x}, \widetilde{y}, \widetilde{z}) \in \mathbb{R}^3 / -\overline{L} < \widetilde{y} < 0\}, \tag{2}$$

which is the domain containing the material. Now we describe precisely the basic cell. For this we first introduce the following cylinder with square base:

$$\widetilde{Z}^e = [-\frac{e}{2}, \frac{e}{2}] \times [-e, 0] \times \mathbb{R}, \tag{3}$$

We consider $\alpha$ such that $0 < \alpha < 1$, and $\widetilde{R}^e = \alpha\frac{e}{2}$. We set

$$\widetilde{D}^e = \{(\widetilde{x}, \widetilde{y}) \in \mathbb{R}^2 / (\widetilde{x}^2 + (\widetilde{y} + \frac{e}{2})^2) < (\widetilde{R}^e)^2\}. \tag{4}$$

We define the cylinder containing the fiber as (see fig 1):

$$\widetilde{C}^e = \widetilde{D}^e \times \mathbb{R}. \tag{5}$$

Then the part of the basic cell containing the matrix is

$$\widetilde{Y}^e_R = \widetilde{Z}^e \setminus \widetilde{C}^e, \tag{6}$$

and by definition, the basic cell $\widetilde{Y}^e$ is the couple

$$(\widetilde{Y}^e_R, \widetilde{C}^e). \tag{7}$$

The composite material results from a periodic extension of the basic cell. More precisely the part of the material that contains the carbon fibers is

$$\widetilde{\Omega}_c = \widetilde{P} \cap \{(ie, je, 0) + \widetilde{C}^e, i \in \mathbb{Z}, j \in \mathbb{Z}^-\}, \tag{8}$$

where the intersection with $\widetilde{P}$ limits the periodic extension to the area where stands the material. Set $\{(ie, je, 0) + \widetilde{C}^e, i \in \mathbb{Z}, j \in \mathbb{Z}^-\}$ is a short notation for

$$\{(\widetilde{x}, \widetilde{y}, \widetilde{z}) \in \mathbb{R}^3, \exists i \in \mathbb{Z}, \exists j \in \mathbb{Z}^-, \exists (x_b, y_b, z_b) \in \widetilde{C}^e; \widetilde{x} = x_b + ie, \widetilde{y} = y_b + je, \widetilde{z} = z_b\}. \tag{9}$$

In the same way the part of the material that contains the resin is

$$\widetilde{\Omega}_r = \widetilde{P} \cap \{(ie, je, 0) + \widetilde{Y}^e_R\}, \tag{10}$$

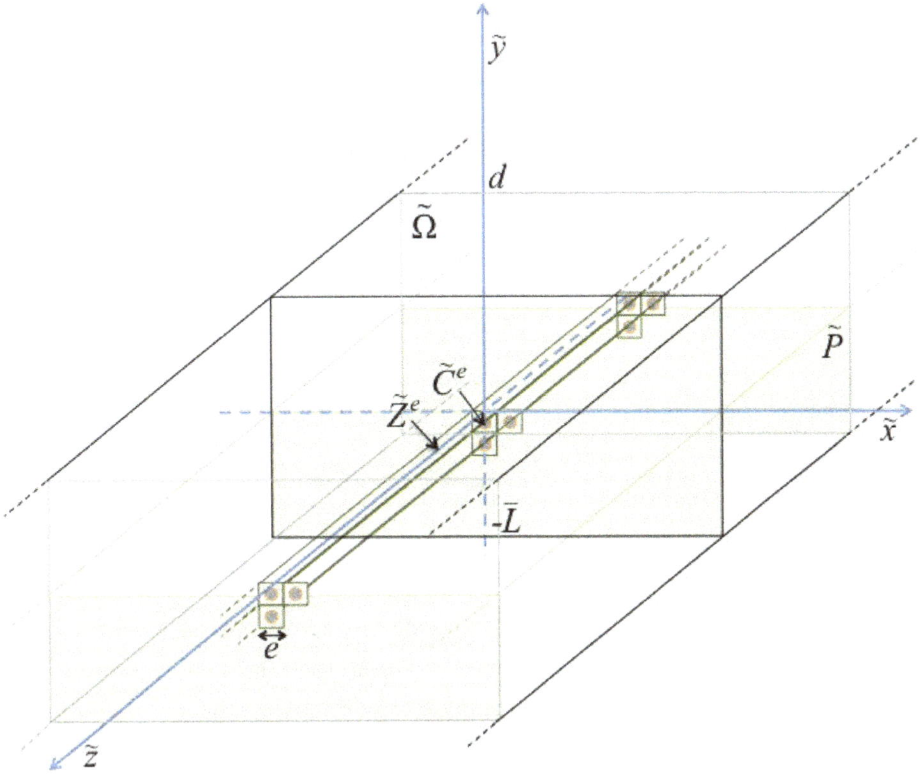


**Figure 1. The global domain**

or equivalently

$$\widetilde{\Omega}_r = \widetilde{P} \cap \{(ie, je, 0) + \widetilde{Z}^e \setminus \widetilde{C}^e\} = (\mathbb{R} \times (-\widetilde{L}, 0) \times \mathbb{R}) \setminus \widetilde{\Omega}_c. \tag{11}$$

So the geometrical model of our composite material is couple $(\widetilde{\Omega}_c, \widetilde{\Omega}_r)$. Now, it remains to set the domain that contains the air:

$$\widetilde{\Omega}_a = \{(\widetilde{x}, \widetilde{y}, \widetilde{z}) / 0 \leq \widetilde{y} < d\}. \tag{12}$$

We consider that $d$ is of the same order as $\widetilde{L}$ and we introduce the upper frontier $\widetilde{\Gamma}_d = \{(\widetilde{x}, \widetilde{y}, \widetilde{z}) / \widetilde{y} = d\}$ of domain $\widetilde{\Omega}$. On this frontier we will consider that the electric field and magnetic field are given. We also introduce the lower frontier $\widetilde{\Gamma}_L = \{(\widetilde{x}, \widetilde{y}, \widetilde{z}) / \widetilde{y} = -\widetilde{L}\}$ with those definitions we have $\widetilde{\Omega}_a \cap \widetilde{P} = \emptyset$, $\widetilde{\Omega}_c \cap \widetilde{\Omega}_r = \emptyset$, $\widetilde{P} = \widetilde{\Omega}_r \cup \widetilde{\Omega}_c$, $\widetilde{\Omega} = \widetilde{\Omega}_a \cup \widetilde{P} = \widetilde{\Omega}_a \cup \widetilde{\Omega}_r \cup \widetilde{\Omega}_c$, and for any $(\widetilde{x}, \widetilde{y}, \widetilde{z}) \in \partial\widetilde{\Omega} = \widetilde{\Gamma}_d \cup \widetilde{\Gamma}_L$ and, we write $\widetilde{n}$, the unit vector, orthogonal to $\partial\widetilde{\Omega}$ and pointing outside $\widetilde{\Omega}$. We have :

$$\begin{aligned} \widetilde{n} &= e_2 \ \text{ on } \ \widetilde{\Gamma}_d \\ \widetilde{n} &= -e_2 \ \text{ on } \ \widetilde{\Gamma}_L. \end{aligned} \tag{13}$$

In the following we need to describe what happens at the interfaces between resin and carbon fibers, and resin and air. So we define $\Gamma_{ra} = \{(\widetilde{x}, \widetilde{y}, \widetilde{z}) \ / \ \widetilde{y} = 0\}$ and $\Gamma_{cr}$ the boundary of the set defined by (9).

## 2.2. *Maxwell equations*

In $\widetilde{\Omega}$, we now write a PDE model that has to do with electromagnetic waves radiated from return stroke. We are well aware that the model we write is a simplified one. Nonetheless, it seems to be well dimensioned for our problem which consists in making homogenization. It is well known see Maxwell [17] the propagation of the electromagnetic field is described by the Maxwell equations which write:

$$-\frac{\partial \widetilde{D}^{\star}}{\partial t} + \nabla\times\widetilde{H}^{\star} = \widetilde{J}^{\star}, \tag{14}$$

$$\frac{\partial \widetilde{B}^{\star}}{\partial t} + \nabla\times\widetilde{E}^{\star} = 0, \tag{15}$$

$$\nabla\cdot\widetilde{D}^{\star} = \widetilde{\rho}^{\star}, \tag{16}$$

$$\nabla\cdot\widetilde{B}^{\star} = 0, \tag{17}$$

in $\mathbb{R}\times\widetilde{\Omega}$.

In (14)–(17), $\nabla\times$ and $\nabla\cdot$ are the curl and divergence operators. $\widetilde{E}^{\star}(t,x,y,z)$ is the electric field, $\widetilde{H}^{\star}(t,x,y,z)$ the magnetic field, $\widetilde{D}^{\star}(t,x,y,z)$ the electric induction, $\widetilde{B}^{\star}(t,x,y,z)$ the magnetic induction and $\widetilde{\rho}^{\star}(t,x,y,z)$ is the charges density see T. Abboud and I. Terrasse [1].

System of Maxwell equations ((14) – (17)) is completed by the constitutive laws which are given in $\mathbb{R}\times\widetilde{\Omega}$ by :

$$\widetilde{D}^{\star} = \epsilon_0\epsilon^{\star}\widetilde{E}^{\star}, \tag{18}$$

$$\widetilde{B}^{\star} = \mu_0\widetilde{H}^{\star}. \tag{19}$$

where $\mu_0$ and $\epsilon_0$ are the permeability and permittivity of free space. $\epsilon^{\star}$ is the relative permittivity of the domains defined by

$$\epsilon^{\star}_{|\widetilde{\Omega_a}} = 1, \epsilon^{\star}_{|\widetilde{\Omega_r}} = \epsilon_r, \epsilon^{\star}_{|\widetilde{\Omega_c}} = \epsilon_c, \tag{20}$$

where $\epsilon_r$ and $\epsilon_c$ are positives constants. In order to account for energy transfer between the electromagnetic compartment and the propagation of the electric charges, we take for granted the Ohmic law, in $\mathbb{R}\times\widetilde{\Omega}$

$$\widetilde{J}^{\star} = \sigma\widetilde{E}^{\star}, \tag{21}$$

where $\sigma$ is the electric conductivity. Its value depends on the location:

$$\sigma_{|\widetilde{\Omega_a}} = \sigma_a, \sigma_{|\widetilde{\Omega_r}} = \sigma_r, \sigma_{|\widetilde{\Omega_c}} = \sigma_c, \tag{22}$$

where $\widetilde{\Omega_a}$, $\widetilde{\Omega_r}$ and $\widetilde{\Omega_c}$ were defined in (12), (10) and (8).

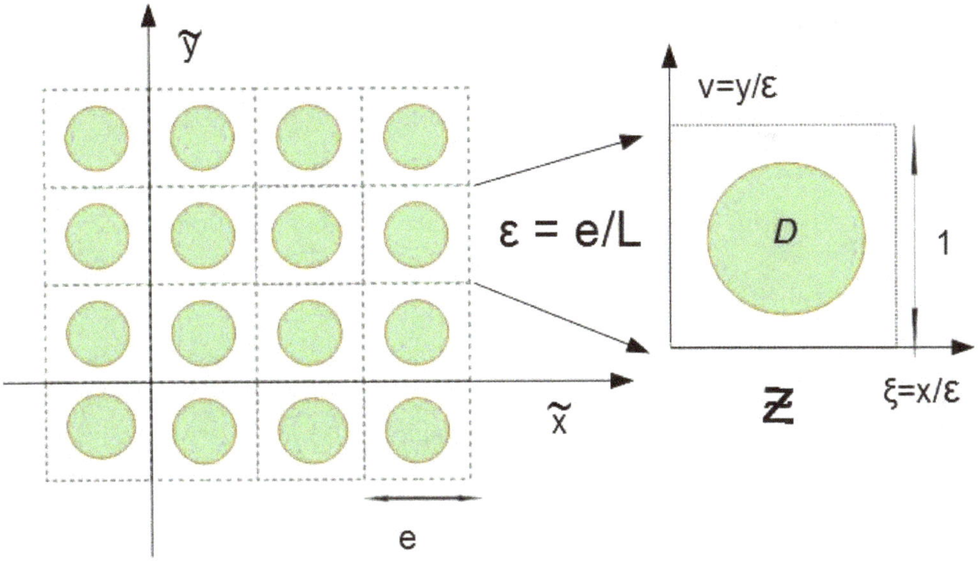


**Figure 2. Left: The global microstructure in 2D. Right: $\mathcal{Z}$-cell of the periodic structure.**

*2.3. Boundary conditions*

For mathematical as well as physical reasons we have to set boundary conditions on $\widetilde{\Gamma}_d$ and $\widetilde{\Gamma}_L$. On $\widetilde{\Gamma}_d$ we will write conditions that translate that $\widetilde{E}^\star$ and $\widetilde{H}^\star$ are given by the source located in $\widetilde{y} = d$. The way we chose consists in setting:

$$\widetilde{E}^\star \times \widetilde{n} = \widetilde{E}_d^\star \times \widetilde{n}; \quad \widetilde{H}^\star \times \widetilde{n} = \widetilde{H}_d^\star \times \widetilde{n} \quad \text{on} \quad \mathbb{R} \times \widetilde{\Gamma}_d, \tag{23}$$

where $\widetilde{E}_d^\star$, $\widetilde{H}_d^\star$ are functions defined on $\widetilde{\Gamma}_d$ for any $t \in \mathbb{R}$. On $\widetilde{\Gamma}_L$, we chose something simple, i.e :

$$\nabla \times \widetilde{E}^\star \times \widetilde{n} = 0 \quad \text{on} \quad \mathbb{R} \times \widetilde{\Gamma}_L, \tag{24}$$

that translate that $\widetilde{E}^\star$ does not vary in the $\widetilde{y}$-direction near $\widetilde{\Gamma}_L$.

Problem (14)–(21) supplemented with (23) and (24), is considered as containing all physics we want to account for. In the following we will consider simplifications of it.

*2.4. Time-harmonic Maxwell equations*

The first simplification we make, consists in considering the harmonic version of the Maxwell equations (14)–(22). This simplification is used in many references studying electromagnetic phenomena and especially for lightning applications [14], in spite of the fact that it considers implicitly that every fields and currents are waves of the form, for all $\widetilde{\omega} \in \mathbb{R}$ :

$$a(\widetilde{x}, \widetilde{y}, \widetilde{z}) \cos(-\widetilde{\omega} t + \phi(\widetilde{x}, \widetilde{y}, \widetilde{z})) = \Re e[a(\widetilde{x}, \widetilde{y}, \widetilde{z}) \exp^{i\widetilde{\omega} t} \exp^{i\phi(\widetilde{x}, \widetilde{y}, \widetilde{z})}], \tag{25}$$

where $\widetilde{\omega}$ is the pulsation, $\phi(\widetilde{x}, \widetilde{y}, \widetilde{z})$ is the phase shift of the wave and $a(\widetilde{x}, \widetilde{y}, \widetilde{z})$ is its amplitude. In particular, it supposes $\widetilde{E}_d^\star$, $\widetilde{H}_d^\star$ in (23) are of the form, for all $\widetilde{w} \in \mathbb{R}$:

$$\widetilde{E}_d^\star(t, \widetilde{x}, \widetilde{z}) = \Re e(\widetilde{E}_d(\widetilde{x}, \widetilde{z}) \exp^{i\widetilde{\omega} t}), \tag{26}$$

$$\widetilde{H}_d^\star(t,\widetilde{x},\widetilde{z}) = \Re e(\widetilde{H}_d(\widetilde{x},\widetilde{z})\exp^{i\widetilde{\omega}t}), \tag{27}$$

where $\widetilde{E}_d$ and $\widetilde{H}_d$ take into account the amplitude and the phase shift of their corresponding fields. Taking (21) into account, the time-harmonic Maxwell equations, which describe the electromagnetic radiation, are written:

$$\nabla\times\widetilde{H} - i\widetilde{\omega}\epsilon_0\epsilon^\star\widetilde{E} = \sigma\widetilde{E}, \text{ Maxwell - Ampere equation} \tag{28}$$

$$\nabla\times\widetilde{E} + i\widetilde{\omega}\mu_0\widetilde{H} = 0, \text{ Maxwell - Faraday equation} \tag{29}$$

$$\nabla\cdot(\epsilon_0\epsilon^\star\widetilde{E}) = \widetilde{\rho}, \tag{30}$$

$$\nabla\cdot(\mu_0\widetilde{H}) = 0, \tag{31}$$

where $\widetilde{E}^\star(t,\widetilde{x},\widetilde{y},\widetilde{z}) = \Re e(\widetilde{E}(\widetilde{x},\widetilde{y},\widetilde{z})\exp^{i\widetilde{\omega}t})$ and $\widetilde{H}^\star(t,\widetilde{x},\widetilde{y},\widetilde{z}) = \Re e(\widetilde{H}(\widetilde{x},\widetilde{y},\widetilde{z})\exp^{i\widetilde{\omega}t})$, $(\widetilde{x},\widetilde{y},\widetilde{z}) \in \widetilde{\Omega}$. The magnetic field $\widetilde{H}$ can be directly computed from the electric field $\widetilde{E}$

$$\widetilde{H} = -\frac{1}{i\omega\mu_0}\nabla\times\widetilde{E}. \tag{32}$$

Now, for the electric approach, taking the curl of equation (32) yields an expression of $\nabla\times\widetilde{H}$ in term of $\nabla\times\nabla\times\widetilde{E}$. Inserting $\nabla\times\widetilde{H}$ in (28) we get the following equation for the electric field:

$$\nabla\times\nabla\times\widetilde{E} + (-\widetilde{\omega}^2\mu_0\epsilon_0\epsilon^\star + i\widetilde{\omega}\mu_0\sigma)\widetilde{E} = 0 \quad \text{in} \quad \widetilde{\Omega}. \tag{33}$$

Taking the divergence of the equation (28) yields the natural gauge condition:

$$\nabla\cdot[(i\widetilde{\omega}\epsilon_0\epsilon^\star + \sigma)\widetilde{E}] = 0 \quad \text{in} \quad \widetilde{\Omega}. \tag{34}$$

Notice that $i\widetilde{\omega}\epsilon_0 + \sigma$ is equal to $i\widetilde{\omega}\epsilon_0 + \sigma_a$ in $\widetilde{\Omega_a}$, to $i\widetilde{\omega}\epsilon_0\epsilon_r + \sigma_r$ in $\widetilde{\Omega_r}$ and to $i\widetilde{\omega}\epsilon_0\epsilon_c + \sigma_c$ in $\widetilde{\Omega_c}$, those quantities being all nonzero. Then (34) is equivalent to:

$$\nabla\cdot\widetilde{E}_{|\Omega_a} = 0 \text{ in } \widetilde{\Omega_a}, \quad \nabla\cdot\widetilde{E}_{|\Omega_r} = 0 \text{ in } \widetilde{\Omega_r}, \quad \nabla\cdot\widetilde{E}_{|\Omega_c} = 0 \text{ in } \widetilde{\Omega_c}. \tag{35}$$

with the transmission conditions

$$\begin{aligned} (i\widetilde{\omega}\epsilon_0 + \sigma_a)\widetilde{E}_{|\widetilde{\Omega_a}}.\widetilde{n} &= (i\widetilde{\omega}\epsilon_0\epsilon_r + \sigma_r)\widetilde{E}_{|\widetilde{\Omega_r}}.\widetilde{n} \text{ on } \widetilde{\Gamma_{ra}}, \\ (i\widetilde{\omega}\epsilon_0\epsilon_r + \sigma_r)\widetilde{E}_{|\widetilde{\Omega_r}}.\widetilde{n} &= (i\widetilde{\omega}\epsilon_0\epsilon_c + \sigma_c)\widetilde{E}_{|\widetilde{\Omega_c}}.\widetilde{n} \text{ on } \widetilde{\Gamma_{cr}}. \end{aligned} \tag{36}$$

Summarizing, we finally obtain the PDE model:

$$\nabla\times\nabla\times\widetilde{E} + (-\widetilde{\omega}^2\mu_0\epsilon_0\epsilon^\star + i\widetilde{\omega}\mu_0\sigma)\widetilde{E} = 0 \quad \text{in} \quad \widetilde{\Omega}. \tag{37}$$

According to the tangential trace of the Maxwell-Faraday equation (29) we obviously obtain that using boundary condition (23), is equivalent to using:

$$\nabla\times\widetilde{E}\times e_2 = -i\widetilde{\omega}\mu_0\widetilde{H}_d(\widetilde{x},\widetilde{z})\times e_2 \quad \text{on} \quad \widetilde{\Gamma_d} \tag{38}$$

where $\widetilde{H}_d$ is defined in (27) and where we used (13). And on $\widetilde{\Gamma_L}$ we have the following boundary condition:

$$\nabla\times\widetilde{E}\times e_2 = 0 \quad \text{on} \quad \widetilde{\Gamma_L}. \tag{39}$$

### 2.5. Scaling

In this subsection we propose a rescaling of system ((37)–(39)), we will consider a set of characteristic sizes related to our problem. Physical factors are then rewritten using those values leading to a new set of dimensionless and unitless variables and fields in which the system is rewritten. The considered characteristic sizes are : $\overline{\omega}$ the characteristic pulsation, $\overline{\sigma}$ the characteristic electric conductivity, $\overline{E}$ the characteristic electric magnitude, $\overline{H}$ the characteristic magnetic magnitude. We also use the already introduced thickness $\overline{L}$ of the plate $\widetilde{P}$. We then introduce the dimensionless variables: $\mathbf{x} = (x, y, z)$ with $x = \frac{\widetilde{x}}{\overline{L}}$, $y = \frac{\widetilde{y}}{\overline{L}}$, $z = \frac{\widetilde{z}}{\overline{L}}$ and fields $E$, $H$ and $\sigma$ that are such that

$$\begin{cases} E(\omega, \mathbf{x}) = \dfrac{1}{\overline{E}}\widetilde{E}(\overline{\omega}\omega, \overline{L}x, \overline{L}y, \overline{L}z), \\ H(\omega, \mathbf{x}) = \dfrac{1}{\overline{H}}\widetilde{H}(\overline{\omega}\omega, \overline{L}x, \overline{L}y, \overline{L}z), \\ \sigma(\mathbf{x}) = \dfrac{1}{\overline{\sigma}}\widetilde{\sigma}(\overline{L}x, \overline{L}y, \overline{L}z), \end{cases} \tag{40}$$

Taking (22) into account, $\sigma$ also reads:

$$\begin{aligned} \sigma(\mathbf{x}) &= \frac{\sigma_a}{\overline{\sigma}} \quad \text{if} \quad 0 \leq \overline{L}y \leq d, \\ \sigma(\mathbf{x}) &= \frac{\sigma_r}{\overline{\sigma}} \quad \text{if} \quad (\overline{L}x, \overline{L}y, \overline{L}z) \in \widetilde{\Omega}_r, \\ \sigma(\mathbf{x}) &= \frac{\sigma_c}{\overline{\sigma}} \quad \text{if} \quad (\overline{L}x, \overline{L}y, \overline{L}z) \in \widetilde{\Omega}_c. \end{aligned} \tag{41}$$

Doing this gives the status of units to the characteristic sizes. Since, for instance:

$$\frac{\partial E}{\partial x}(\omega, \mathbf{x}) = \frac{\overline{L}}{\overline{E}}\frac{\partial \widetilde{E}}{\partial \widetilde{x}}(\overline{\omega}\omega, \overline{L}x, \overline{L}y, \overline{L}z), \tag{42}$$

using those dimensionless variables and fields and taking (41) into account, equation (37) gives:

$$\overline{E}\,\nabla\times\nabla\times E(\omega, \mathbf{x}) - \Big(\frac{\overline{L}^2\overline{\omega}^2}{c^2}\epsilon^{\star}\omega^2 + i\overline{\sigma}\,\overline{\omega}\,\omega\overline{L}^2\mu_0\sigma(\mathbf{x}, \omega)\Big)\overline{E}E(\omega, x, y, z) = 0, \tag{43}$$

for any $(\omega, \mathbf{x})$ such that $(\overline{\omega}\omega, \overline{L}x, \overline{L}y, \overline{L}z) \in \widetilde{\Omega}$. Now we exhibit

$$\overline{\lambda} = \frac{2\pi c}{\overline{\omega}}, \tag{44}$$

which is the characteristic wave length and

$$\overline{\delta} = \frac{1}{\sqrt{\overline{\omega}\,\overline{\sigma}\mu_0}}, \tag{45}$$

which is the characteristic skin thickness. Using those quantities equation (43) reads, for any $(\omega, \mathbf{x}) \in$

$\Omega$:

$$
\begin{aligned}
&\nabla\times\nabla\times E(\omega,\mathbf{x}) + \left(-\frac{4\pi^2\overline{L}^2}{\overline{\lambda}^2}\omega^2 + i\frac{\overline{L}^2}{\overline{\delta}^2}\frac{\sigma_a}{\overline{\sigma}}\omega\right)E(\omega,\mathbf{x}) = 0 \quad \text{when} \quad 0 \le \overline{L}y \le d,\\
&\nabla\times\nabla\times E(\omega,\mathbf{x}) + \left(-\frac{4\pi^2\overline{L}^2}{\overline{\lambda}^2}\epsilon_r\,\omega^2 + i\frac{\overline{L}^2}{\overline{\delta}^2}\frac{\sigma_r}{\overline{\sigma}}\omega\right)E(\omega,\mathbf{x}) = 0 \quad \text{when} \quad (\overline{L}x,\overline{L}y,\overline{L}z) \in \widetilde{\Omega}_r,\\
&\nabla\times\nabla\times E(\omega,\mathbf{x}) + \left(-\frac{4\pi^2\overline{L}^2}{\overline{\lambda}^2}\epsilon_c\,\omega^2 + i\frac{\overline{L}^2}{\overline{\delta}^2}\frac{\sigma_c}{\overline{\sigma}}\omega\right)E(\omega,\mathbf{x}) = 0 \quad \text{when} \quad (\overline{L}x,\overline{L}y,\overline{L}z) \in \widetilde{\Omega}_c.
\end{aligned} \tag{46}
$$

In the following expressions, $\frac{\overline{L}}{\overline{\lambda}}$ and $\frac{\overline{L}}{\overline{\delta}}$ appearing in the equations above will be rewritten in terms of a small parameter $\varepsilon$.

The boundary conditions are written

$$
\begin{aligned}
&\nabla\times E(\omega,\mathbf{x}) \times e_2 = -i\omega\overline{\omega}\mu_0\frac{\overline{L}}{\overline{E}}\widetilde{H_d}(\overline{L}x,\overline{L}z) \times e_2 \quad \text{when} \quad (\overline{L}x,\overline{L}y,\overline{L}z) \in \widetilde{\Gamma}_d,\\
&\nabla\times E(\omega,\mathbf{x}) \times e_2 = 0 \quad \text{when} \quad (\overline{L}x,\overline{L}y,\overline{L}z) \in \widetilde{\Gamma}_L.
\end{aligned} \tag{47}
$$

The characteristic thickness of the plate $\overline{L}$ is about $10^{-3}$m and the size of the basic cell $e$ is about $10^{-5}$m. Since $e$ is much smaller than the thickness of the plate $\overline{L}$, it is pertinent to assume the ratio $\frac{e}{\overline{L}}$ equals a small parameter $\varepsilon$:

$$
\frac{e}{\overline{L}} \sim 10^{-2} = \varepsilon. \tag{48}
$$

Then, in what concerns the characteristic pulsation $\overline{\omega}$, in the tables below we consider several values. The lightning is seen as a low frequency phenomenon. Indeed, energy associated with radiation tracers and return stroke are mainly burn by low and very low frequencies (from 1 kHz to 300 kHz). Components of the frequency spectrum are however observed beyond 1GHz (see [16]). So, in the case when we want to catch low frequency ie we consider $\overline{\omega} = 100$ rad/s, (in our study we will consider $\overline{\omega} = 10^6$ rad/s), for medium frequency we set $\overline{\omega} = 10^{10}$ rad/s and for high frequency phenomena $\overline{\omega} = 10^{12}$ rad/s. Then, concerning the characteristic electric conductivity it seems to be reasonable to take for $\overline{\sigma}$ the value of the effective electric conductivity of the composite material. Yet this choice implies to compute a coarse estimate of this effective conductivity at this level.

For this we take into account that the composite material is composed of carbon fibers and epoxy resin. The resin can be doped, which increases strongly its conductivity, or not. The tables below summarize the cases when the resin is doped and also when the resin is not doped. We also account for the fact there is not only one effective electric conductivity but a first one in the fiber direction : the effective longitudinal electric conductivity (in cases 1, 2, 5 and 6 of the tables below), and a second effective electric conductivity, in the direction transverse to the fibers (considered in cases 3, 4, 7 and 8). In this context, we consider the basic model which is based on the electrical analogy and the law of mixtures. It corresponds to the Wiener limits: the harmonic average and the arithmetic average. The effective values are the extreme limits of the conductivity of the composite introduced by Wiener in 1912 see S. Berthier p 76 [7].

| | case 1 | case 2 | case 3 | case 4 | case 5 | case 6 | case 7 | case 8 |
|---|---|---|---|---|---|---|---|---|
| $\overline{L}(m)$ | $10^{-3}$ | $10^{-3}$ | $10^{-3}$ | $10^{-3}$ | $10^{-3}$ | $10^{-3}$ | $10^{-3}$ | $10^{-3}$ |
| $e(m)$ | $10^{-5}$ | $10^{-5}$ | $10^{-5}$ | $10^{-5}$ | $10^{-5}$ | $10^{-5}$ | $10^{-5}$ | $10^{-5}$ |
| $\overline{\lambda}(m)$ | $10^{6}$ | $10^{6}$ | $10^{6}$ | $10^{6}$ | $10^{6}$ | $10^{6}$ | $10^{6}$ | $10^{6}$ |
| $\overline{\sigma}(S.m^{-1})$ | 40000 | 40000 | $10^{-10}$ | $10^{-3}$ | 40000 | $10^{4}$ | $10^{-10}$ | $10^{-3}$ |
| $\overline{\delta}(m)$ | 0, 1 | 0, 1 | $10^{7}$ | $10^{3}$ | 0, 1 | 0, 1 | $10^{7}$ | $10^{3}$ |
| $\sigma_c(S.m^{-1})$ | $\overline{\sigma}$ | $\overline{\sigma}$ | $\frac{\overline{\sigma}}{\varepsilon^7}$ | $\frac{\overline{\sigma}}{\varepsilon^4}$ | $\overline{\sigma}$ | $\overline{\sigma}$ | $\frac{\overline{\sigma}}{\varepsilon^7}$ | $\frac{\overline{\sigma}}{\varepsilon^4}$ |
| $\sigma_r(S.m^{-1})$ | $\varepsilon^7\overline{\sigma}$ | $\varepsilon^4\overline{\sigma}$ | $\overline{\sigma}$ | $\overline{\sigma}$ | $\varepsilon^7\overline{\sigma}$ | $\varepsilon^4\overline{\sigma}$ | $\overline{\sigma}$ | $\overline{\sigma}$ |
| $\sigma_a(S.m^{-1})$ | $\varepsilon^9\overline{\sigma}$ | $\varepsilon^9\overline{\sigma}$ | $\varepsilon^2\overline{\sigma}$ | $\varepsilon^6\overline{\sigma}$ | $\varepsilon\overline{\sigma}$ | $\varepsilon\overline{\sigma}$ | $\frac{\overline{\sigma}}{\varepsilon^5}$ | $\frac{\overline{\sigma}}{\varepsilon^2}$ |
| $\frac{4\pi\overline{L}^2}{\overline{\lambda}^2}$ | $\varepsilon^9$ | $\varepsilon^9$ | $\varepsilon^9$ | $\varepsilon^9$ | $\varepsilon^9$ | $\varepsilon^9$ | $\varepsilon^9$ | $\varepsilon^9$ |
| $\frac{\overline{L}^2}{\overline{\delta}^2}$ | $\varepsilon^2$ | $\varepsilon^2$ | $\varepsilon^{10}$ | $\varepsilon^7$ | $\varepsilon^2$ | $\varepsilon^2$ | $\varepsilon^{10}$ | $\varepsilon^7$ |

**Table 1. for $\overline{\omega} = 100 rad.s^{-1}$.**

The effective longitudinal electric conductivity corresponding of the upper Wiener limit is expressed by the equation:

$$\overline{\sigma} = \sigma_{\text{long}} = f_c\, \sigma_c + (1 - f_c)\, \sigma_r, \tag{49}$$

where $f_c = \pi\frac{\alpha^2}{4}$ is the volume fraction of the carbon fiber.

The effective transverse electric conductivity corresponding of the lower Wiener limit is expressed by

$$\overline{\sigma} = \sigma_{\text{trans}} = \frac{1}{\frac{f_c}{\sigma_c} + \frac{(1-f_c)}{\sigma_r}}. \tag{50}$$

For the computation, we take values close to reality. We consider composite materials with similar proportions of carbon and resin, this means that $\alpha$ is close to $\frac{1}{2}$. When the resin is not doped $\sigma_r \sim 10^{-10}\ S.m^{-1}$ is much smaller than $\sigma_c \sim 40000\ S.m^{-1}$. Then, $\overline{\sigma} = \sigma_{\text{long}}$ is close to $\pi\frac{\alpha^2}{4}\ \sigma_c \sim \sigma_c$ and $\overline{\sigma} = \sigma_{\text{trans}}$ is close to $\frac{\sigma_r}{(1-\pi\frac{\alpha^2}{4})} \sim \sigma_r$.

Now, we express the electric conductivity of the air in terms of $\overline{\sigma}$, we consider two possibilities. The first one is relevant for a situation with a ionized channel. The second one of situation with a strong atmospheric electric field but without a ionized channel. In this situation air is not ionized and has a low conductivity. All possible situations are gathered in the tables below. Cases 5 to 8 are associated with the first situation with air conductivity $\sigma_a$ being $\sigma_{lightning} = 4242\ S.m^{-1}$ for an ionized lightning channel see [13]. Cases 1 to 4, to the second one, with $\sigma_a = 10^{-14}\ S.m^{-1}$.

All calculations of the different cases of the tables are detailed in Annex A. In our study we consider the case 6 for $\omega = 10^6\ rad.s^{-1}$, which corresponds to the air ionized, a resin doped and the effective longitudinal electric conductivity of the carbon fibers.

As it is well known the tables confirm that at high frequencies the thickness of the plate is much greater than the skin depth. This one depends on $\overline{\sigma}$ and $\overline{\omega}$ and decreases strongly for high conductivity or high frequencies. For $\overline{\omega} = 10^{10}\ rad.s^{-1}$ and $\overline{\sigma} = 4 * 10^4\ S.m^{-1}$, the effective conductivity in the

| | case 1 | case 2 | case 3 | case 4 | case 5 | case 6 | case 7 | case 8 |
|---|---|---|---|---|---|---|---|---|
| $\overline{L}(m)$ | $10^{-3}$ | $10^{-3}$ | $10^{-3}$ | $10^{-3}$ | $10^{-3}$ | $10^{-3}$ | $10^{-3}$ | $10^{-3}$ |
| $e(m)$ | $10^{-5}$ | $10^{-5}$ | $10^{-5}$ | $10^{-5}$ | $10^{-5}$ | $10^{-5}$ | $10^{-5}$ | $10^{-5}$ |
| $\overline{\lambda}(m)$ | $10^3$ | $10^3$ | $10^3$ | $10^3$ | $10^3$ | $10^3$ | $10^3$ | $10^3$ |
| $\overline{\sigma}(S.m^{-1})$ | 40000 | 40000 | $10^{-10}$ | $10^{-3}$ | 40000 | 40000 | $10^{-10}$ | $10^{-3}$ |
| $\overline{\delta}(m)$ | $10^{-3}$ | $10^{-3}$ | $10^5$ | 10 | $10^{-3}$ | $10^{-3}$ | $10^5$ | 10 |
| $\sigma_c(S.m^{-1})$ | $\overline{\sigma}$ | $\overline{\sigma}$ | $\frac{\overline{\sigma}}{\varepsilon^7}$ | $\frac{\overline{\sigma}}{\varepsilon^5}$ | $\overline{\sigma}$ | $\overline{\sigma}$ | $\frac{\overline{\sigma}}{\varepsilon^7}$ | $\frac{\overline{\sigma}}{\varepsilon^5}$ |
| $\sigma_r(S.m^{-1})$ | $\varepsilon^7\overline{\sigma}$ | $\varepsilon^4\overline{\sigma}$ | $\overline{\sigma}$ | $\overline{\sigma}$ | $\varepsilon^7\overline{\sigma}$ | $\varepsilon^4\overline{\sigma}$ | $\overline{\sigma}$ | $\overline{\sigma}$ |
| $\sigma_a(S.m^{-1})$ | $\varepsilon^9\overline{\sigma}$ | $\varepsilon^9\overline{\sigma}$ | $\varepsilon^2\overline{\sigma}$ | $\varepsilon^6\overline{\sigma}$ | $\varepsilon\overline{\sigma}$ | $\varepsilon\overline{\sigma}$ | $\frac{\overline{\sigma}}{\varepsilon^5}$ | $\frac{\overline{\sigma}}{\varepsilon^2}$ |
| $\frac{4\pi\overline{L}^2}{\overline{\lambda}^2}$ | $\varepsilon^5$ | $\varepsilon^5$ | $\varepsilon^5$ | $\varepsilon^5$ | $\varepsilon^5$ | $\varepsilon^5$ | $\varepsilon^5$ | $\varepsilon^5$ |
| $\frac{\overline{L}^2}{\overline{\delta}^2}$ | 1 | 1 | $\varepsilon^8$ | $\varepsilon^5$ | 1 | 1 | $\varepsilon^8$ | $\varepsilon^5$ |

**Table 2. for $\overline{\omega} = 10^6 rad.s^{-1}$.**

| | case 1 | case 2 | case 3 | case 4 | case 5 | case 6 | case 7 | case 8 |
|---|---|---|---|---|---|---|---|---|
| $\overline{L}(m)$ | $10^{-3}$ | $10^{-3}$ | $10^{-3}$ | $10^{-3}$ | $10^{-3}$ | $10^{-3}$ | $10^{-3}$ | $10^{-3}$ |
| $e(m)$ | $10^{-5}$ | $10^{-5}$ | $10^{-5}$ | $10^{-5}$ | $10^{-5}$ | $10^{-5}$ | $10^{-5}$ | $10^{-5}$ |
| $\overline{\lambda}(m)$ | $10^{-1}$ | $10^{-1}$ | $10^{-1}$ | $10^{-1}$ | $10^{-1}$ | $10^{-1}$ | $10^{-1}$ | $10^{-1}$ |
| $\overline{\sigma}(S.m^{-1})$ | 40000 | 40000 | $10^{-10}$ | $10^{-3}$ | 40000 | 40000 | $10^{-10}$ | $10^{-3}$ |
| $\overline{\delta}(m)$ | $10^{-5}$ | $10^{-5}$ | $10^3$ | $10^{-1/2}$ | $10^{-5}$ | $10^{-5}$ | $10^3$ | $10^{-1/2}$ |
| $\sigma_c(S.m^{-1})$ | $\overline{\sigma}$ | $\overline{\sigma}$ | $\frac{\overline{\sigma}}{\varepsilon^7}$ | $\frac{\overline{\sigma}}{\varepsilon^4}$ | $\overline{\sigma}$ | $\overline{\sigma}$ | $\frac{\overline{\sigma}}{\varepsilon^7}$ | $\frac{\overline{\sigma}}{\varepsilon^4}$ |
| $\sigma_r(S.m^{-1})$ | $\varepsilon^7\overline{\sigma}$ | $\varepsilon^4\overline{\sigma}$ | $\overline{\sigma}$ | $\overline{\sigma}$ | $\varepsilon^7\overline{\sigma}$ | $\varepsilon^4\overline{\sigma}$ | $\overline{\sigma}$ | $\overline{\sigma}$ |
| $\sigma_a(S.m^{-1})$ | $\varepsilon^9\overline{\sigma}$ | $\varepsilon^9\overline{\sigma}$ | $\varepsilon^2\overline{\sigma}$ | $\varepsilon^6\overline{\sigma}$ | $\varepsilon\overline{\sigma}$ | $\varepsilon\overline{\sigma}$ | $\frac{\overline{\sigma}}{\varepsilon^5}$ | $\frac{\overline{\sigma}}{\varepsilon^2}$ |
| $\frac{4\pi\overline{L}^2}{\overline{\lambda}^2}$ | $\varepsilon$ | $\varepsilon$ | $\varepsilon$ | $\varepsilon$ | $\varepsilon$ | $\varepsilon$ | $\varepsilon$ | $\varepsilon$ |
| $\frac{\overline{L}^2}{\overline{\delta}^2}$ | $\frac{1}{\varepsilon^2}$ | $\frac{1}{\varepsilon^2}$ | $\varepsilon^6$ | $\varepsilon^3$ | $\frac{1}{\varepsilon^2}$ | $\frac{1}{\varepsilon^2}$ | $\epsilon^6$ | $\epsilon^3$ |

**Table 3. for $\overline{\omega} = 10^{10} rad.s^{-1}$.**

| | case 1 | case 2 | case 3 | case 4 | case 5 | case 6 | case 7 | case 8 |
|---|---|---|---|---|---|---|---|---|
| $\overline{L}(m)$ | $10^{-3}$ | $10^{-3}$ | $10^{-3}$ | $10^{-3}$ | $10^{-3}$ | $10^{-3}$ | $10^{-3}$ | $10^{-3}$ |
| $e(m)$ | $10^{-5}$ | $10^{-5}$ | $10^{-5}$ | $10^{-5}$ | $10^{-5}$ | $10^{-5}$ | $10^{-5}$ | $10^{-5}$ |
| $\overline{\lambda}(m)$ | $10^{-3}$ | $10^{-3}$ | $10^{-3}$ | $10^{-3}$ | $10^{-3}$ | $10^{-3}$ | $10^{-3}$ | $10^{-3}$ |
| $\overline{\sigma}(S.m^{-1})$ | 40000 | 40000 | $10^{-10}$ | $10^{-3}$ | 40000 | 40000 | $10^{-10}$ | $10^{-3}$ |
| $\overline{\delta}(m)$ | $10^{-6}$ | $10^{-6}$ | 1 | $10^{-3/2}$ | $10^{-6}$ | $10^{-6}$ | 1 | $10^{-3/2}$ |
| $\sigma_c(S.m^{-1})$ | $\overline{\sigma}$ | $\overline{\sigma}$ | $\frac{\overline{\sigma}}{\varepsilon^7}$ | $\frac{\overline{\sigma}}{\varepsilon^4}$ | $\overline{\sigma}$ | $\overline{\sigma}$ | $\frac{\overline{\sigma}}{\varepsilon^7}$ | $\frac{\overline{\sigma}}{\varepsilon^4}$ |
| $\sigma_r(S.m^{-1})$ | $\varepsilon^7\overline{\sigma}$ | $\varepsilon^4\overline{\sigma}$ | $\overline{\sigma}$ | $\overline{\sigma}$ | $\varepsilon^7\overline{\sigma}$ | $\varepsilon^4\overline{\sigma}$ | $\overline{\sigma}$ | $\overline{\sigma}$ |
| $\sigma_a(S.m^{-1})$ | $\varepsilon^9\overline{\sigma}$ | $\varepsilon^9\overline{\sigma}$ | $\varepsilon^2\overline{\sigma}$ | $\varepsilon^6\overline{\sigma}$ | $\varepsilon\overline{\sigma}$ | $\varepsilon\overline{\sigma}$ | $\frac{\overline{\sigma}}{\varepsilon^5}$ | $\frac{\overline{\sigma}}{\varepsilon^2}$ |
| $\frac{4\pi\overline{L}^2}{\overline{\lambda}^2}$ | 1 | 1 | 1 | 1 | 1 | 1 | 1 | 1 |
| $\frac{\overline{L}^2}{\overline{\delta}^2}$ | $\frac{1}{\varepsilon^3}$ | $\frac{1}{\varepsilon^3}$ | $\varepsilon^3$ | $\varepsilon$ | $\frac{1}{\varepsilon^3}$ | $\frac{1}{\varepsilon^3}$ | $\epsilon^3$ | $\varepsilon$ |

**Table 4. for $\overline{\omega} = 10^{12} rad.s^{-1}$.**

direction of the carbon fibers, which the skin effect phenomenon appears. Indeed, for high frequencies $\omega = 10^{12} rad.s^{-1}$ and when $\overline{\sigma}$ is the effective conductivity is in direction of the carbon fibers *i.e.* in high conductivity, $\overline{\delta} = 10^{-4}$ m. In low frequencies and low conductivity $\delta$ is large so the electromagnetic wave can penetrate the composite material. The high conductivity limits the penetration of the electromagnetic wave to a boundary layer whose depth is about $\overline{\delta}$.

Now, we will discuss on the values of $\overline{E}$ and $\overline{\rho}$. It seems that the density of electrons in a ionized channel is about $10^{10}$ part.$m^{-3}$. Hence we take $\overline{\rho} = 10^{10}$. When the air is not ionized, the charge density is much smaller, and we choose: $\overline{\rho} = 1$.

For the boundary conditions, in the context of the case 6 and $\overline{\omega} = 10^6$ rad/s, we consider the peak of the current of the first return stroke. Then the magnetic field magnitude $\overline{H}$ is $\overline{H_d}$ given by (1).

Then the dimensionless boundary conditions (38) writes:

$$\nabla \times E(\mathbf{x}, \omega) \times e_2 = -i\omega\overline{\omega}\mu_0 \frac{\overline{L}}{\overline{E}} \overline{H_d} H_d(x, z) \times e_2, \tag{51}$$

where $\overline{H_d} H_d(x, z) = \widetilde{H_d}(\overline{L}x, \overline{L}z)$ and where $\overline{\omega}\mu_0 \frac{\overline{L}}{\overline{E}} \overline{H_d}$ being of order 1, with the characteristic electric field $\overline{E} = 20$ kV/m.

From the physical spatial coordinates $(\widetilde{x}, \widetilde{y}, \widetilde{z}) \in \widetilde{\Omega}$ we define $\mathbf{y} = (\xi, \nu, \zeta)$ with $\xi = \frac{\widetilde{x}}{e}, \nu = \frac{\widetilde{y}}{e}, \zeta = \frac{\widetilde{z}}{e}$ or equivalently $\xi = \frac{x}{\varepsilon}, \nu = \frac{y}{\varepsilon}, \zeta = \frac{z}{\varepsilon}$. And we now introduce $Y$, the basic cell. It is built from: $Z = [-\frac{1}{2}, \frac{1}{2}] \times [-1, 0] \times \mathbb{R}$ and the set $C = D \times \mathbb{R}$ with the disc $D$ defined by:

$$D = \{(\xi, \nu) \in \mathbb{R}^2 \, / \xi^2 + (\nu + \frac{1}{2})^2 < R^2\}, \tag{52}$$

and $R = \frac{\alpha}{2}$. The set $\Omega_c$ is then defined as:

$$\Omega_c = \{(i, j, 0) + C, i \in \mathbb{Z}, j \in \mathbb{Z}^-\}. \tag{53}$$

We denote $Y_r$ as $Y_r = Z \backslash C$ and then the set

$$\Omega_r = \{(i, j, 0) + Y_r, i \in \mathbb{Z}, j \in \mathbb{Z}^-\}. \tag{54}$$

Then unit cell $Y$ is defined as $Y = (Y_r, C)$. Finally, we define the domain $\Omega_a$:

$$\Omega_a = \{\mathbf{y} = (\xi, \nu, \zeta) \ / \ \nu > 0\}. \tag{55}$$

Using this, we will give a new expression of the sets in which the variables range in equations (46). We see the following:

$$(\overline{L}x, \overline{L}y, \overline{L}z) \in \widetilde{\Omega}_r \Leftrightarrow \begin{cases} (\overline{L}x, \overline{L}y, \overline{L}z) \in \widetilde{P}, \\ (\frac{\overline{L}}{e}x, \frac{\overline{L}}{e}y, \frac{\overline{L}}{e}z) \in \Omega_r, \end{cases} \tag{56}$$

*i.e.*

$$(\overline{L}x, \overline{L}y, \overline{L}z) \in \widetilde{\Omega}_r \Leftrightarrow \begin{cases} (\overline{L}x, \overline{L}y, \overline{L}z) \in \widetilde{P}, \\ (\frac{x}{\varepsilon}, \frac{y}{\varepsilon}, \frac{z}{\varepsilon}) \in \Omega_r. \end{cases} \tag{57}$$

In the same way:

$$(\overline{L}x, \overline{L}y, \overline{L}z) \in \widetilde{\Omega}_c \Leftrightarrow \begin{cases} (\overline{L}x, \overline{L}y, \overline{L}z) \in \widetilde{P}, \\ (\frac{x}{\varepsilon}, \frac{y}{\varepsilon}, \frac{z}{\varepsilon}) \in \Omega_c, \end{cases} \tag{58}$$

and:

$$0 \le \overline{L}y \le d \Leftrightarrow \begin{cases} y \in \mathbb{R}^2 \\ \overline{L}y \le d, \end{cases} \tag{59}$$

or

$$(\overline{L}x, \overline{L}y, \overline{L}z) \in \widetilde{\Omega}_a \Leftrightarrow \begin{cases} \overline{L}y \le d \\ (\frac{x}{\varepsilon}, \frac{y}{\varepsilon}, \frac{z}{\varepsilon}) \in \Omega_a. \end{cases} \tag{60}$$

We define:

$$\Sigma^\varepsilon(\mathbf{y}) = \Sigma^\varepsilon(\xi, \nu, \zeta) = \begin{cases} \Sigma_a^\varepsilon & \text{in } \Omega_a, \\ \Sigma_r^\varepsilon & \text{in } \Omega_r, \\ \Sigma_c^\varepsilon & \text{in } \Omega_c, \end{cases} \tag{61}$$

where $\Sigma_a^\varepsilon = \frac{\sigma_a}{\overline{\sigma}} \frac{\overline{L}^2}{\overline{\delta}^2}$, $\Sigma_r^\varepsilon = \frac{\sigma_r}{\overline{\sigma}} \frac{\overline{L}^2}{\overline{\delta}^2}$ and $\Sigma_c^\varepsilon = \frac{\sigma_c}{\overline{\sigma}} \frac{\overline{L}^2}{\overline{\delta}^2}$ have their expressions in term of $\varepsilon$ given from Tables above depending on the case we are interested in. The detail of this expressions are in appendix B. The model that we present is the case $\omega = 10^6\ rad.s^{-1}$, $\eta = 5$, $\Sigma_a^\varepsilon = \varepsilon$, $\Sigma_r^\varepsilon = \varepsilon^4$ and $\Sigma_c^\varepsilon = 1$.

Defining also mapping

$$\begin{aligned} \psi_\varepsilon : \ & \mathbb{R}^3 \to \mathbb{R}^3 \\ & (x, y, z) \mapsto (\frac{x}{\varepsilon}, \frac{y}{\varepsilon}, \frac{z}{\varepsilon}), \end{aligned} \tag{62}$$

we can set $\Omega_a^\varepsilon$ as $\psi_\varepsilon^{-1}(\Omega_a) \cap (\mathbb{R} \times [0, \frac{d}{L}] \times \mathbb{R})$, $\Omega_r^\varepsilon$ as $\psi_\varepsilon^{-1}(\Omega_r) \cap \widetilde{P}$ and $\Omega_c^\varepsilon$ as $\psi_\varepsilon^{-1}(\Omega_c) \cap \widetilde{P}$. We also define the boundaries $\Gamma_d = \{\mathbf{x} \in \mathbb{R}^3,\ \ y = \frac{d}{L}\}$ and $\Gamma_L = \{\mathbf{x} \in \mathbb{R}^3,\ \ y = -\overline{L}\}$ and interfaces $\Gamma_{ra} = \{\mathbf{x} \in \mathbb{R}^3,\ \ y = 0\}$ and $\Gamma_{cr}^\varepsilon = \partial\Omega_c$. Hence equation (46) reads:

$$\nabla\times\nabla\times E^{\varepsilon} + (-\omega^2\varepsilon^{\eta}\epsilon^{\star} + i\,\omega\,\sigma^{\varepsilon}(x,y,z))E^{\varepsilon} = 0 \quad \text{in } \Omega, \tag{63}$$

where $\Omega = \Omega_a^{\varepsilon} \cup \Omega_r^{\varepsilon} \cup \Omega_c^{\varepsilon} = \{\mathbf{x} \in \mathbb{R}^3, -1 < y < \frac{d}{L}\}$ does not depend on $\varepsilon$. Only its partition in $\Omega_a^{\varepsilon}$, $\Omega_r^{\varepsilon}$ and $\Omega_c^{\varepsilon}$ is $\varepsilon$-dependent where

$$\sigma^{\varepsilon}(x,y,z) = \Sigma^{\varepsilon}(\frac{x}{\varepsilon}, \frac{y}{\varepsilon}, \frac{z}{\varepsilon}), \tag{64}$$

with $\Sigma^{\varepsilon}$ given by (61) and

$$\varepsilon^{\eta} = \frac{4\pi^2\overline{L}^{-2}}{\lambda^2}, \tag{65}$$

with the value of $\eta \geq 0$ extracted from Tables, and where we replace $E$ by $E^{\varepsilon}$, to clearly state that it depends on $\varepsilon$.

Equation (63) is provided with the following boundary conditions:

$$\nabla\times E^{\varepsilon} \times e_2 = -i\omega H_d(x,z) \times e_2 \text{ on } \Gamma_d, \tag{66}$$

coming from (51). And, coming from (47),

$$\nabla\times E^{\varepsilon} \times e_2 = 0 \text{ on } \Gamma_L. \tag{67}$$

From (63) we can deduce the condition on the divergence of $E^{\varepsilon}$ which can be written in two ways. As previously in (34), (35) and (36) we obtain:

$$\nabla\cdot[(-\omega^2\varepsilon^{\eta}\epsilon^{\star} + i\omega\sigma^{\varepsilon})E^{\varepsilon}] = 0 \quad \text{in } \Omega, \tag{68}$$

which will be preferentially used with (63) and its second one is

$$\nabla\cdot E^{\varepsilon}_{|\Omega_a^{\varepsilon}} = 0 \text{ in } \Omega_a^{\varepsilon},\quad \nabla\cdot E^{\varepsilon}_{|\Omega_r^{\varepsilon}} = 0 \text{ in } \Omega_r^{\varepsilon},\quad \nabla\cdot E^{\varepsilon}_{|\Omega_c^{\varepsilon}} = 0 \text{ in } \Omega_c^{\varepsilon}, \tag{69}$$

with the transmission conditions on the interfaces $\Gamma_{ra}$ and $\Gamma_{cr}^{\varepsilon}$

$$\begin{aligned} (-\omega^2\varepsilon^{\eta} + i\omega\Sigma_a^{\varepsilon})\, E^{\varepsilon}_{|\Omega_a^{\varepsilon}} \cdot n_{|\Omega_a^{\varepsilon}} &= (-\omega^2\varepsilon^{\eta}\epsilon_r + i\omega\Sigma_r^{\varepsilon})\, E^{\varepsilon}_{|\Omega_r^{\varepsilon}} \cdot n_{|\Omega_r^{\varepsilon}} \text{ on } \Gamma_{ra},\\ (-\omega^2\varepsilon^{\eta}\epsilon_r + i\omega\Sigma_r^{\varepsilon})\, E^{\varepsilon}_{|\Omega_r^{\varepsilon}} \cdot n_{|\Omega_r^{\varepsilon}} &= (-\omega^2\varepsilon^{\eta}\epsilon_c + i\omega\Sigma_c^{\varepsilon})\, E^{\varepsilon}_{|\Omega_c^{\varepsilon}} \cdot n_{|\Omega_c^{\varepsilon}} \text{ on } \Gamma_{cr}^{\varepsilon}. \end{aligned} \tag{70}$$

Before treating mathematically the question we are interested in, we make a last simplification. Since it seems clear that physical relevant phenomena occur in the upper part of the plate. The boundary condition on the lower boundary of the plate has very little influence on the physics of what happens in the upper part, we consider that the lower boundary of $\Omega$ is located in $y = -\infty$ in place of $y = -1$, making the second boundary condition useless. Besides, as $\overline{L}$ and $d$ are of the same order it seems reasonable to set $\Gamma_d = \{\mathbf{x} \in \mathbb{R}^3,\ y = 1\}$ and consequently

$$\begin{cases} \Omega = \{\mathbf{x} \in \mathbb{R}^3, y < 1\} = \Omega_a^{\varepsilon} \cup \Omega_r^{\varepsilon} \cup \Omega_c^{\varepsilon}, \text{ with,}\\ \Omega_a^{\varepsilon} = \psi_{\varepsilon}^{-1}(\Omega_a),\\ \Omega_r^{\varepsilon} = \psi_{\varepsilon}^{-1}(\Omega_r),\\ \Omega_c^{\varepsilon} = \psi_{\varepsilon}^{-1}(\Omega_c), \end{cases} \tag{71}$$

with $\psi_{\varepsilon}$ defined in (62). We have that the border of $\Omega$ is $\Gamma_d$. In the following section we will establish existence and uniqueness results.

## 3. Mathematical analysis of the models

### *3.1. Preliminaries*

We are going to make precise the variational formulation. First of all, we need to introduce the following functional spaces. We have the standard function spaces $\mathbf{L}^2(\Omega^\varepsilon) = [L^2(\Omega^\varepsilon)]^3$

$$\begin{aligned} \mathbf{H}(\text{curl}, \Omega) &= \{u \in \mathbf{L}^2(\Omega) : \ \nabla \times u \in \mathbf{L}^2(\Omega)\}, \\ \mathbf{H}(\text{div}, \Omega) &= \{u \in \mathbf{L}^2(\Omega) : \ \nabla \cdot u \in L^2(\Omega)\}, \end{aligned} \tag{72}$$

with the usual norms:

$$\begin{aligned} \|u\|^2_{\mathbf{H}(\text{curl},\Omega)} &= \|u\|^2_{\mathbf{L}^2(\Omega)} + \|\nabla \times u\|^2_{\mathbf{L}^2(\Omega)}, \\ \|u\|^2_{\mathbf{H}(\text{div},\Omega)} &= \|u\|^2_{\mathbf{L}^2(\Omega)} + \|\nabla \cdot u\|^2_{L^2(\Omega)}. \end{aligned} \tag{73}$$

They are well known Hilbert spaces.

We use in this paper, the trace spaces $H^{-\frac{1}{2}}(\text{curl}, \Gamma_d)$ and $H^{-\frac{1}{2}}(\text{div}, \Gamma_d)$ defined by

$$H^{-\frac{1}{2}}(\text{curl}, \Gamma_d) = \{u \in H^{-\frac{1}{2}}(\Gamma_d, \mathbb{R}^3),\ (n \cdot u)_{|\Gamma_d} = 0,\ \text{curl}_{\Gamma_d} u \in H^{-\frac{1}{2}}(\Gamma_d, \mathbb{R}^3)\}, \tag{74}$$

$$H^{-\frac{1}{2}}(\text{div}, \Gamma_d) = \{u \in H^{-\frac{1}{2}}(\Gamma_d, \mathbb{R}^3),\ (n \cdot u)_{|\Gamma_d} = 0,\ \text{div}_{\Gamma_d} u \in H^{-\frac{1}{2}}(\Gamma_d, \mathbb{R}^3)\} \tag{75}$$

where the surface divergence $\text{div}_{\Gamma_d} u$ and the surface rotation $\text{curl}_{\Gamma_d} u$ are defined by

$$\begin{aligned} &(\text{div}_{\Gamma_d} u, V)_{L^2(\Gamma_d)} = -(u, \nabla_{\Gamma_d} V)_{L^2(\Gamma_d, \mathbb{R}^3)}, \ \ \forall\, V \in C^1(\Gamma_d) \\ &\text{curl}_{\Gamma_d} u = n \cdot (\nabla \times u_{|\Gamma_d}) \end{aligned} \tag{76}$$

and the surface gradient $\nabla_{\Gamma_d} V$ is defined by the orthogonal projection of $\nabla$ on $\Gamma_d$, n denotes the outward unit vector normal to $\Gamma_d$.

Finally we recall the trace theorems, see J.C Nédélec [19] for the demonstration, stating that the traces mappings

$\gamma_T : \mathbf{H}(\text{curl}, \Omega) \longrightarrow H^{-\frac{1}{2}}(\text{curl}, \Gamma_d)$, that assigns any $u \in \mathbf{H}(\text{curl}, \Omega)$ its tangential components $n \times (u \times n)$, is continuous and surjective, that is:

$$\|\gamma_T(u)\|_{H^{-\frac{1}{2}}(\text{curl},\Gamma_d)} \leq C_{\gamma_T} \|u\|_{\mathbf{H}(\text{curl},\Omega)}, \quad \forall u \in \mathbf{H}(\text{curl}, \Omega)$$

$\gamma_t : \mathbf{H}(\text{curl}, \Omega) \longrightarrow H^{-\frac{1}{2}}(\text{div}, \Gamma_d)$, that assigns any $u \in \mathbf{H}(\text{curl}, \Omega)$ its tangential components $u \times n$, is continuous and surjective:

$$\|\gamma_t(u)\|_{H^{-\frac{1}{2}}(\text{div},\Gamma_d)} \leq C_{\gamma_t} \|u\|_{\mathbf{H}(\text{curl},\Omega)}, \quad \forall u \in \mathbf{H}(\text{curl}, \Omega).$$

Moreover, $H^{-\frac{1}{2}}(\text{div}, \Gamma_d)$ is the dual of $H^{-\frac{1}{2}}(\text{curl}, \Gamma_d)$ and one has the Green's formula:

$$\int_\Omega (\nabla \times u \cdot V - u \cdot \nabla \times V) dx = \langle u \times n, V_T \rangle_{\Gamma_d} \ \forall (u, V) \in \mathbf{H}(\text{curl}, \Omega). \tag{77}$$

We define the next space:

$$\mathbf{X}(\Omega) = \{u \in \mathbf{H}(\mathrm{curl}, \Omega) \mid \nabla\cdot u_{|\Omega_a^\varepsilon} \in L^2(\Omega_a^\varepsilon), \nabla\cdot u_{|\Omega_r^\varepsilon} \in L^2(\Omega_r^\varepsilon),\ \nabla\cdot u_{|\Omega_c^\varepsilon} \in L^2(\Omega_c^\varepsilon)\}. \tag{78}$$

Our variational space is:

$$\begin{aligned}\mathbf{X}^\varepsilon(\Omega) = \{u \in \mathbf{X}(\Omega) \mid (-\omega^2\varepsilon^\eta + i\omega\sigma^\varepsilon_{|\Omega_a^\varepsilon})u_{|\Omega_a^\varepsilon}\cdot e_2 = (-\omega^2\varepsilon^\eta\varepsilon_r + i\omega\sigma^\varepsilon_{|\Omega_r^\varepsilon})u_{|\Omega_r^\varepsilon}\cdot e_2,\\ (-\omega^2\varepsilon^\eta\epsilon_r + i\omega\sigma^\varepsilon_{|\Omega_r^\varepsilon})u_{|\Omega_r^\varepsilon}\cdot n^\varepsilon_{|\Omega_r^\varepsilon} = (-\omega^2\varepsilon^\eta\varepsilon_c + i\omega\sigma^\varepsilon_{|\Omega_c^\varepsilon})u_{|\Omega_c^\varepsilon}\cdot n^\varepsilon_{|\Omega_c^\varepsilon}.\end{aligned} \tag{79}$$

Finally

$$\begin{aligned}\mathbf{X}^\varepsilon(\Omega) = \{u \in \mathbf{X}(\Omega) \mid (-\omega^2\varepsilon^\eta + i\omega\Sigma_a^\varepsilon)u_{|\Omega_a^\varepsilon}\cdot e_2 = (-\omega^2\varepsilon^\eta\varepsilon_r + i\omega\Sigma_r^\varepsilon)u_{|\Omega_r^\varepsilon}\cdot e_2,\\ (-\omega^2\varepsilon^\eta\epsilon_r + i\omega\Sigma_r^\varepsilon)u_{|\Omega_r^\varepsilon}\cdot n^\varepsilon_{|\Omega_r^\varepsilon} = (-\omega^2\varepsilon^\eta\varepsilon_c + i\omega\Sigma_c^\varepsilon)u_{|\Omega_c^\varepsilon}\cdot n^\varepsilon_{|\Omega_c^\varepsilon}\}.\end{aligned} \tag{80}$$

This space is equipped with the norm

$$||u||^2_{\mathbf{X}^\varepsilon(\Omega)} = ||u||^2_{\mathbf{L}^2(\Omega)} + ||\nabla\cdot u_{|\Omega_a^\varepsilon}||^2_{L^2(\Omega_a^\varepsilon)} + ||\nabla\cdot u_{|\Omega_r^\varepsilon}||^2_{L^2(\Omega_r^\varepsilon)} + ||\nabla\cdot u_{|\Omega_c^\varepsilon}||^2_{L^2(\Omega_c^\varepsilon)} + ||\nabla\times u||^2_{\mathbf{L}^2(\Omega)}.$$

### 3.2. *Weak formulation*

Now, we introduce the variational formulation of our problem (63), (66) and (67) for the electric field. Integrating (63) over $\Omega$ and using the Green's formula and (66) we obtain

$$\begin{aligned}&\int_\Omega \nabla\times E^\varepsilon\cdot\nabla\times\overline{V}\,d\mathbf{x} + \int_{\Omega_a^\varepsilon}(-\omega^2\varepsilon^\eta + i\omega\Sigma_a^\varepsilon)E^\varepsilon\cdot\overline{V}\,d\mathbf{x}\\ &+\int_{\Omega_c^\varepsilon}(-\omega^2\varepsilon^\eta\epsilon_c + i\omega\Sigma_c^\varepsilon)E^\varepsilon\cdot\overline{V}\,d\mathbf{x} + \int_{\Omega_r^\varepsilon}(-\omega^2\varepsilon^\eta\epsilon_r + i\omega\Sigma_r^\varepsilon)E^\varepsilon\cdot\overline{V}\,d\mathbf{x}\\ &= \int_{\Gamma_d}(\nabla\times E^\varepsilon\times e_2)\cdot\overline{V}_T\,d\sigma\\ &= \int_{\Gamma_d} -i\omega H_d\times e_2\cdot\overline{V}_T\,d\sigma\end{aligned} \tag{81}$$

where $\overline{V}$ is the complex conjugate of $V$ and $V_T = (e_2\times V)\times e_2$. We introduce the sesquilinear form depending on parameters $\eta$ and $\varepsilon$:

$$\begin{cases}\text{For } E^\varepsilon, V \in \mathbf{X}^\varepsilon(\Omega),\\ a^{\varepsilon,\eta}(E^\varepsilon, V) = \displaystyle\int_\Omega \nabla\times E^\varepsilon\cdot\nabla\times\overline{V}\,d\mathbf{x} + \sum_{i=a,r,c}\int_{\Omega_i^\varepsilon}(-\omega^2\varepsilon^\eta\epsilon_i + i\omega\Sigma_i^\varepsilon)\,E^\varepsilon\cdot\overline{V}\,d\mathbf{x}.\end{cases} \tag{82}$$

Hence, the weak formulation of (63), (66) and (67) that we will use is the following:

$$\begin{cases}\text{Find } E^\varepsilon \in \mathbf{X}^\varepsilon(\Omega) \text{ such as } \forall\ V \in \mathbf{X}^\varepsilon(\Omega) \text{ we have :}\\ a^{\varepsilon,\eta}(E^\varepsilon, V) = -i\omega\displaystyle\int_{\Gamma_d} H_d\times e_2\cdot\overline{V}_T\,d\sigma.\end{cases} \tag{83}$$

Integrating by parts in the variational formulation (81), we find the following transmission problem:

$$\left\{\begin{array}{ll} \nabla\times\nabla\times E^\varepsilon + (-\omega^2\varepsilon^\eta + i\,\omega\,\Sigma_a^\varepsilon)E^\varepsilon = 0 & \text{in } \Omega_a^\varepsilon, \\ \nabla\times\nabla\times E^\varepsilon + (-\omega^2\varepsilon^\eta\epsilon_r + i\,\omega\,\Sigma_r^\varepsilon)E^\varepsilon = 0 & \text{in } \Omega_r^\varepsilon, \\ \nabla\times\nabla\times E^\varepsilon + (-\omega^2\varepsilon^\eta\epsilon_c + i\,\omega\,\Sigma_c^\varepsilon)E^\varepsilon = 0 & \text{in } \Omega_c^\varepsilon. \\ E^\varepsilon_{|\Omega_a^\varepsilon}\times e_2 = E^\varepsilon_{|\Omega_r^\varepsilon}\times n_{|\Omega_r^\varepsilon} \text{ on } \Gamma_{ra}, & \\ E^\varepsilon_{|\Omega_r^\varepsilon}\times n_{|\Omega_r^\varepsilon} = E^\varepsilon_{|\Omega_c^\varepsilon}\times n_{|\Omega_c^\varepsilon} \text{ on } \Gamma_{cr}^\varepsilon, & \\ \nabla\times E^\varepsilon_{|\Omega_a^\varepsilon}\times e_2 = \nabla\times E^\varepsilon_{|\Omega_r^\varepsilon}\times n_{|\Omega_r^\varepsilon} \text{ on } \Gamma_{ra}, & \\ \nabla\times E^\varepsilon_{|\Omega_r^\varepsilon}\times n_{|\Omega_r^\varepsilon} = \nabla\times E^\varepsilon_{|\Omega_c^\varepsilon}\times n_{|\Omega_c^\varepsilon} \text{ on } \Gamma_{cr}^\varepsilon, & \end{array}\right. \tag{84}$$

where $e_2$ is the unit outward normal to $\Omega_a^\varepsilon$, $n_{|\Omega_r^\varepsilon}$ is the unit outward normal to $\Omega_r^\varepsilon$ and $n_{|\Omega_c^\varepsilon}$ is the unit outward normal to $\Omega_c^\varepsilon$. We refer to Annex C for the proof that transmission problem (84) is equivalent to ((63), (66), (67), (69)).

### 3.3. Regularized Maxwell equations for the electric field

The sesquilinear form $a^{\varepsilon,\eta}$ is not coercive on $\mathbf{X}^\varepsilon(\Omega)$, so we regularize it adding terms involving the divergence of $E^\varepsilon$ in $\Omega_a^\varepsilon$, $\Omega_r^\varepsilon$ and $\Omega_c^\varepsilon$. Thanks to the additional terms, existence and uniqueness of the regularized variational formulation solution will be established by the Lax-Milgram theory. Let $s$ be an arbitrary positive number, we define the regularized formulation of problem (83):

$$\left\{\begin{array}{l} \text{Find } E^\varepsilon \in \mathbf{X}^\varepsilon(\Omega) \text{ such that for any } V \in \mathbf{X}^\varepsilon(\Omega) \\ a_R^{\varepsilon,\eta}(E^\varepsilon, V) = a^{\varepsilon,\eta}(E^\varepsilon, V) + s\displaystyle\int_{\Omega_a^\varepsilon} \nabla\cdot E^\varepsilon \nabla\cdot\overline{V}\, d\mathbf{x} \\ + s\displaystyle\int_{\Omega_r^\varepsilon} \nabla\cdot E^\varepsilon \nabla\cdot\overline{V}\, d\mathbf{x} + s\int_{\Omega_c^\varepsilon} \nabla\cdot E^\varepsilon \nabla\cdot\overline{V}\, d\mathbf{x} \\ = -i\omega\displaystyle\int_{\Gamma_d} H_d\times e_2\cdot\overline{V}_T\, d\sigma. \end{array}\right. \tag{85}$$

For any $\varepsilon > 0$ and any $\eta \geq 0$, sesquilinear form $a_R^{\varepsilon,\eta}(.,.)$ is continuous over $\mathbf{X}^\varepsilon(\Omega)$ thanks to the continuity conditions. We will show that it is also coercive. The following proposition was inspired by article [9] Lemma 1.1.

**Proposition 3.1.** *For any $\varepsilon > 0$, for any $\eta \geq 0$ and for any $s > 0$, there exists a positive constant $\omega_0$ which does not depend on $\varepsilon$ and such that for all $\omega \in (0, \omega_0)$, there exists a positive constant $C_0$ depending on $\varepsilon_r, \varepsilon_c, s, \omega$ but not on $\varepsilon$ such that:*

$$\forall\, E^\varepsilon \in \mathbf{X}^\varepsilon(\Omega),\quad \Re[\exp(-i\frac{\pi}{4})\, a_R^{\varepsilon,\eta}(E^\varepsilon, E^\varepsilon)] \geq C_0\|E^\varepsilon\|_{\mathbf{X}^\varepsilon(\Omega)} \tag{86}$$

*Proof.* We have:

$$\begin{aligned} \Re[\exp(-i\frac{\pi}{4})\, a_R^{\varepsilon,\eta}(E^\varepsilon, E^\varepsilon)] &= a_R^\varepsilon(E^\varepsilon, E^\varepsilon) - \int_{\Omega_a^\varepsilon} \omega^2\varepsilon^\eta|E^\varepsilon|^2\, d\mathbf{x} \\ &- \int_{\Omega_r^\varepsilon} \omega^2\varepsilon^\eta\epsilon_r|E^\varepsilon|^2\, d\mathbf{x} - \int_{\Omega_c^\varepsilon} \omega^2\varepsilon^\eta\epsilon_c|E^\varepsilon|^2\, d\mathbf{x}. \end{aligned} \tag{87}$$

with

$$
\begin{aligned}
a_R^\varepsilon(E^\varepsilon, E^\varepsilon) = &\int_\Omega |\nabla\times E^\varepsilon|^2\, d\mathbf{x} + s\int_{\Omega_a^\varepsilon} |\nabla\cdot E^\varepsilon|^2\, d\mathbf{x} \\
&+ s\int_{\Omega_r^\varepsilon} |\nabla\cdot E^\varepsilon|^2\, d\mathbf{x} + s\int_{\Omega_c^\varepsilon} |\nabla\cdot E^\varepsilon|^2\, d\mathbf{x} \\
&+ \int_{\Omega_a^\varepsilon} \omega\Sigma_a^\varepsilon|E^\varepsilon|^2\, d\mathbf{x} + \int_{\Omega_r^\varepsilon} \omega\Sigma_r^\varepsilon|E^\varepsilon|^2\, d\mathbf{x} \\
&+ \int_{\Omega_c^\varepsilon} \omega\Sigma_c^\varepsilon|E^\varepsilon|^2\, d\mathbf{x}.
\end{aligned}
\tag{88}
$$

We have the following estimate:

$$
\begin{aligned}
|a_R^\varepsilon(E^\varepsilon, E^\varepsilon)| \geq \min\{1,\omega,s\}(&||\nabla\times E^\varepsilon||^2_{\mathbf{L}^2(\Omega)} + ||\nabla\cdot E^\varepsilon||^2_{L^2(\Omega_a^\varepsilon)} + ||\nabla\cdot E^\varepsilon||^2_{L^2(\Omega_r^\varepsilon)} \\
&+ ||\nabla\cdot E^\varepsilon||^2_{L^2(\Omega_c^\varepsilon)} + ||E^\varepsilon||^2_{\mathbf{L}^2(\Omega)}).
\end{aligned}
\tag{89}
$$

Then we have:

$$
|\, a_R^\varepsilon(E^\varepsilon, E^\varepsilon)| \geq \min\{1,\omega,s\}||E^\varepsilon||^2_{\mathbf{X}^\varepsilon(\Omega)}. \tag{90}
$$

Returning to formulation (86), for $\eta \geq 0$, since $\max(\Sigma_a^\varepsilon, \Sigma_r^\varepsilon, \Sigma_c^\varepsilon) > \varepsilon^\eta$, inequality (86) is valid with $C_0 = \min\{1,\omega,s\}$ as soon as $\omega^2\min\{1,\varepsilon_r,\varepsilon_c\} < \min\{1,\omega,s\}$ or $\omega < \sqrt{\frac{\min\{1,\omega,s\}}{\min\{1,\varepsilon_r,\varepsilon_c\}}}$. This ends the proof of Proposition 3.1. □

Thanks to Proposition 3.1 we can state the existence and uniqueness of the solution to regularized problem (85).

**Theorem 3.2.** *Under the assumptions of Proposition 3.1, there exists a unique solution $E^\varepsilon$ to regularized problem (85).*

*Proof.* The sesquilinear form $a_R^{\varepsilon,\eta}$ is continuous, bounded, coercive thanks to Proposition 3.1 and the right hand side is continuous on $X^\varepsilon(\Omega)$, then problem (85) has a unique solution in $X^\varepsilon(\Omega)$ thanks to the Lax-Milgram Lemma. □

### *3.4. Existence, uniqueness and estimate*

**Theorem 3.3.** *For any $\varepsilon > 0$, for any $\eta \geq 0$, there exists a positive constant $\omega_0$ which does not depend on $\varepsilon$ and such that for all $\omega \in (0, \omega_0)$, there exists a unique solution of (84) or ((63), (66), (67), (69)).*

*Proof.* We show that for an appropriate choice of $s$ that $E^\varepsilon$ satisfies all equations (84) or ((63), (66), (67), (69)). It is obvious that any solution of (84) or of ((63), (66), (67),(69)) is also solution to (85). Indeed, since from (84) or from ((63), (66), (67),(69)) we have $\nabla\cdot E^\varepsilon_{|\Omega_a^\varepsilon} = 0$, $\nabla\cdot E^\varepsilon_{|\Omega_r^\varepsilon} = 0$, $\nabla\cdot E^\varepsilon_{|\Omega_c^\varepsilon} = 0$, the additional terms $s\int_{\Omega_a^\varepsilon} \nabla\cdot E^\varepsilon \nabla\cdot\overline{V}\, d\mathbf{x} + s\int_{\Omega_r^\varepsilon} \nabla\cdot E^\varepsilon \nabla\cdot\overline{V}\, d\mathbf{x} + s\int_{\Omega_c^\varepsilon} \nabla\cdot E^\varepsilon \nabla\cdot\overline{V}\, d\mathbf{x}$ cancel in (85).

Uniqueness follows from that if $E_1^\varepsilon$ and $E_2^\varepsilon$ are two solutions to (63) with the boundary condition (67) their difference $e^\varepsilon = E_2^\varepsilon - E_1^\varepsilon$ satisfies the problem (63) with (67). Then it comes

$$\begin{aligned}\int_\Omega |\nabla\times e^\varepsilon|^2\, d\mathbf{x} &+ \int_{\Omega_a^\varepsilon} (-\omega^2\varepsilon^\eta + i\omega\Sigma_a^\varepsilon)|e^\varepsilon|^2\, d\mathbf{x}\\ &+ \int_{\Omega_c^\varepsilon} (-\omega^2\varepsilon^\eta\epsilon_c + i\omega\Sigma_c^\varepsilon)|e^\varepsilon|^2\, d\mathbf{x} + \int_{\Omega_r^\varepsilon} (-\omega^2\varepsilon^\eta\epsilon_r + i\omega\Sigma_r^\varepsilon)|e^\varepsilon|^2\, d\mathbf{x}\\ &= 0.\end{aligned} \tag{91}$$

Taking the imaginary part of the expression we get $\int_{\Omega_a^\varepsilon} \omega\Sigma_a^\varepsilon|e^\varepsilon|^2\, d\mathbf{x} + \int_{\Omega_c^\varepsilon} \omega\Sigma_c^\varepsilon|e^\varepsilon|^2\, d\mathbf{x} + \int_{\Omega_r^\varepsilon} \omega\Sigma_r^\varepsilon|e^\varepsilon|^2\, d\mathbf{x} = 0$ and then $e^\varepsilon = 0$.

Let us consider the reciprocal assertion, according to the same proof of S. Hassani, S. Nicaise, A. Maghnouji in [23], we define $H_0^1(\Omega_c^\varepsilon, \Delta)$ the subspace of $\psi \in H_0^1(\Omega_c^\varepsilon)$ such that $\Delta(\psi) \in L^2(\Omega_c^\varepsilon)$.

Let $E_s^\varepsilon$ be the solution of the regularized formulation (85). In (85) we take a test function $V = \nabla\psi$ where $\psi \in H_0^1(\Omega_c^\varepsilon, \Delta)$, extended by zero outside $\Omega_c^\varepsilon$. We get:

$$\int_{\Omega_c^\varepsilon} s\nabla\cdot E_s^\varepsilon \nabla\cdot(\nabla\psi)\, d\mathbf{x} + \int_{\Omega_c^\varepsilon} (-\omega^2\varepsilon^\eta\epsilon_c + i\omega\Sigma_c^\varepsilon)E_s^\varepsilon \cdot \nabla\psi\, d\mathbf{x} = 0. \tag{92}$$

By Green's formula, $\forall\psi \in H_0^1(\Omega_c^\varepsilon, \Delta)$, we obtain:

$$\int_{\Omega_c^\varepsilon} \nabla\cdot E_s^\varepsilon\Big(\Delta\psi + \frac{\omega^2\varepsilon^\eta\epsilon_c - i\omega\Sigma_c^\varepsilon}{s}\psi\Big)\, d\mathbf{x} = 0. \tag{93}$$

Thus, if we choose $s$ such that $\frac{\omega^2\varepsilon^\eta\epsilon_c - i\omega\Sigma_c^\varepsilon}{s}$ is not an eigenvalue of $(\Delta_{dir}, \Omega_c^\varepsilon)$: the Laplacian operator in $\Omega_c^\varepsilon$ with Dirichlet condition on its boundary, then for all $\varphi \in L(\Omega_c^\varepsilon)^2$ there exists $\psi \in H_0^1(\Omega_c^\varepsilon, \Delta)$ solution of

$$\Delta\psi + \frac{\omega^2\varepsilon^\eta\epsilon_c - i\omega\Sigma_c^\varepsilon}{s}\psi = \varphi. \tag{94}$$

Then, we conclude that

$$\nabla\cdot E^\varepsilon_{s|\Omega_c^\varepsilon} = 0. \tag{95}$$

A similar argument in $\Omega_a^\varepsilon$ yields $\nabla\cdot E^\varepsilon_{s|\Omega_a^\varepsilon} = 0$ for $s$ such that $\frac{\omega^2\varepsilon^\eta - i\omega\Sigma_a^\varepsilon}{s}$ is not an eigenvalue of $(\Delta_{dir}, \Omega_a^\varepsilon)$. In the same way, we obtain in $\Omega_r^\varepsilon$, $\nabla\cdot E^\varepsilon_{s|\Omega_r^\varepsilon} = 0$ with $s$ such that $\frac{\omega^2\varepsilon^\eta\epsilon_r - i\omega\Sigma_r^\varepsilon}{s}$ is not an eigenvalue of $(\Delta_{dir}, \Omega_r^\varepsilon)$.

Hence $\nabla\cdot E_s^\varepsilon = 0$ in $\Omega_c^\varepsilon$, this cancels the additional term $s\int_{\Omega_c^\varepsilon} \nabla\cdot E_s^\varepsilon \nabla\cdot\overline{V}\, d\mathbf{x}$ in (85). In the same way, $\nabla\cdot E_s^\varepsilon = 0$ in $\Omega_r^\varepsilon$ and $\nabla\cdot E_s^\varepsilon = 0$ in $\Omega_a^\varepsilon$ cancel $s\int_{\Omega_r^\varepsilon} \nabla\cdot E_s^\varepsilon\nabla\cdot\overline{V}\, d\mathbf{x}$ and $s\int_{\Omega_a^\varepsilon} \nabla\cdot E_s^\varepsilon\nabla\cdot\overline{V}\, d\mathbf{x}$ in (85). So, (85) becomes (81). Applying Green's formula, we find (63). □

**Theorem 3.4.** *Under the assumptions of Theorem 3.2, $E^\varepsilon \in X^\varepsilon(\Omega)$ solution of (85) satisfies*

$$\|E^\varepsilon\|_{\mathbf{X}^\varepsilon(\Omega)} \leq C \tag{96}$$

*with* $C = \frac{C_{\gamma_t}C_{\gamma_T}}{C_0}\|H_d\|_{H(curl,\Omega)}$

*Proof.* The sesquilinear form $a_R^{\varepsilon,\eta}(E^\varepsilon, V)$ is coercive, weak formulation (85) becomes:

$$\begin{aligned} C_0\|E^\varepsilon\|^2_{\mathbf{X}^\varepsilon(\Omega)} &\leq \Re(\exp(-i\frac{\pi}{4})a_R^{\varepsilon,\eta}(E^\varepsilon, E^\varepsilon)) \\ &\leq |\exp(-i\frac{\pi}{4}) \cdot a_R^{\varepsilon,\eta}(E^\varepsilon, E^\varepsilon)| = |a_R^{\varepsilon,\eta}(E^\varepsilon, E^\varepsilon)| \\ &\leq |\int_{\Gamma_d} -i\omega H_d \times e_2 \cdot E_T^\varepsilon \; d\sigma| \\ &\leq \|H_d \times e_2\|_{H^{-\frac{1}{2}}(\mathrm{div},\Gamma_d)} \|E_T^\varepsilon\|_{H^{-\frac{1}{2}}(\mathrm{curl},\Gamma_d)} \\ &\leq C_{\gamma_t} C_{\gamma_T} \|H_d \times e_2\|_{\mathbf{H}(\mathrm{curl},\Omega)} \|E^\varepsilon\|_{\mathbf{H}(\mathrm{curl},\Omega)} \end{aligned} \tag{97}$$

where $E_T^\varepsilon = e_2 \times (E^\varepsilon \times e_2)$ and the continuous dependence of the trace norm with $C = \frac{C_{\gamma_t} C_{\gamma_T}}{C_0} \|H_d\|_{\mathbf{H}(curl,\Omega)}$ gives:

$$\|E^\varepsilon\|^2_{\mathbf{X}^\varepsilon(\Omega)} \leq C\|E^\varepsilon\|_{\mathbf{H}(\mathrm{curl},\Omega)} \leq C\|E^\varepsilon\|_{\mathbf{X}^\varepsilon(\Omega)}. \tag{98}$$

□

## 4. Homogenization

With the aim to obtain a convergence result for the problem (63), (66) and (67), we propose an approach based on two-scale convergence. This concept was introduced by G. Nguetseng [21, 22] and specified by G. Allaire [3, 4] which studied properties of the two-scale convergence. M. Neuss-Radu in [20] presented an extension of two-scale convergence method to the periodic surfaces. Many authors applied two-scale convergence approach D. Cionarescu and P. Donato [8], N. Crouseilles, E. Frénod, S. Hirstoaga and A. Mouton [10], Y. Amirat, K. Hamdache and A. Ziani [2] and also A. Back, E. Frénod [6]. This mathematical concept were applied to homogenize the time-harmonic Maxwell equations S. Ouchetto, O. Zouhdi and A. Bossavit [24], H.E. Pak [25].

In our model, the parallel carbon cylinders are periodically distributed in direction x and z, as the material is homogenous in the y direction, we can consider that the material is periodic with a three directional cell of periodicity. In other words, introducing $\mathcal{Z} = [-\frac{1}{2}, \frac{1}{2}] \times [-1, 0]^2$, function $\Sigma^\varepsilon$ given by (61) is naturally periodic with respect to $(\xi, \zeta)$ with period $[-\frac{1}{2}, \frac{1}{2}] \times [-1, 0]$ but it is also periodic with respect to $\mathbf{y}$ with period $\mathcal{Z}$.

Now, we review some basis definitions and results about two-scale convergence.

### *4.1. Two-scale convergence*

We first define the function spaces

$$\begin{cases} \mathbf{H}_\#(\mathrm{curl}, \mathcal{Z}) = \{u \in \mathbf{H}(\mathrm{curl}, \mathbb{R}^3) : \;\; u \;\text{ is } \mathcal{Z}\text{-periodic}\} \\ \mathbf{H}_\#(\mathrm{div}, \mathcal{Z}) = \{u \in \mathbf{H}(\mathrm{div}, \mathbb{R}^3) : \;\; u \;\text{ is } \mathcal{Z}\text{-periodic}\} \end{cases} \tag{99}$$

and where $\mathbf{H}(\mathrm{curl}, \mathbb{R}^3)$ and $\mathbf{H}(\mathrm{div}, \mathbb{R}^3)$ are defined by (72) with $\Omega^\varepsilon$ replaced by $\mathbb{R}^3$. We introduce

$$\mathbf{L}^2_\#(\mathcal{Z}) = \{u \in \mathbf{L}^2(\mathbb{R}^3), u \text{ is } \mathcal{Z}\text{-periodic}\}, \tag{100}$$

and

$$\mathbf{H}^1_\#(\mathcal{Z}) = \{u \in \mathbf{H}^1(\mathbb{R}^3), u \text{ is } \mathcal{Z}\text{-periodic}\}, \tag{101}$$

where $\mathbf{H}^1(\mathbb{R}^3)$ is the usual Sobolev space on $\mathbb{R}^3$. First, denoting by $\mathbf{C}^0_\#(\mathcal{Z})$ the space of functions in $\mathbf{C}^0(\mathbb{R}^3)$ and $\mathcal{Z}$-periodic, $\mathbf{C}^0_0(\mathbb{R}^3)$ the space of continuous functions over $\mathbb{R}^3$ with compact support, we have the following definitions:

**Definition 4.1.** *A sequence $u^\varepsilon(\mathbf{x})$ in $\mathbf{L}^2(\Omega)$ two-scale converges to $u_0(\mathbf{x}, \mathbf{y}) \in \mathbf{L}^2(\Omega, \mathbf{L}^2_\#(\mathcal{Z}))$ if for every $V(\mathbf{x}, \mathbf{y}) \in \mathbf{C}^0_0(\Omega, C^0_\#(\mathcal{Z}))$*

$$\lim_{\varepsilon \to 0} \int_\Omega u^\varepsilon(\mathbf{x}) \cdot V(\mathbf{x}, \mathbf{x}/\varepsilon)\, dx = \int_\Omega \int_{\mathcal{Z}} u_0(\mathbf{x}, \mathbf{y}) \cdot V(\mathbf{x}, \mathbf{y})\, d\mathbf{x} d\mathbf{y}. \tag{102}$$

**Proposition 4.2.** *If $u^\varepsilon(\mathbf{x})$ two-scale converges to $u_0(\mathbf{x}, \mathbf{y}) \in \mathbf{L}^2(\Omega, \mathbf{L}^2_\#(\mathcal{Z}))$, we have for all $v(\mathbf{x}) \in C_0(\overline{\Omega})$ and all $w(\mathbf{y}) \in \mathbf{L}^2_\#(\mathcal{Z})$*

$$\lim_{\varepsilon \to 0} \int_\Omega u^\varepsilon(\mathbf{x}) \cdot v(\mathbf{x}) w(\frac{\mathbf{x}}{\varepsilon})\, d\mathbf{x} = \int_\Omega \int_{\mathcal{Z}} u_0(\mathbf{x}, \mathbf{y}) \cdot v(\mathbf{x}) w(\mathbf{y})\, d\mathbf{x} d\mathbf{y}. \tag{103}$$

**Theorem 4.3.** *(Nguetseng). Let $u^\varepsilon(\mathbf{x}) \in \mathbf{L}^2(\Omega)$. Suppose there exists a constant $c > 0$ such that for all $\varepsilon$*

$$\|u^\varepsilon\|_{L^2(\Omega)} \leq c.$$

*Then there exists a subsequence of $\varepsilon$ (still denoted $\varepsilon$) and $u_0(\mathbf{x}, \mathbf{y}) \in \mathbf{L}^2(\Omega, \mathbf{L}^2_\#(\mathcal{Z}))$ such that:*

$$u^\varepsilon(\mathbf{x}) \twoheadrightarrow u_0(\mathbf{x}, \mathbf{y}). \tag{104}$$

**Proposition 4.4.** *Let $u^\varepsilon(\mathbf{x})$ be a sequence of functions in $\mathbf{L}^2(\Omega)$, which two-scale converges to a limit $u_0(\mathbf{x}, \mathbf{y}) \in \mathbf{L}^2(\Omega, \mathbf{L}^2_\#(\mathcal{Z}))$. Then $u^\varepsilon(\mathbf{x})$ converges also to $u(\mathbf{x}) = \int_{\mathcal{Z}} u_0(\mathbf{x}, \mathbf{y}) d\mathbf{y}$ in $\mathbf{L}^2(\Omega)$ weakly. Furthermore, we have*

$$\lim_{\varepsilon \to 0} \|u^\varepsilon\|_{\mathbf{L}^2(\Omega)} \geq \|u_0\|_{\mathbf{L}^2(\Omega \times Y)} \geq \|u\|_{\mathbf{L}^2(\Omega)}. \tag{105}$$

**Remark 4.5.** *: - For any smooth function $u(\mathbf{x}, \mathbf{y})$, being $\mathcal{Z}$-periodic in $\mathbf{y}$, the associated sequence $u^\varepsilon(\mathbf{x}) = u(\mathbf{x}, \frac{\mathbf{x}}{\varepsilon})$ two-scale converges to $u(\mathbf{x}, \mathbf{y})$.*

*- Any sequence $u^\varepsilon$ that converges strongly in $L^2(\Omega)$ to a limit $u(\mathbf{x})$, two-scale converges to the same limit $u(\mathbf{x})$.*

*- If $u^\varepsilon$ admits an asymptotic expansion of the type $u^\varepsilon(\mathbf{x}) = u_0(\mathbf{x}, \mathbf{x}/\varepsilon) + \varepsilon u_1(\mathbf{x}, \mathbf{x}/\varepsilon) + \varepsilon^2 u_2(\mathbf{x}, \mathbf{x}/\varepsilon) + \ldots$ , where the functions $u_i(\mathbf{x}, \mathbf{y})$ are smooth and $\mathcal{Z}$-periodic in $\mathbf{y}$, two-scale convergence allows to identify the first term of the expansion $u_0(\mathbf{x}, \mathbf{y})$ with the two-scale limit of $u^\varepsilon$ and the two-scale limit of $\frac{u^\varepsilon(\mathbf{x}) - u_0(\mathbf{x}, \frac{\mathbf{x}}{\varepsilon})}{\varepsilon}$ with $u_1(\mathbf{x}, \mathbf{y})$ see (Frénod, Raviart and Sonnendrucker [11]).*

**Proposition 4.6.** *Let $u^\varepsilon(\mathbf{x})$ in $\mathbf{L}^2(\Omega)$. Suppose there exists a constant $c > 0$ such that for all $\varepsilon$*

$$\|u^\varepsilon\|_{L^2(\Omega)} \leq c.$$

*Up to a subsequence,* $u^\varepsilon(\mathbf{x})$ *two-scale converges to* $u_0(\mathbf{x},\mathbf{y}) \in \mathbf{L}^2(\Omega, \mathbf{L}^2_\#(\mathcal{Z}))$ *such that:*

$$u_0(\mathbf{x},\mathbf{y}) = u(\mathbf{x}) + \widetilde{u_0}(\mathbf{x},\mathbf{y}), \tag{106}$$

*where* $\widetilde{u_0}(\mathbf{x},\mathbf{y}) \in \mathbf{L}^2(\Omega, \mathbf{L}^2_\#(\mathcal{Z}))$ *satisfies*

$$\int_{\mathcal{Z}} \widetilde{u_0}(\mathbf{x},\mathbf{y})\, d\mathbf{y} = 0, \tag{107}$$

*and* $u(\mathbf{x}) = \displaystyle\int_{\mathcal{Z}} u_0(\mathbf{x},\mathbf{y})\, d\mathbf{y}$ *is a weak limit in* $\mathbf{L}^2(\Omega)$.

*Proof.* $u^\varepsilon(\mathbf{x})$ is bounded in $\mathbf{L}^2(\Omega)$, then by application of Theorem 4.3, we get the first part of the proposition. Furthermore by defining $\widetilde{u_0}$ as

$$\widetilde{u_0}(\mathbf{x},\mathbf{y}) = u_0(\mathbf{x},\mathbf{y}) - \int_{\mathcal{Z}} u_0(\mathbf{x},\mathbf{y}) d\mathbf{y}, \tag{108}$$

we obtain the decomposition of $u_0$. □

Defining $\nabla_{\mathbf{x}} = (\frac{\partial}{\partial x}; \frac{\partial}{\partial y}; \frac{\partial}{\partial z})$, $\nabla_{\mathbf{y}} = (\frac{\partial}{\partial \xi}; \frac{\partial}{\partial v}; \frac{\partial}{\partial \zeta})$, we have

**Proposition 4.7.** *Let* $u^\varepsilon(\mathbf{x})$ *be bounded in* $\mathbf{H}(curl, \Omega)$. *Then, up to a subsequence, there exists a function* $u_1 \in \mathbf{L}^2(\Omega, H_\#(curl, \mathcal{Z}))$ *such that*

$$\nabla \times u^\varepsilon(\mathbf{x}) \twoheadrightarrow \nabla_{\mathbf{x}} \times u_0(\mathbf{x},\mathbf{y}) + \nabla_{\mathbf{y}} \times u_1(\mathbf{x},\mathbf{y}), \tag{109}$$

*where* $u_0$ *is given by Proposition 4.6.*

*Proof.* From Theorem 4.3, since $u^\varepsilon$ and $\nabla \times u^\varepsilon$ are bounded in $\mathbf{L}^2(\Omega)$ then, up to a subsequence, they two-scale converge to $u_0(\mathbf{x},\mathbf{y}) \in \mathbf{L}^2(\Omega, \mathbf{L}^2_\#(\mathcal{Z}))$ and $\eta_0(\mathbf{x},\mathbf{y}) \in \mathbf{L}^2(\Omega, \mathbf{L}^2_\#(\mathcal{Z}))$. So we have for all $V(\mathbf{x},\mathbf{y}) \in \mathbf{C}^0_0(\Omega; C^0_\#(\mathcal{Z}))$:

$$\lim_{\varepsilon\to 0} \int_\Omega u^\varepsilon(\mathbf{x}) \cdot V(\mathbf{x}, \mathbf{x}/\varepsilon)\, d\mathbf{x} = \int_\Omega \int_{\mathcal{Z}} u_0(\mathbf{x},\mathbf{y}) \cdot V(\mathbf{x},\mathbf{y}) d\mathbf{x} d\mathbf{y}, \tag{110}$$

$$\lim_{\varepsilon\to 0} \int_\Omega \nabla \times u^\varepsilon(\mathbf{x}) \cdot V(\mathbf{x}, \mathbf{x}/\varepsilon)\, d\mathbf{x} = \int_\Omega \int_{\mathcal{Z}} \eta_0(\mathbf{x},\mathbf{y}) \cdot V(\mathbf{x},\mathbf{y}) d\mathbf{x} d\mathbf{y}. \tag{111}$$

Next, by integration by parts, we have:

$$\int_\Omega \nabla \times u^\varepsilon(\mathbf{x}) \cdot V(\mathbf{x}, \mathbf{x}/\varepsilon)\, d\mathbf{x} = \int_\Omega u^\varepsilon(\mathbf{x}) \cdot (\nabla_{\mathbf{x}} \times V(\mathbf{x}, \mathbf{x}/\varepsilon) + \frac{1}{\varepsilon} \nabla_{\mathbf{y}} \times V(\mathbf{x}, \mathbf{x}/\varepsilon))\, d\mathbf{x}. \tag{112}$$

If we choose a test function $V \in \mathbf{C}^0_0(\Omega, \mathbf{C}^0_\#(\mathcal{Z}))$ such that $\nabla_{\mathbf{y}} \times V = 0$, passing to the limit in the left-hand side (111) we get

$$\begin{aligned} \int_\Omega \nabla_{\mathbf{x}} \times u^\varepsilon(\mathbf{x}) \cdot V(\mathbf{x}, \mathbf{x}/\varepsilon)\, d\mathbf{x} &\to \int_\Omega \int_{\mathcal{Z}} u_0(\mathbf{x},\mathbf{y}) \cdot \nabla_{\mathbf{x}} \times V(\mathbf{x},\mathbf{y})\, d\mathbf{x} d\mathbf{y} \\ &= \int_\Omega \int_{\mathcal{Z}} \nabla_{\mathbf{x}} \times u_0(\mathbf{x},\mathbf{y}) \cdot V(\mathbf{x},\mathbf{y})\, d\mathbf{x} d\mathbf{y}. \end{aligned} \tag{113}$$

This means that with the difference between (111) and (113):

$$\int_\Omega\int_\mathcal{Z}[\eta_0(\mathbf{x},\mathbf{y})-\nabla_\mathbf{x}\times u_0(\mathbf{x},\mathbf{y})]\cdot V(\mathbf{x},\mathbf{y})\,d\mathbf{x}d\mathbf{y}=0, \tag{114}$$

for all functions $V\in\mathbf{C}_0^1(\Omega)$ with $\nabla_\mathbf{y}\times V=0$. It follows that function $\eta_0(\mathbf{x},\mathbf{y})-\nabla_\mathbf{x}\times u_0(\mathbf{x},\mathbf{y})$ is orthogonal to functions with zero rotational in $\mathbf{L}^2(\Omega,\mathbf{H}_\#(\text{curl}),\mathcal{Z})$. This implies that there exists a function $u_1\in\mathbf{L}^2(\Omega,\mathbf{H}_\#(\text{curl},\mathcal{Z}))$ such that

$$\nabla_\mathbf{y}\times u_1(\mathbf{x},\mathbf{y})=\eta_0(\mathbf{x},\mathbf{y})-\nabla_\mathbf{x}\times u_0(\mathbf{x},\mathbf{y}). \tag{115}$$

Thus

$$\nabla\times u^\varepsilon(\mathbf{x})\twoheadrightarrow\nabla_\mathbf{x}\times u_0(\mathbf{x},\mathbf{y})+\nabla_\mathbf{y}\times u_1(\mathbf{x},\mathbf{y}). \tag{116}$$

□

**Proposition 4.8.** *Let $u^\varepsilon$ be a bounded sequence in $\mathbf{H}(curl,\Omega)$. Then a subsequence $u^\varepsilon$ can be extrated from $\varepsilon$ such that, letting $\varepsilon\to 0$*

$$u^\varepsilon(\mathbf{x})\twoheadrightarrow u(\mathbf{x})+\nabla_\mathbf{y}\Phi(\mathbf{x},\mathbf{y}). \tag{117}$$

*where $\Phi\in\mathbf{L}^2(\Omega,\mathbf{H}_\#^1(\mathcal{Z}))$ is a scalar-valued function and where $u\in\mathbf{L}^2(\Omega)$. And we have*

$$\nabla\times u^\varepsilon(\mathbf{x})\rightharpoonup\nabla_\mathbf{x}\times u(\mathbf{x})\ \ \textit{weakly in}\ \ \boldsymbol{L}^2(\Omega). \tag{118}$$

*where $u(\mathbf{x})$ is given by Proposition 4.6.*

*Proof.* Proof of (117), for any $V(\mathbf{x},\mathbf{y})\in\mathbf{C}_0^1(\Omega,\mathbf{C}_\#^1(\mathcal{Z}))$, we have

$$\int_\Omega\nabla\times u^\varepsilon(\mathbf{x})\cdot V(\mathbf{x},\frac{\mathbf{x}}{\varepsilon})\,d\mathbf{x}=\int_\Omega u^\varepsilon(\mathbf{x})\{\nabla_\mathbf{x}\times V(\mathbf{x},\frac{\mathbf{x}}{\varepsilon})+\frac{1}{\varepsilon}\nabla_\mathbf{y}\times V(\mathbf{x},\frac{\mathbf{x}}{\varepsilon})\}\,d\mathbf{x}. \tag{119}$$

Multiplying by $\varepsilon$ we have

$$\varepsilon\int_\Omega\nabla\times u^\varepsilon(\mathbf{x})\cdot V(\mathbf{x},\frac{\mathbf{x}}{\varepsilon})\,d\mathbf{x}=\int_\Omega u^\varepsilon(\mathbf{x})\{\varepsilon\nabla_\mathbf{x}\times V(\mathbf{x},\frac{\mathbf{x}}{\varepsilon})+\nabla_\mathbf{y}\times V(\mathbf{x},\frac{\mathbf{x}}{\varepsilon})\}\,d\mathbf{x}. \tag{120}$$

Taking the two-scale limit as $\varepsilon\to 0$ we obtain

$$0=\int_\Omega\int_\mathcal{Z}u_0(\mathbf{x},\mathbf{y})\cdot\nabla_\mathbf{y}\times V(\mathbf{x},\mathbf{y})\,d\mathbf{x}d\mathbf{y}, \tag{121}$$

which implies that $\nabla_\mathbf{y}\times u_0(\mathbf{x},\mathbf{y})=0$. Thus $u_0(\mathbf{x},\mathbf{y})$ is a gradient with respect to the variable $\mathbf{y}$ for some scalar function $\Phi(\mathbf{x},\mathbf{y})$. And according to Proposition (4.6) $u_0(\mathbf{x},\mathbf{y})$ can be written as $u_0(\mathbf{x},\mathbf{y})=u(\mathbf{x})+\nabla_\mathbf{y}\Phi(\mathbf{x},\mathbf{y})$, where $u(\mathbf{x})=\int_\mathcal{Z}u_0(\mathbf{x},\mathbf{y})d\mathbf{y}$ for some scalar function $\Phi(\mathbf{x},\mathbf{y})$.

Next, we choose a test function $V(\mathbf{x})\in\mathbf{L}^2(\Omega)$. Integration by parts yields:

$$\begin{aligned}\lim_{\varepsilon\to0}\int_\Omega\nabla\times u^\varepsilon(\mathbf{x})\cdot V(\mathbf{x})\,d\mathbf{x}&=\lim_{\varepsilon\to0}\int_\Omega u^\varepsilon(\mathbf{x})\cdot\nabla\times V(\mathbf{x})\,d\mathbf{x}\\&=\int_\Omega\int_\mathcal{Z}u_0(\mathbf{x},\mathbf{y})\,d\mathbf{y}\cdot\nabla\times V(\mathbf{x})\,d\mathbf{x}\\&=\int_\Omega\nabla\times u(\mathbf{x})\cdot V(\mathbf{x})\,d\mathbf{x}.\end{aligned} \tag{122}$$

□

These results are important properties of the two-scales convergence. We note that the usual concepts of convergence do not preserve information concerning the micro-scale of the function. However, the two-scale convergence preserves information on the micro-scale.

### *4.2. Homogenized problem*

We will explore in this section the behavior of electromagnetic field $E^\varepsilon$ using the two-scale convergence to determine the homogenized problem. We place in the context of the case 6 with $\delta > L$ and $\overline{\omega} = 10^6 rad.s^{-1}$, then we have $\eta = 5$ and $\Sigma_a^\varepsilon = \varepsilon$, $\Sigma_r^\varepsilon = \varepsilon^4$, $\Sigma_c^\varepsilon = 1$ which gives the following equation:

$$\nabla\times\nabla\times E^\varepsilon - \omega^2\varepsilon^5 k(\epsilon)E^\varepsilon + i\omega[(\mathbf{1}_C^\varepsilon(\frac{\mathbf{x}}{\varepsilon}) + \varepsilon^4\mathbf{1}_R^\varepsilon(\frac{\mathbf{x}}{\varepsilon}))\mathbf{1}_{\{y<0\}} + \varepsilon\mathbf{1}_{\{y>0\}}]E^\varepsilon = 0, \tag{123}$$

where for a given set $\mathcal{A}$, $\mathbf{1}_\mathcal{A}$ stands for the characteristic function of $\mathcal{A}$ and where $\mathbf{1}_\mathcal{A}^\varepsilon(\mathbf{x}) = \mathbf{1}_\mathcal{A}(\frac{\mathbf{x}}{\varepsilon})$, hence $\mathbf{1}_C^\varepsilon$ and $\mathbf{1}_R^\varepsilon$ are the characteristic functions of the sets filled by carbon fibers and by resin. And where $k(\epsilon) = (\epsilon_c\mathbf{1}_C^\varepsilon(\mathbf{x}) + \epsilon_r\mathbf{1}_R^\varepsilon(\mathbf{x}))\mathbf{1}_{\{y<0\}} + \mathbf{1}_{\{y>0\}}$.

**Remark 4.9.** *We recall that $\epsilon_c$ and $\epsilon_r$ are respectively the relative permittivity of the carbon fibers and the resin. You should not confused with the microscopic scale $\varepsilon$.*

On this purpose, we have the following Theorem:

**Theorem 4.10.** *Under assumptions of Theorem 3.4, sequence $E^\varepsilon$ solution of (85) or (84) or ((63), (66), (67), (69)) converges to $E(\mathbf{x}) \in \mathbf{L}^2(\Omega)$ which is the unique solution of the homogenized problem:*

$$\begin{cases} \theta_1\nabla_\mathbf{x}\times\nabla_\mathbf{x}\times E(\mathbf{x}) + i\omega\theta_2 E(\mathbf{x}) = 0 \quad in\ \Omega, \\ \theta_1\nabla_\mathbf{x}\times E(\mathbf{x})\times e_2 = -i\omega H_d\times e_2 \quad on\ \Gamma_d, \\ \nabla_\mathbf{x}\times E(\mathbf{x})\times e_2 = 0 \quad on\ \Gamma_L, \end{cases} \tag{124}$$

*with $\theta_1 = \int_\mathcal{Z} \mathrm{Id} - \nabla_y\chi(y)\, d\mathbf{y}$ and $\theta_2 = \int_\mathcal{Z} \mathbf{1}_C(\mathbf{y})(\mathrm{Id} - \nabla_y\chi(\mathbf{y}))\, d\mathbf{y}$.*

*And where the scalar function $\chi$ is the unique solution, up to an additive constant in the Hilbert space of $\mathcal{Z}$ periodic functions $H^1_\#(\mathcal{Z})$, of the following boundary value problem*

$$\begin{cases} \Delta_\mathbf{y}(\chi(\mathbf{y})) = 0 \quad in\ \mathcal{Z}\backslash\partial\Omega_C, \\ [\frac{\partial\chi}{\partial n}] = -n_j \quad on\ \partial\Omega_C, \\ [\chi] = 0 \quad on\ \partial\Omega_C, \end{cases} \tag{125}$$

*where $[f]$ is the jump across the surface of $\partial\Omega_C$, $n_j$, $j = \{1,2,3\}$ is the projection on the axis $e_j$ of the normal of $\partial\Omega_C$.*

*Proof.* **Step 1: Two-scale convergence.** Due to the estimate (96), $E^\varepsilon$ is bounded in $\mathbf{L}^2(\Omega)$. Hence, up to a subsequence, $E^\varepsilon$ two-scale converges to $E_0(\mathbf{x},\mathbf{y})$ belonging to $\mathbf{L}^2(\Omega, \mathbf{L}^2_\#(\mathcal{Z}))$. That means for any $V(\mathbf{x},\mathbf{y}) \in \mathbf{C}^1_0(\Omega, \mathbf{C}^1_\#(\mathcal{Z}))$, we have:

$$\lim_{\varepsilon\to 0}\int_\Omega E^\varepsilon(\mathbf{x})\cdot V(\mathbf{x},\frac{\mathbf{x}}{\varepsilon})\, d\mathbf{x} = \int_\Omega\int_\mathcal{Z} E_0(\mathbf{x},\mathbf{y})\cdot V(\mathbf{x},\mathbf{y})\, d\mathbf{y}d\mathbf{x}. \tag{126}$$

**Step 2: Deduction of the constraint equation.** We multiply the equation (123) by oscillating test function $V^\varepsilon(\mathbf{x}) = V(\mathbf{x}, \frac{\mathbf{x}}{\varepsilon})$ where $V(\mathbf{x}, \mathbf{y}) \in \mathbf{C}_0^1(\Omega, \mathbf{C}_\#^1(\mathcal{Z}))$:

$$\begin{aligned}
&\int_\Omega \nabla\times E^\varepsilon(\mathbf{x}) \cdot (\nabla_\mathbf{x} \times V^\varepsilon(\mathbf{x}, \frac{\mathbf{x}}{\varepsilon}) + \frac{1}{\varepsilon}\nabla_\mathbf{y} \times V^\varepsilon(\mathbf{x}, \frac{\mathbf{x}}{\varepsilon})) + [-\omega^2\varepsilon^5 k(\epsilon) \\
&\quad + i\omega((\mathbf{1}_C^\varepsilon(\frac{\mathbf{x}}{\varepsilon}) + \varepsilon^4\mathbf{1}_R^\varepsilon(\frac{\mathbf{x}}{\varepsilon}))\mathbf{1}_{\{y<0\}} + \varepsilon\mathbf{1}_{\{y>0\}})]E^\varepsilon \cdot V^\varepsilon(\mathbf{x}, \frac{\mathbf{x}}{\varepsilon})\, d\mathbf{x} \\
&\quad = -i\omega \int_{\Gamma_d} H_d \times e_2 \cdot (e_2 \times V(x, 1, z, \xi, \frac{1}{\varepsilon}, \zeta)) \times e_2\, d\sigma.
\end{aligned} \tag{127}$$

Integrating by parts, we get:

$$\begin{aligned}
&\int_\Omega E^\varepsilon(\mathbf{x}) \cdot (\nabla_\mathbf{x} \times \nabla_\mathbf{x} \times V^\varepsilon(\mathbf{x}, \frac{\mathbf{x}}{\varepsilon}) + \frac{1}{\varepsilon}\nabla_\mathbf{y} \times \nabla_\mathbf{x} \times V^\varepsilon(\mathbf{x}, \frac{\mathbf{x}}{\varepsilon}) \\
&+ \frac{1}{\varepsilon}\nabla_\mathbf{x} \times \nabla_\mathbf{y} \times V^\varepsilon(\mathbf{x}, \frac{\mathbf{x}}{\varepsilon}) + \frac{1}{\varepsilon^2}\nabla_\mathbf{y} \times \nabla_\mathbf{y} \times V^\varepsilon(\mathbf{x}, \frac{\mathbf{x}}{\varepsilon})) + [-\omega^2\varepsilon^5 k(\epsilon) \\
&+ i\omega(\mathbf{1}_C^\varepsilon(\frac{\mathbf{x}}{\varepsilon}) + \varepsilon^4\mathbf{1}_R^\varepsilon(\frac{\mathbf{x}}{\varepsilon}))\mathbf{1}_{\{y<0\}} + \varepsilon\mathbf{1}_{\{y>0\}}]E^\varepsilon(\mathbf{x}) \cdot V^\varepsilon(\mathbf{x}, \frac{\mathbf{x}}{\varepsilon})\, d\mathbf{x} \\
&= -i\omega \int_{\Gamma_d} H_d \times e_2 \cdot (e_2 \times V(x, 1, z, \xi, \frac{1}{\varepsilon}, \zeta)) \times e_2\, d\sigma.
\end{aligned} \tag{128}$$

Now we multiply (128) by $\varepsilon^2$ and we pass to the two-scale limit, applying Theorem 4.3 we obtain:

$$\int_\Omega \int_\mathcal{Z} E_0(\mathbf{x}, \mathbf{y})(\nabla_\mathbf{y} \times \nabla_\mathbf{y} \times V(\mathbf{x}, \mathbf{y}))\, d\mathbf{y}d\mathbf{x} = 0. \tag{129}$$

We deduce the constraint equation for the profile $E_0$:

$$\nabla_\mathbf{y} \times \nabla_\mathbf{y} \times E_0(\mathbf{x}, \mathbf{y}) = 0. \tag{130}$$

**Step 3. Looking for the solutions to the constraint equation.** Multiplying Equation (130) by $E_0$ and integrating by parts over $\mathcal{Z}$ leads to

$$\int_\mathcal{Z} \nabla_\mathbf{y} \times \nabla_\mathbf{y} \times E_0(\mathbf{x}, \mathbf{y})E_0(\mathbf{x}, \mathbf{y})\ d\mathbf{y} = \int_\mathcal{Z} |\nabla_\mathbf{y} \times E_0(\mathbf{x}, \mathbf{y})|^2\ d\mathbf{y} = 0. \tag{131}$$

We deduce that equation (131) is equivalent to

$$\nabla_\mathbf{y} \times E_0(\mathbf{x}, \mathbf{y}) = 0. \tag{132}$$

Moreover a solution of (132) is also solution of (130). So (130) and (132) are equivalent.

Hence, from Proposition (117) we conclude that $E_0(\mathbf{x}, \mathbf{y})$ can be decomposed as

$$E_0(\mathbf{x}, \mathbf{y}) = E(\mathbf{x}) + \nabla_\mathbf{y}\Phi_0(\mathbf{x}, \mathbf{y}). \tag{133}$$

**Step 4. Equations for $E(\mathbf{x})$ and $\Phi_0(\mathbf{x}, \mathbf{y})$.** The divergence equation of (123) is multiplied with $V(\mathbf{x}, \frac{\mathbf{x}}{\varepsilon}) = \varepsilon v(\mathbf{x})\psi(\frac{\mathbf{x}}{\varepsilon})$, where $v \in \mathbf{C}_0^1(\Omega)$ and $\psi \in \mathbf{H}_\#^1(\mathcal{Z})$. Theorem 4.3 and integration by parts yields for all $\psi \in \mathbf{H}_\#^1(\mathcal{Z})$ and $v \in \mathbf{C}_0^1(\Omega)$

$$
\begin{aligned}
&\lim_{\varepsilon\to 0}\int_\Omega \nabla\cdot\{-\omega^2\varepsilon^5 k(\epsilon)E^\varepsilon(\mathbf{x}) + i\omega[(\mathbf{1}_C^\varepsilon(\frac{\mathbf{x}}{\varepsilon}) + \varepsilon^4\mathbf{1}_R^\varepsilon(\frac{\mathbf{x}}{\varepsilon}))\mathbf{1}_{\{y<0\}} + \varepsilon\mathbf{1}_{\{y>0\}}]E^\varepsilon(\mathbf{x})\}\varepsilon v(\mathbf{x})\psi(\frac{\mathbf{x}}{\varepsilon})\,d\mathbf{x} \\
&= -\lim_{\varepsilon\to 0}\int_\Omega \{-\omega^2\varepsilon^5 k(\epsilon)E^\varepsilon(\mathbf{x}) + i\omega[\mathbf{1}_C^\varepsilon(\frac{\mathbf{x}}{\varepsilon}) + \varepsilon^4\mathbf{1}_R^\varepsilon(\frac{\mathbf{x}}{\varepsilon}))\mathbf{1}_{\{y<0\}} \\
&+ \varepsilon\mathbf{1}_{\{y>0\}}]E^\varepsilon\}\cdot(\varepsilon v(\mathbf{x})\psi(\frac{\mathbf{x}}{\varepsilon}) + v(\mathbf{x})\nabla_\mathbf{y}\psi(\frac{\mathbf{x}}{\varepsilon}))\,d\mathbf{x} \\
&= -\int_\Omega\int_\mathcal{Z} v(\mathbf{x})\nabla_\mathbf{y}\psi(\mathbf{y})\cdot[i\omega\mathbf{1}_C(\mathbf{y})E_0(\mathbf{x},\mathbf{y})]\,d\mathbf{y}d\mathbf{x} = 0,
\end{aligned} \tag{134}
$$

from which it follows that

$$\nabla_\mathbf{y}\cdot[i\omega\mathbf{1}_C(\mathbf{y})E_0(\mathbf{x},\mathbf{y})] = 0, \tag{135}$$

with $E_0$ given by the decomposition (117). So we obtain the local equation

$$\nabla_\mathbf{y}\cdot[i\omega\mathbf{1}_C(\mathbf{y})\{E(\mathbf{x}) + \nabla_\mathbf{y}\Phi_0(\mathbf{x},\mathbf{y})\}]\,d\mathbf{y} = 0. \tag{136}$$

The potential $\Phi_0$ may be written on the form

$$\Phi_0(\mathbf{x},\mathbf{y}) = \sum_{j=1}^{3}\chi_j(\mathbf{y})e_j\cdot E(\mathbf{x}) = \chi(\mathbf{y})\cdot E(\mathbf{x}). \tag{137}$$

From (133) and (137), we get:

$$E_0(\mathbf{x},\mathbf{y}) = (\mathrm{Id} + \nabla_\mathbf{y}\chi(\mathbf{y}))E(\mathbf{x}). \tag{138}$$

Inserting $E_0$ in (136) we obtain

$$\nabla_\mathbf{y}\cdot[i\omega\mathbf{1}_C(\mathbf{y})(\mathrm{Id} + \nabla_\mathbf{y}\chi(\mathbf{y})] = 0. \tag{139}$$

Now, we build oscillating test functions satisfying constraint (133) and use them in weak formulation (128). We define test function $V(\mathbf{x},\mathbf{y}) = \alpha(\mathbf{x}) + \nabla_\mathbf{y}\beta(\mathbf{x},\mathbf{y})$, $V(\mathbf{x},\mathbf{y}) \in \mathbf{C}_0^1(\Omega, \mathbf{C}_\#^1(\mathcal{Z}))$ and we inject in (128) test function $V^\varepsilon = V(\mathbf{x}, \frac{\mathbf{x}}{\varepsilon})$, which gives:

$$
\begin{aligned}
&\int_\Omega E^\varepsilon(\mathbf{x})\cdot(\nabla_\mathbf{x}\times\nabla_\mathbf{x}\times V(\mathbf{x},\frac{\mathbf{x}}{\varepsilon}) + \frac{2}{\varepsilon}\nabla_\mathbf{x}\times\nabla_\mathbf{y}\times V(\mathbf{x},\frac{\mathbf{x}}{\varepsilon}) \\
&+ \frac{1}{\varepsilon^2}\nabla_\mathbf{y}\times\nabla_\mathbf{y}\times V(\mathbf{x},\frac{\mathbf{x}}{\varepsilon})) + [-\omega^2\varepsilon^5 k(\epsilon) + i\omega((\mathbf{1}_C^\varepsilon(\frac{\mathbf{x}}{\varepsilon}) \\
&+ \varepsilon^4\mathbf{1}_R^\varepsilon(\frac{\mathbf{x}}{\varepsilon}))\mathbf{1}_{\{y<0\}} + \varepsilon\mathbf{1}_{\{y>0\}})]E^\varepsilon(\mathbf{x})\cdot V(\mathbf{x},\frac{\mathbf{x}}{\varepsilon})\,d\mathbf{x} \\
&= -i\omega\int_{\Gamma_d} H_d\times e_2\cdot(e_2\times V^\ddagger(x,1,z,\xi,\zeta))\times e_2\,d\sigma,
\end{aligned} \tag{140}
$$

with $V(x,1,z,\xi,\nu,\zeta) = V^\ddagger(x,1,z,\xi,\zeta)$ the restriction on $V$ which does not depend on $\nu$. The term containing the constraint, the third one, disappears. Passing to the limit $\varepsilon\to 0$ and replacing the expression of $V$ by the term $\alpha(\mathbf{x}) + \nabla_{\mathbf{y}}\beta(\mathbf{x},\mathbf{y})$, we have

$$\begin{aligned}\nabla_{\mathbf{x}} \times \nabla_{\mathbf{y}} \times V(\mathbf{x},\mathbf{y}) &= \nabla_{\mathbf{x}} \times \nabla_{\mathbf{y}} \times [\alpha(\mathbf{x}) + \nabla_{\mathbf{y}}\beta(\mathbf{x},\mathbf{y})] \\ &= \nabla_{\mathbf{x}} \times \nabla_{\mathbf{y}} \times (\alpha(\mathbf{x})) + \nabla_{\mathbf{x}} \times \nabla_{\mathbf{y}} \times (\nabla_{\mathbf{y}}\beta(\mathbf{x},\mathbf{y})) \\ &= \nabla_{\mathbf{x}} \times \nabla_{\mathbf{y}} \times (\nabla_{\mathbf{y}}\beta(\mathbf{x},\mathbf{y})).\end{aligned} \tag{141}$$

Since $\nabla_{\mathbf{y}} \times (\nabla_{\mathbf{y}}) = 0$, the term $\frac{2}{\varepsilon}\nabla_{\mathbf{x}} \times \nabla_{\mathbf{y}} \times \nabla_{\mathbf{y}}\beta(\mathbf{x},\mathbf{y}))$ vanishes. Therefore, (140) becomes:

$$\begin{aligned}&\int_\Omega \int_{\mathcal{Z}} E_0(\mathbf{x},\mathbf{y}) \cdot \nabla_{\mathbf{x}} \times \nabla_{\mathbf{x}} \times (\alpha(\mathbf{x}) + \nabla_{\mathbf{y}}\beta(\mathbf{x},\mathbf{y})) \\ &\quad + i\omega \mathbf{1}_C(\mathbf{y}) E_0(\mathbf{x},\mathbf{y}) \cdot (\alpha(\mathbf{x}) + \nabla_{\mathbf{y}}\beta(\mathbf{x},\mathbf{y})\, d\mathbf{y} d\mathbf{x} \\ &\quad = -i\omega \int_{\Gamma_d} H_d \times e_2 \cdot (e_2 \times (\alpha(x,1,z) + \nabla_{\mathbf{y}}\beta(x,1,z,\xi,\zeta))) \times e_2\, d\sigma.\end{aligned} \tag{142}$$

Now in (142) we replace expression $E_0$ giving by (138). We obtain

$$\begin{aligned}&\int_\Omega \int_{\mathcal{Z}} (\mathrm{Id} + \nabla_{\mathbf{y}}\chi(\mathbf{y}))E(\mathbf{x}) \cdot (\nabla_{\mathbf{x}} \times \nabla_{\mathbf{x}} \times (\alpha(\mathbf{x}) + \nabla_{\mathbf{y}}\beta(\mathbf{x},\mathbf{y})) \\ &\quad + i\omega \mathbf{1}_C(\mathbf{y})(\mathrm{Id} + \nabla_{\mathbf{y}}\chi(\mathbf{y}))E(\mathbf{x})) \cdot (\alpha(\mathbf{x}) + \nabla_{\mathbf{y}}\beta(\mathbf{x},\mathbf{y}))\, d\mathbf{y} d\mathbf{x} \\ &\quad = -i\omega \int_{\Gamma_d} H_d \times e_2 \cdot (e_2 \times (\alpha(x,1,z) + \nabla_{\mathbf{y}}\beta(x,1,z,\xi,\zeta))) \times e_2\, d\sigma.\end{aligned} \tag{143}$$

Taking $\alpha(\mathbf{x}) = 0$ in (143), we obtain

$$\begin{aligned}&\int_\Omega \int_{\mathcal{Z}} (\mathrm{Id}+\nabla_{\mathbf{y}}\chi(\mathbf{y}))\nabla_{\mathbf{x}} \times \nabla_{\mathbf{x}} \times E(\mathbf{x})\nabla_{\mathbf{y}}\beta(\mathbf{x},\mathbf{y}) \\ &\qquad + i\omega \mathbf{1}_C(\mathbf{y})(\mathrm{Id} + \nabla_{\mathbf{y}}\chi(\mathbf{y}))E(\mathbf{x}) \cdot \nabla_{\mathbf{y}}\beta(\mathbf{x},\mathbf{y}) d\mathbf{y} d\mathbf{x} = 0.\end{aligned} \tag{144}$$

Integrating by parts

$$\begin{aligned}&\int_\Omega \int_{\mathcal{Z}} -\nabla_{\mathbf{y}} \cdot \{(\mathrm{Id} - \nabla_{\mathbf{y}}\chi(\mathbf{y}))\nabla_{\mathbf{x}} \times \nabla_{\mathbf{x}} \times E(\mathbf{x})\}\beta(\mathbf{x},\mathbf{y}) \\ &\qquad - i\omega \nabla_{\mathbf{y}} \cdot \{\mathbf{1}_C(\mathbf{y})(\mathrm{Id} - \nabla_{\mathbf{y}}\chi(\mathbf{y}))E(\mathbf{x})\}\beta(\mathbf{x},\mathbf{y})\, d\mathbf{y} d\mathbf{x} = 0.\end{aligned} \tag{145}$$

And since $\nabla_{\mathbf{y}} \cdot \{\mathbf{1}_C(\mathbf{y})(\mathrm{Id} + \nabla_{\mathbf{y}}\chi(\mathbf{y}))E(\mathbf{x})\} = 0$ we obtain

$$\int_\Omega \int_{\mathcal{Z}} -\nabla_{\mathbf{y}} \cdot \{(\mathrm{Id} + \nabla_{\mathbf{y}}\chi(\mathbf{y}))\nabla_{\mathbf{x}} \times \nabla_{\mathbf{x}} \times E(\mathbf{x})\}\beta(\mathbf{x},\mathbf{y})\, d\mathbf{y} d\mathbf{x} = 0, \tag{146}$$

which gives the cell problem

$$\nabla_{\mathbf{y}} \cdot [\mathrm{Id} + \nabla_{\mathbf{y}}\chi(\mathbf{y})] = 0. \tag{147}$$

From (139) and (147), the scalar function $\chi$ is the unique solution, thanks to Lax-Milgram Lemma, up to an additive constant in the Hilbert space of $\mathcal{Z}$ periodic function $H^1_\#(\mathcal{Z})$ of the following boundary value problem

$$\begin{cases} \Delta_{\mathbf{y}}(\chi(\mathbf{y})) = 0 \ \text{ in } \mathcal{Z}\backslash\partial\Omega_C, \\ [\dfrac{\partial\chi}{\partial n}] = -n_j \ \text{ on } \partial\Omega_C, \\ [\chi] = 0 \ \text{ on } \partial\Omega_C, \end{cases} \tag{148}$$

where $[f]$ is the jump across the surface of $\partial\Omega_C$ and $n_j$, $j = \{1, 2, 3\}$ is the projection on the axis $e_j$ of the normal of $\partial\Omega_C$.

**Remark 4.11.** *(148) can be seen as an electrostatic problem. Solving (139) and (147) reduces to look for a potential induced by surface density of charges. Then $\chi$ is this potential induced by the charges on the interface of carbon fiber.*

Setting $\beta(\mathbf{x}, \mathbf{y}) = 0$ in (143) and integrating by parts, we get

$$\begin{aligned}\int_\Omega \int_Z &(\mathrm{Id} + \nabla_{\mathbf{y}}\chi(\mathbf{y}))\nabla_{\mathbf{x}} \times \nabla_{\mathbf{x}} \times E(\mathbf{x}) \cdot \alpha(\mathbf{x}) \\ &+ i\omega \mathbf{1}_C(\mathbf{y})(\mathrm{Id} + \nabla_{\mathbf{y}}\chi(\mathbf{y}))E(\mathbf{x})\alpha(\mathbf{x})\, d\mathbf{y}d\mathbf{x} \\ &= -i\omega \int_{\Gamma_d} H_d \times e_2 \cdot (e_2 \times \alpha(x, 1, z)) \times e_2 \, d\sigma,\end{aligned} \tag{149}$$

which gives the following well posed problem for $E(\mathbf{x})$

$$\begin{cases} \theta_1 \nabla_{\mathbf{x}} \times \nabla_{\mathbf{x}} \times E(\mathbf{x}) + i\omega\theta_2 E(\mathbf{x}) = 0 \text{ in } \Omega, \\ \theta_1 \nabla_{\mathbf{x}} \times E(\mathbf{x}) \times e_2 = -i\omega H_d \times e_2 \text{ on } \Gamma_d, \\ \nabla_{\mathbf{x}} \times E(\mathbf{x}) \times e_2 = 0 \text{ on } \Gamma_L, \end{cases} \tag{150}$$

with $\theta_1 = \int_Z \mathrm{Id} + \nabla_{\mathbf{y}}\chi(\mathbf{y})\, d\mathbf{y}$ and $\theta_2 = \int_Z \mathbf{1}_C(\mathbf{y})(\mathrm{Id} + \nabla_{\mathbf{y}}\chi(\mathbf{y}))\, d\mathbf{y}$.
This concludes the proof of Theorem (124). □

## 5. Conclusion

We presented in this paper the homogenization of time harmonic Maxwell equation by the method of two-scale convergence. We started by studying the time harmonic Maxwell equations with coefficients depending of $\varepsilon$. We remind that $\lambda$ is the wave length, $\delta$ is the skin length, $L$ is thickness of the medium and $e$ the size of the basic cell and then $\varepsilon = \frac{e}{L}$ is the small parameter. We find for low frequencies the macroscopic homogenized Maxwell equations depending on the volume fraction of the carbon fibers and we find also the microscopic equation.

## 6. Annexes

## A. Presentation of all cases of tables 1, 2, 3 and 4

- The case 1 corresponds to the air not ionized, a resin not doped and $\overline{\sigma}$ is the effective electric conductivity in the direction of the carbon fibers. We have for the effective electric conductivity $\overline{\sigma} = \sigma_c \sim 40000\ S.m^{-1}$, the resin conductivity is about $\sigma_r \sim 10^{-10}\ S.m^{-1}$ and the conductivity in the air is about $10^{-14}\ S.m^{-1}$. So when we want to calculate the ratio in (**??**)-(41) depending on $\varepsilon$ we get: $\frac{\sigma_r}{\overline{\sigma}} \sim \varepsilon^7$ and $\frac{\sigma_a}{\overline{\sigma}} \sim \varepsilon^9$.

- In case 2, the air is not ionized, the resin is doped and $\overline{\sigma}$ is the effective conductivity is in direction of carbon fibers. We have like the case 1 $\overline{\sigma} = \sigma_c \sim 40000\ S.m^{-1}$. The resin conductivity is about $\sigma_r \sim 10^{-3}\ S.m^{-1}$ and the conductivity in the air is about $10^{-14}\ S.m^{-1}$. So $\frac{\sigma_r}{\overline{\sigma}} \sim \varepsilon^4$ and $\frac{\sigma_a}{\overline{\sigma}} \sim \varepsilon^9$.

- In case 3, the air is not ionized, the resin is not doped and $\overline{\sigma}$ is the effective conductivity is orthogonal to the fibers. $\overline{\sigma} = \sigma_r \sim 10^{-10}\ S.m^{-1}$. The carbon fiber conductivity is about $\sigma_c \sim 10^4\ S.m^{-1}$ and the conductivity in the air is about $10^{-14}\ S.m^{-1}$. $\frac{\sigma_c}{\overline{\sigma}} \sim \frac{1}{\varepsilon^7}$ and $\frac{\sigma_a}{\overline{\sigma}} \sim \varepsilon^2$.

- Case 4 corresponds to the air non ionized, the resin doped and $\overline{\sigma}$ is the effective conductivity orthogonal to the fibers. The effective electric conductivity is $\overline{\sigma} = \sigma_r \sim 10^{-3}\ S.m^{-1}$. The carbon fiber conductivity is about $\sigma_c \sim 40000\ S.m^{-1}$ and the conductivity in the air is about $10^{-14}\ S.m^{-1}$. $\frac{\sigma_c}{\overline{\sigma}} \sim \frac{1}{\varepsilon^4}$ and $\frac{\sigma_a}{\overline{\sigma}} \sim \varepsilon^6$.

- In case 5, the air is ionized, the resin is not doped and $\overline{\sigma}$ is the effective conductivity is in the direction of the carbon fibers. This one is equal $\overline{\sigma} = \sigma_c \sim 40000\ S.m^{-1}$, the resin conductivity is about $\sigma_r \sim 10^{-10}\ S.m^{-1}$ and the conductivity in the air is now about $4242\ S.m^{-1}$. $\frac{\sigma_r}{\overline{\sigma}} \sim \varepsilon^7$ and $\frac{\sigma_a}{\overline{\sigma}} \sim \varepsilon$.

- Case 6 corresponds to the air ionized, the resin doped and $\overline{\sigma}$ is the effective conductivity in direction of the carbon fibers. This one is equal $\overline{\sigma} = \sigma_c \sim 40000\ S.m^{-1}$, the resin conductivity is about $\sigma_r \sim 10^3\ S.m^{-1}$ and the conductivity in the air is now about $4242\ S.m^{-1}$. $\frac{\sigma_r}{\overline{\sigma}} \sim \varepsilon^4$ and $\frac{\sigma_a}{\overline{\sigma}} \sim \varepsilon$.

- Case 7 corresponds to the air ionized, the resin not doped and $\overline{\sigma}$ is the effective conductivity orthogonal to the fibers. The effective conductivity is $\overline{\sigma} = \sigma_r \sim 10^{-10}\ S.m^{-1}$, the carbon fibers conductivity is about $\sigma_c \sim 40000\ S.m^{-1}$ and the conductivity in the air is now about $4242\ S.m^{-1}$. $\frac{\sigma_c}{\overline{\sigma}} \sim \frac{1}{\varepsilon^7}$ and $\frac{\sigma_a}{\overline{\sigma}} \sim \frac{1}{\varepsilon^6}$.

- Case 8 corresponds to the air ionized, the resin doped and $\overline{\sigma}$ is the effective conductivity orthogonal to the fibers. The effective conductivity is $\overline{\sigma} = \sigma_r \sim 10^{-3}\ S.m^{-1}$, the carbon fibers conductivity is about $\sigma_c \sim 40000\ S.m^{-1}$ and the conductivity in the air is now about $4242\ S.m^{-1}$. $\frac{\sigma_c}{\overline{\sigma}} \sim \frac{1}{\varepsilon^4}$ and $\frac{\sigma_a}{\overline{\sigma}} \sim \frac{1}{\varepsilon^2}$.

## B. Structure of the equations depending of $\varepsilon$

For $\overline{\omega} = 100\ rad.s^{-1}$, we have

Case 1

$$\eta = 9 \text{ and } \Sigma_a^\varepsilon = \varepsilon^{11},\ \Sigma_r^\varepsilon = \varepsilon^9,\ \Sigma_c^\varepsilon = \varepsilon^2. \tag{151}$$

Case 2

$$\eta = 9 \text{ and } \Sigma_a^\varepsilon = \varepsilon^{11},\ \Sigma_r^\varepsilon = \varepsilon^6,\ \Sigma_c^\varepsilon = \varepsilon^2. \tag{152}$$

Case 3

$$\eta = 9 \text{ and } \Sigma_a^\varepsilon = \varepsilon^{12},\ \Sigma_r^\varepsilon = \varepsilon^{10},\ \Sigma_c^\varepsilon = \varepsilon^3. \tag{153}$$

Case 4

$$\eta = 9 \text{ and } \Sigma_a^\varepsilon = \varepsilon^{13},\ \Sigma_r^\varepsilon = \varepsilon^7,\ \Sigma_c^\varepsilon = \varepsilon^3. \tag{154}$$

Case 5

$$\eta = 9 \text{ and } \Sigma_a^\varepsilon = \varepsilon^3,\ \Sigma_r^\varepsilon = \varepsilon^9,\ \Sigma_c^\varepsilon = \varepsilon^2. \tag{155}$$

Case 6

$$\eta = 9 \text{ and } \Sigma_a^\varepsilon = \varepsilon^3, \Sigma_r^\varepsilon = \varepsilon^6, \Sigma_c^\varepsilon = \varepsilon^2. \quad (156)$$

Case 7

$$\eta = 9 \text{ and } \Sigma_a^\varepsilon = \varepsilon^5, \Sigma_r^\varepsilon = \varepsilon^{10}, \Sigma_c^\varepsilon = \varepsilon^3. \quad (157)$$

Case 8

$$\eta = 9 \text{ and } \Sigma_a^\varepsilon = \varepsilon^5, \Sigma_r^\varepsilon = \varepsilon^7, \Sigma_c^\varepsilon = \varepsilon^3. \quad (158)$$

For $\overline{\omega} = 10^6 \; rad.s^{-1}$

Case 1

$$\eta = 5 \text{ and } \Sigma_a^\varepsilon = \varepsilon^9, \Sigma_r^\varepsilon = \varepsilon^7, \Sigma_c^\varepsilon = 1. \quad (159)$$

Case 2

$$\eta = 5 \text{ and } \Sigma_a^\varepsilon = \varepsilon^9, \Sigma_r^\varepsilon = \varepsilon^4, \Sigma_c^\varepsilon = 1. \quad (160)$$

Case 3

$$\eta = 5 \text{ and } \Sigma_a^\varepsilon = \varepsilon^{10}, \Sigma_r^\varepsilon = \varepsilon^8, \Sigma_c^\varepsilon = \varepsilon. \quad (161)$$

Case 4

$$\eta = 5 \text{ and } \Sigma_a^\varepsilon = \varepsilon^{11}, \Sigma_r^\varepsilon = \varepsilon^5, \Sigma_c^\varepsilon = 1. \quad (162)$$

Case 5

$$\eta = 5 \text{ and } \Sigma_a^\varepsilon = \varepsilon, \Sigma_r^\varepsilon = \varepsilon^7, \Sigma_c^\varepsilon = 1. \quad (163)$$

Case 6

$$\eta = 5 \text{ and } \Sigma_a^\varepsilon = \varepsilon, \Sigma_r^\varepsilon = \varepsilon^4, \Sigma_c^\varepsilon = 1. \quad (164)$$

Case 7

$$\eta = 5 \text{ and } \Sigma_a^\varepsilon = \varepsilon^3, \Sigma_r^\varepsilon = \varepsilon^8, \Sigma_c^\varepsilon = \varepsilon. \quad (165)$$

Case 8

$$\eta = 5 \text{ and } \Sigma_a^\varepsilon = \varepsilon^3, \Sigma_r^\varepsilon = \varepsilon^5, \Sigma_c^\varepsilon = 1. \quad (166)$$

For $\overline{\omega} = 10^{10} \; rad.s^{-1}$

Case 1

$$\eta = 1 \text{ and } \Sigma_a^\varepsilon = \varepsilon^7, \Sigma_r^\varepsilon = \varepsilon^5, \Sigma_c^\varepsilon = \tfrac{1}{\varepsilon^2}. \quad (167)$$

Case 2

$$\eta = 1 \text{ and } \Sigma_a^\varepsilon = \varepsilon^7,\ \Sigma_r^\varepsilon = \varepsilon^2,\ \Sigma_c^\varepsilon = \tfrac{1}{\varepsilon^2}. \tag{168}$$

Case 3

$$\eta = 1 \text{ and } \Sigma_a^\varepsilon = \varepsilon^8,\ \Sigma_r^\varepsilon = \varepsilon^6,\ \Sigma_c^\varepsilon = \tfrac{1}{\varepsilon}. \tag{169}$$

Case 4

$$\eta = 1 \text{ and } \Sigma_a^\varepsilon = \varepsilon^9,\ \Sigma_r^\varepsilon = \varepsilon^3,\ \Sigma_c^\varepsilon = \tfrac{1}{\varepsilon}. \tag{170}$$

Case 5

$$\eta = 1 \text{ and } \Sigma_a^\varepsilon = \tfrac{1}{\varepsilon},\ \Sigma_r^\varepsilon = \varepsilon^5,\ \Sigma_c^\varepsilon = \tfrac{1}{\varepsilon^2}. \tag{171}$$

Case 6

$$\eta = 1 \text{ and } \Sigma_a^\varepsilon = \tfrac{1}{\varepsilon},\ \Sigma_r^\varepsilon = \varepsilon^2,\ \Sigma_c^\varepsilon = \tfrac{1}{\varepsilon^2}. \tag{172}$$

Case 7

$$\eta = 1 \text{ and } \Sigma_a^\varepsilon = \tfrac{1}{\varepsilon},\ \Sigma_r^\varepsilon = \varepsilon^6,\ \Sigma_c^\varepsilon = \tfrac{1}{\varepsilon}. \tag{173}$$

Case 8

$$\eta = 1 \text{ and } \Sigma_a^\varepsilon = \varepsilon,\ \Sigma_r^\varepsilon = \varepsilon^3,\ \Sigma_c^\varepsilon = \tfrac{1}{\varepsilon}. \tag{174}$$

For $\overline{\omega} = 10^{12}\ rad.s^{-1}$

Case 1

$$\eta = 0 \text{ and } \Sigma_a^\varepsilon = \varepsilon^6,\ \Sigma_r^\varepsilon = \varepsilon^4,\ \Sigma_c^\varepsilon = \tfrac{1}{\varepsilon^3}. \tag{175}$$

Case 2

$$\eta = 0 \text{ and } \Sigma_a^\varepsilon = \varepsilon^6,\ \Sigma_r^\varepsilon = \varepsilon,\ \Sigma_c^\varepsilon = \tfrac{1}{\varepsilon^3}. \tag{176}$$

Case 3

$$\eta = 0 \text{ and } \Sigma_a^\varepsilon = \varepsilon^5,\ \Sigma_r^\varepsilon = \varepsilon^3,\ \Sigma_c^\varepsilon = \tfrac{1}{\varepsilon^4}. \tag{177}$$

Case 4

$$\eta = 0 \text{ and } \Sigma_a^\varepsilon = \varepsilon^7,\ \Sigma_r^\varepsilon = \varepsilon,\ \Sigma_c^\varepsilon = \tfrac{1}{\varepsilon^3}. \tag{178}$$

Case 5

$$\eta = 0 \text{ and } \Sigma_a^\varepsilon = \tfrac{1}{\varepsilon^2},\ \Sigma_r^\varepsilon = \varepsilon^4,\ \Sigma_c^\varepsilon = \tfrac{1}{\varepsilon^3}. \tag{179}$$

Case 6

$$\eta = 0 \text{ and } \Sigma_a^\varepsilon = \tfrac{1}{\varepsilon^2},\ \Sigma_r^\varepsilon = \varepsilon,\ \Sigma_c^\varepsilon = \tfrac{1}{\varepsilon^3}. \tag{180}$$

Case 7

$$\eta = 0 \text{ and } \Sigma_a^\varepsilon = \tfrac{1}{\varepsilon^2},\ \Sigma_r^\varepsilon = \varepsilon^3,\ \Sigma_c^\varepsilon = \tfrac{1}{\varepsilon^4}. \tag{181}$$

Case 8

$$\eta = 0 \text{ and } \Sigma_a^\varepsilon = \tfrac{1}{\varepsilon},\ \Sigma_r^\varepsilon = \varepsilon,\ \Sigma_c^\varepsilon = \tfrac{1}{\varepsilon^3}. \tag{182}$$

## C. The transmission Maxwell problem

Taking a test function $V \in C^1(\Omega)$ with compact support in $\Omega_c^\varepsilon$, in weak formulation (83) associated with the problem ((63), (66), (67)). Since

$$\int_\Omega \nabla\times E^\varepsilon_{|\Omega_c^\varepsilon} \cdot \nabla\times\overline{V} \ d\mathbf{x} = \langle \nabla\times\nabla\times E^\varepsilon_{|\Omega_c^\varepsilon}, \overline{V}\rangle_{\Omega_c^\varepsilon}, \tag{183}$$

we deduce the third equation in (84). Similarly, taking $V \in C^1(\Omega)$ with compact support respectively in $\Omega_r^\varepsilon$ and $\Omega_a^\varepsilon$, we obtain the first and the second equation in (84). Now, since $E_{|\Omega_a^\varepsilon} \in \mathbf{H}(\mathrm{curl}, \Omega_a^\varepsilon)$ and $E_{|\Omega_r^\varepsilon} \in \mathbf{H}(\mathrm{curl}, \Omega_r^\varepsilon)$, let $V \in C_0^1(\Omega_a^\varepsilon \bigcup \Omega_r^\varepsilon)$ integrating by parts we get

$$\begin{aligned}\int_{\Omega_a^\varepsilon \bigcup \Omega_r^\varepsilon} E \cdot \nabla\times\overline{V} \ d\mathbf{x} &= \int_{\Omega_a^\varepsilon} E_{|\Omega_a^\varepsilon} \cdot \nabla\times\overline{V} \ d\mathbf{x} + \int_{\Omega_r^\varepsilon} E_{|\Omega_r^\varepsilon} \cdot \nabla\times\overline{V} \ d\mathbf{x} \\ &= \int_{\Omega_a^\varepsilon} \nabla\times E_{|\Omega_a^\varepsilon} \cdot \overline{V} \ d\mathbf{x} + \int_{\Omega_r^\varepsilon} \nabla\times E_{|\Omega_r^\varepsilon} \cdot \overline{V} \ d\mathbf{x} \\ &+ \int_{\Gamma_{ra}} (E_{|\Omega_a^\varepsilon} \times e_2 - E_{|\Omega_r^\varepsilon} \times n_{|\Omega_r^\varepsilon}) \cdot \overline{V} \, ds.\end{aligned} \tag{184}$$

Since on every point of $\Gamma_{ra}$ $e_2 = -n_{|\Omega_r^\varepsilon}$ the assumed continuity require

$$E_{|\Omega_a^\varepsilon} \times e_2 = E_{|\Omega_r^\varepsilon} \times n_{|\Omega_r^\varepsilon}, \tag{185}$$

we obtain the fourth relation in (84). With the same argument on $\Gamma_{cr}^\varepsilon$, we obtain the last relation in (84). This shows that (83) implies (84). And, if $E^\varepsilon$ is solution to (84) following that for any regular set $\widehat{\Omega}$ in $\Omega$ the Stokes's formula gives, for more details see p 57, 58 of P. Monk's book [18]:

$$\forall \ E, V \in \mathbf{H}(\mathrm{curl}, \widehat{\Omega}) \quad \int_{\widehat{\Omega}} \nabla\times E \cdot \overline{V} - E \cdot \nabla\times\overline{V} \ d\mathbf{x} = \langle E \times n_{\widehat{\Omega}}, \overline{V_T}\rangle_{\partial\widehat{\Omega}} \tag{186}$$

$\mathbf{H}(\mathrm{curl}, \widehat{\Omega})$ has the same definition as $\mathbf{H}(\mathrm{curl}, \Omega)$ with $\Omega$ replaced by $\widehat{\Omega}$ and where $V_T = (n \times V) \times n$, and $n_{\widehat{\Omega}}$ is the unit outward normal of $\partial\widehat{\Omega}$. For all $V \in \mathbf{H}(\mathrm{curl}, \Omega)$, $V_{|\Omega_r^\varepsilon} \in \mathbf{H}(\mathrm{curl}, \Omega_r^\varepsilon)$, $V_{|\Omega_a^\varepsilon} \in \mathbf{H}(\mathrm{curl}, \Omega_a^\varepsilon)$ and $V_{|\Omega_c^\varepsilon} \in \mathbf{H}(\mathrm{curl}, \Omega_c^\varepsilon)$. Hence, fixing any $E' \in \mathbf{H}(\mathrm{curl}, \Omega)$ according to the second equation in (84), we have $\nabla\times E^\varepsilon_{|\Omega_r^\varepsilon} \in \mathbf{H}(\mathrm{curl}, \Omega_r^\varepsilon)$ then applying (186) in $\Omega_r^\varepsilon$ with $E = \nabla\times E^\varepsilon_{|\Omega_r^\varepsilon}$ and $V$ we get

$$\begin{aligned}\int_{\Omega_r^\varepsilon} \nabla\times E^\varepsilon_{|\Omega_r^\varepsilon} \cdot \nabla\times\overline{V} \ d\mathbf{x} = \int_{\Omega_r^\varepsilon} \nabla\times\nabla\times E^\varepsilon_{|\Omega_r^\varepsilon} \cdot \overline{V} \, d\mathbf{x} + \langle \nabla\times E^\varepsilon_{|\Omega_r^\varepsilon} \times n_{|\Omega_r^\varepsilon}, \overline{V_T}\rangle_{\Gamma_{ra}} \\ + \langle \nabla\times E^\varepsilon_{|\Omega_c^\varepsilon} \times n_{|\Omega_c^\varepsilon}, \overline{V_T}\rangle_{\Gamma_{cr}^\varepsilon}.\end{aligned} \tag{187}$$

Doing the same for $\Omega_c^\varepsilon$, we have

$$\int_{\Omega_c^\varepsilon} \nabla\times E^\varepsilon_{|\Omega_c^\varepsilon} \cdot \nabla\times\overline{V} \ d\mathbf{x} = \int_{\Omega_c^\varepsilon} \nabla\times\nabla\times E^\varepsilon_{|\Omega_c^\varepsilon} \cdot \overline{V} \, d\mathbf{x} + \langle \nabla\times E^\varepsilon_{|\Omega_c^\varepsilon} \times n^\varepsilon_{|\Omega_c^\varepsilon}, \overline{V_T}\rangle_{\Gamma_{cr}^\varepsilon}. \tag{188}$$

Finally for $\Omega_a^\varepsilon$, we have

$$\begin{aligned}\int_{\Omega_a^\varepsilon} \nabla\times E^\varepsilon_{|\Omega_a^\varepsilon} \cdot \nabla\times\overline{V} \ d\mathbf{x} = \int_{\Omega_a^\varepsilon} \nabla\times\nabla\times E^\varepsilon_{|\Omega_a^\varepsilon} \cdot \overline{V} \, d\mathbf{x} + \langle \nabla\times E^\varepsilon_{|\Omega_a^\varepsilon} \times e_2, \overline{V_T}\rangle_{\Gamma_d} \\ - \langle \nabla\times E^\varepsilon_{|\Omega_a^\varepsilon} \times e_2, \overline{V_T}\rangle_{\Gamma_{ra}^\varepsilon}.\end{aligned} \tag{189}$$

Summing the relations above since in every point of $\Gamma_{ra}$ $n_{|\Omega_r^\varepsilon} = -e_2$ and in every point of $\Gamma_{cr}^\varepsilon$ $n_{|\Omega_c^\varepsilon} = -n_{|\Omega_r^\varepsilon}$, it comes

$$\begin{aligned}\int_\Omega \nabla\times E^\varepsilon \cdot \nabla\times\overline{V}\ d\mathbf{x} &= \int_\Omega \nabla\times\nabla\times E^\varepsilon \cdot \overline{V}\ d\mathbf{x} + < [\nabla\times E^\varepsilon \times n], \overline{V_T} >_{\Gamma_{ra}} \\ &+ \langle[\nabla\times E^\varepsilon \times n], \overline{V_T}\rangle_{\Gamma_{cr}} - i\omega \int_{\Gamma_d} H_d \times n^\varepsilon \cdot \overline{V}_T\ d\sigma.\end{aligned} \tag{190}$$

According to (83) and the first, second and third equations in (84) we have

$$\begin{aligned}&\langle\nabla\times E^\varepsilon_{|\Omega_c^\varepsilon} \times n_{|\Omega_c^\varepsilon}, \overline{V_T}\rangle_{\Gamma_{cr}^\varepsilon} - \langle\nabla\times E^\varepsilon_{|\Omega_r^\varepsilon} \times n_{|\Omega_r^\varepsilon}, \overline{V_T}\rangle_{\Gamma_{cr}^\varepsilon} \\ &+ \langle\nabla\times E^\varepsilon_{|\Omega_r^\varepsilon} \times n_{|\Omega_r^\varepsilon}, \overline{V_T}\rangle_{\Gamma_{ra}} + \langle\nabla\times E^\varepsilon_{|\Omega_a^\varepsilon} \times e_2, \overline{V_T}\rangle_{\Gamma_{ra}} = 0,\end{aligned} \tag{191}$$

for all $V \in \mathbf{H}(\mathrm{curl}, \Omega)$ which causes the last two equalities in (84) and concludes the first part of the proof.

Reciprocally, integrating by parts (84) we have:

$$\begin{aligned}\forall\ V \in \mathbf{X}^\varepsilon(\Omega),\quad &\int_\Omega \nabla\times E^\varepsilon \cdot \nabla\times\overline{V}\, d\mathbf{x} + \int_{\Omega_a^\varepsilon} (-\omega^2\varepsilon^\eta + i\omega\Sigma_a^\varepsilon)E^\varepsilon \cdot \overline{V}\, d\mathbf{x} \\ &= -i\omega \int_{\Gamma_d} H_d \times n^\varepsilon \cdot \overline{V}_T\ d\sigma,\end{aligned} \tag{192}$$

and

$$\forall\ V \in \mathbf{X}^\varepsilon(\Omega),\quad \int_\Omega \nabla\times E^\varepsilon \cdot \nabla\times\overline{V}\, d\mathbf{x} + \int_{\Omega_r^\varepsilon} (-\omega^2\varepsilon^\eta\epsilon_r + i\,\omega\Sigma_r^\varepsilon)E^\varepsilon \cdot \overline{V}\, d\mathbf{x} = 0, \tag{193}$$

and

$$\forall\ V \in \mathbf{X}^\varepsilon(\Omega),\quad \int_\Omega \nabla\times E^\varepsilon \cdot \nabla\times\overline{V}\, d\mathbf{x} + \int_{\Omega_c^\varepsilon} (-\omega^2\varepsilon^\eta\epsilon_c + i\,\omega\Sigma_c^\varepsilon)E^\varepsilon \cdot \overline{V}\, d\mathbf{x} = 0. \tag{194}$$

By adding these three integrals, we get the variational formulation (83) associated with the problem ((63), (66), (67)).

Taking the divergence of the first three equations of (84) we get (69).

## Conflicts of Interest

All authors declare no conflicts of interest in this paper.

## References

1. T. Abboud and I. Terrasse, *Modélisation des phénomènes de propagation d'ondes,* Centre Poly-Média de l'école Polytechnique, 2007.
2. Y. Amirat, K. Hamdache and A. Ziani, *Homogénéisation d'équations hyperboliques du premier ordre et application aux écoulements missibles en milieux poreux,* Ann. Inst. H. Poincaré, **6** (1989), 397-417.

3. G. Allaire, *Homogenization and Two-scale Convergence,* SIAM Journal on Mathematical Analysis, **23** (1992), 1482-1518.
4. G. Allaire and M. Briand, *Multiscale convergence and reiterated homogenization,* Roy.Soc.Edinburgh, **126** (1996), 297-342.
5. Y. Amirat and V. Shelukhin, *Homogenization of time-harmonic Maxwell equations and the frequency dispersion effect,* J.Maths.Pures.Appl., **95** (2011), 420-443.
6. A. Back and E. Frenod, *Geometric Two-Scale Convergence on Manifold and Applications to the Vlasov Equation Discrete and Continuous Dynamical Systems - Serie S. Special Issue on Numerical Methods based on Homogenization and Two-Scale Convergence,* **8** (2015), 223-241.
7. S. Berthier, *Optique des milieux composites,* Ed. Polytechnicia, 1993.
8. D. Cionarescu and P. Donato, *An introduction to homogenization,* Oxford University Press., 1999.
9. M. Costabel, M. Dauge and S. Nicaise, *Corner Singularities of Maxwell interface and Eddy current problems,* Advances and Applications, **147** (2004), 241-256.
10. N. Crouseilles, E. Frenod, S. Hirstoaga and A. Mouton, *Two-Scale Macro-Micro decomposition of the Vlasov equation with a strong magnetic field,* Mathematical Models and Methods in Applied Sciences, **23** (2012), 1527-1559.
11. E. Frénod, P. A. Raviart and E. Sonnendrücker, *Asymptotic Expansion of the Vlasov Equation in a Large External Magnetic Field*, J. Math. Pures et Appl. **80**, (2001), 815-843.
12. S. Guenneau, F. Zolla and A. Nicolet, *Homogenization of 3D finite photonic crystals with heterogeneous permittivity and permeability,* Waves in Random and Complex Media, **17** (2007), 653-697.
13. P.R.P. Hoole and S.R.H. Hoole, *Guided waves along an unmagnetized lightning plasma channel,* IEEE Transactions on Magnetics, **24** (1998), 3165-3167.
14. P.R.P. Hoole, S.R.H. Hoole, S. Thirukumaran, R. Harikrishnan, K. Jeevan and K. Pirapaharan, *Aircraft-lightning electrodynamics using the transmission line model part I: review of the transmission line model,* Progress In Electromagnetics Research M, **31**, (2013), 85-101.
15. P. Laroche, P. Blanchet, A. Delannoy, and F. Issac, *Experimental Studies of Lightning Strikes to Aircraft,* JOURNAL AEROSPACELAB, **112** (2012).
16. M. Leboulch, *Analyse spectrale VHF, UHF du rayonnement deséclairs,* Hamelin, CENT.
17. J.C. Maxwell, *A dynamical theory of the Electromagnetic Field,* Phisophical transacting of the Royal Society of London, (1885), 459-512.
18. P. Monk, *Finite Element Methods for Maxwell's Equations,* Oxford Science publication, Numerical Mathematics and scientific computation, Clarendon Press - Oxford, 2003.
19. J.C. Nédélec, *Acoustic and electromagnetic equations; integral representations for harmonic problems,* Springer-Verlag, Berlin, 2001.
20. M. Neuss-Radu, *Some extensions of two-scale convergence,* Comptes rendus de l'Academie des sciences, **322** (1996), 899-904.
21. G. Nguetseng. *A General Convergence Result for a Functional Related to the Theory of Homogenization,* **20** (1989), 608-623.

22. G. Nguetseng, *Asymptotic Analysis for a Stiff Variational Problem Arising in Mechanics,* SIAM Journal on Mathematical Analysis, **21** (1990), 1394-1414.
23. S. Nicaise, S. Hassani and A. Maghnouji, *Limit behaviors of some boundary value problems with high and/or low valued parameters,* Advances in differential equations, **14** (2009), 875-910.
24. O. Ouchetto, S. Zouhdi and A. Bossavit et al., *Effective constitutive parameters of periodic composites,* Microwave conference, European, **2** (2005).
25. H.E. Pak, *Geometric two-scale convergence on forms and its applications to Maxwell's equations,* Proceedings of the Royal Society of Edinburgh, European, **135A** (2005), 133-147.
26. N. Wellander, *Homogenization of the Maxwell equations: Case I. Linear theory,* Appl Math, **46** (2001), 29-51.
27. N. Wellander, *Homogenization of the Maxwell equations: Case II. Nonlinear conductivity,* Appl Math, **47** (2002), 255-283.
28. N. Wellander and B. Kristensson, *Homogenization of the Maxwell equations at fixed frequency,* Technical Report, (2002), 1-37.
29. Pr. Welter, *Cours : Matériaux diélectriques,* Master Matériaux, Institut Le Bel.

# 14

# On the Sum of Unitary Divisors Maximum Function

**Bhabesh Das**[1,*] **and Helen K. Saikia**[2]

[1] Department of Mathematics, B.P.Chaliha College, Assam-781127, India
[2] Department of Mathematics, Gauhati University, Assam-781014, India

[*] **Correspondence:** Email: mtbdas99@gmail.com

**Abstract:** It is well-known that a positive integer $d$ is called a unitary divisor of an integer $n$ if $d|n$ and $\gcd\left(d, \frac{n}{d}\right) = 1$. Divisor function $\sigma^*(n)$ denote the sum of all such unitary divisors of $n$. In this paper we consider the maximum function $U^*(n) = \max\{k \in \mathbb{N} : \sigma^*(k)|n\}$ and study the function $U^*(n)$ for $n = p^m$, where $p$ is a prime and $m \geq 1$.

**Keywords:** Unitary Divisor function; Smarandache function; Fermat prime

## 1. Introduction

Any function whose domain of definition is the set of positive integers is said to be an arithmetic function. Let $f : \mathbb{N} \to \mathbb{N}$ be an arithmetic function with the property that for each $n \in \mathbb{N}$ there exists at least one $k \in \mathbb{N}$ such that $n|f(k)$. Let

$$F_f(n) = \min\{k \in \mathbb{N} : n|f(k)\} \tag{1.1}$$

This function generalizes some particular functions. If $f(k) = k!$, then one gets the well known Smarandache function, while for $f(k) = \frac{k(k+1)}{2}$ one has the Pseudo Smarandache function [1, 5, 6]. The dual of these two functions are defined by J. Sandor [5, 7]. If $g$ is an arithmetic function having the property that for each $n \in \mathbb{N}$ , there exists at least one $k \in \mathbb{N}$ such that $g(k)|n$, then the dual of $F_f(n)$ is defined as

$$G_g(n) = \max\{k \in \mathbb{N} : g(k)|n\} \tag{1.2}$$

The dual Smarandache function is obtained for $g(k) = k!$ and for $g(k) = \frac{k(k+1)}{2}$ one gets the dual Pseudo-Smarandache function. The Euler minimum function has been first studied by P. Moree and H. Roskam [4] and it was independently studied by Sandor [11]. Sandor also studied the maximum and minimum functions for the various arithmetic functions like unitary toitent function $\varphi^*(n)$ [9], sum of divisors function $\sigma(n)$, product of divisors function $T(n)$ [10] , the exponential totient function $\varphi^e(n)$ [8].

## 2. Preliminary

A positive integer $d$ is called a unitary divisor of $n$ if $d|n$ and $\gcd\left(d, \frac{n}{d}\right) = 1$ . The notion of unitary divisor related to arithmetical function was introduced by E.Cohen[3]. If the integer $n > 1$ has the prime factorization $n = p_1^{\alpha_1} p_2^{\alpha_2}.....p_r^{\alpha_r}$ ,then $d$ is a unitary divisor of $n$ if and only if $d = p_1^{\beta_1} p_2^{\beta_2}.....p_r^{\beta_r}$ ,where $\beta_i = 0$ or $\beta_i = \alpha_i$ for every $i \in \{1, 2, 3....r\}$ . The unitary divisor function, denoted by $\sigma^*(n)$ , is the sum of all positive unitary divisors of $n$. It is to noted that $\sigma^*(n)$ is a multiplicative function. Thus $\sigma^*(n)$ satisfies the functional condition $\sigma^*(nm) = \sigma^*(n)\sigma^*(m)$ for $\gcd(m, n) = 1$. If $n > 1$ has the prime factorization $n = p_1^{\alpha_1} p_2^{\alpha_2}.....p_r^{\alpha_r}$, then we have $\sigma^*(n) = \sigma^*(p_1^{\alpha_1})\sigma^*(p_2^{\alpha_2})...\sigma^*(p_r^{\alpha_r}) = (p_1^{\alpha_1} + 1)(p_2^{\alpha_2} + 1)....(p_r^{\alpha_r} + 1)$ In this paper, we consider the case (1.2) for the unitary divisor function $\sigma^*(n)$ and investigate various characteristics of this function. In (1.2), taking $g(k) = \sigma^*(k)$ we define maximum function as follows

$$U^*(n) = \max\{k \in \mathbb{N} : \sigma^*(k)|n\}$$

First we discuss some preliminary results related to the function $\sigma^*(n)$.

**Lemma 2.1.** Let $n \geq 2$ be a positive integer and let $r$ denote the number of distinct prime factors of $n$. Then

$$\sigma^*(n) \geq (1 + n^{\frac{1}{r}})^r \geq 1 + n$$

**Proof.** Let $n = p_1^{\alpha_1} p_2^{\alpha_2}.....p_r^{\alpha_r}$ be the prime factorization of the natural number $n \geq 2$ , where $p_i$ are distinct primes and $\alpha_i \geq 0$ . For any positive numbers $x_1, x_2, x_3..., x_r$ by Huyggens inequality,we have $((1 + x_1)(1 + x_2)(1 + x_r))^{\frac{1}{r}} \geq 1 + (x_1 x_2..x_r)^{\frac{1}{r}}$ For $i = 1, 2, ..r$, putting $x_i = p_i^{\alpha_i}$ in the above inequality, we obtain $\sigma^*(n)^{\frac{1}{r}} \geq 1 + n^{\frac{1}{r}}$, giving $\sigma^*(n) \geq (1 + n^{\frac{1}{r}})^r$ . Again for any numbers $a, b \geq 0, r \geq 1$, from binomial theory we have $(a + b)^r \geq a^r + b^r$. Therefore we obtain $\sigma^*(n) \geq (1 + n^{\frac{1}{r}})^r \geq 1 + n$. Thus for all $n \geq 2$, we have $\sigma^*(n) \geq 1 + n$. The equality holds only when $n$ is a prime or $n$ is power of a prime.

**Remark 2.1.** From the lemma 2.1,for all $k \geq 2$ we have $\sigma^*(k) \geq k + 1$ and from $\sigma^*(k)|n$ it follows that $\sigma^*(k) \leq n$, so $k + 1 \leq n$ . Thus $U^*(n) \leq n - 1$. Therefore the maximum function $U^*(n)$ is finite and well defined.

**Lemma 2.2.** Let $p$ be a prime. The equation $\sigma^*(x) = p$ has solution if and only if $p$ is a Fermat prime.

**Proof.** If $x$ is a composite number with at least two distinct prime factors, then $\sigma^*(x)$ is also a composite number. Therefore, for any composite number $x$ with at least two distinct prime factors, $\sigma^*(x) \neq p$, a prime. So $x$ must be of the form $x = q^\alpha$ for some prime $q$. Thus $x = q^\alpha$ gives $\sigma^*(x) = q^\alpha + 1 = p$ if and only if $q^\alpha = p - 1$. If $p = 2$, then $q = 1$ and $\alpha = 1$ , which is impossible , so $p$ must be an odd prime . If $p \geq 3$, then $p - 1$ is even, so we must have $q = 2$, i.e., $p = 2^\alpha + 1$. It is clear that such prime exists when $\alpha$ is a power of 2 giving thereby that $p$ is Fermat prime (see [2], page-236).

**Lemma 2.3.** Let $p$ be a prime. The equation $\sigma^*(x) = p^2$ only has the following two solutions: $x = 3, p = 2$ and $x = 8, p = 3$.

**Proof.** Let $x = p_1^{\alpha_1} p_2^{\alpha_2}.....p_r^{\alpha_r}$ be solution of $\sigma^*(x) = p^2$, then $(1 + p_1^{\alpha_1})(1 + p_2^{\alpha_2})....(1 + p_r^{\alpha_r}) = p^2$ if and only if

(a) $p_1^{\alpha_1} + 1 = p^2$
(b)$p_1^{\alpha_1} + 1 = 1,\ p_2^{\alpha_2} + 1 = p^2$
(c)$p_1^{\alpha_1} + 1 = 1,\ p_2^{\alpha_2} + 1 = p\ ,\ p_3^{\alpha_3} + 1 = p$
(d)$p_1^{\alpha_1} + 1 = p,\ p_2^{\alpha_2} + 1 = p$

Since $p_i$ are distinct primes, so the cases (b), (c) and (d) are impossible. Therefore only possible case is (a). If $x$ is odd, then from the case (a) we must have only $p = 2$. In this case we have $p_1 = 3$, $\alpha_1 = 1$, so $x = 3$. If $x$ is even then only possibility is $p_1 = 2$, so from the case (a),we have $2^{\alpha_1} = p^2 - 1$ , then $2^{\alpha_1} = (p-1)(p+1)$, giving the equations $2^a = p - 1$, $2^b = p + 1$, where $a + b = \alpha_1$. Solving we get $2^b = 2(1 + 2^{a-1})$ and $p = 2^{a-1} + 2^{b-1}$. Since $2^b = 2(1 + 2^{a-1})$ is possible only when $b = 2$ and $a = 1$, therefore $p = 2^{a-1} + 2^{b-1}$ gives $p = 3$. Thus $\alpha_1 = 3$ and hence $x = 8$.

**Lemma 2.4.** Let $p$ be a prime. The equation $\sigma^*(x) = p^3$ has a unique solution: $x = 7$, $p = 2$.

**Proof.** Proceeding as the lemma 2.3, we are to find the solution of the equation $p_1^{\alpha_1} + 1 = p^3$. If $p_1$ is odd, then only possible value of $p$ is 2 and in that case the solution is $x = 7$. If $p_1$ is even, then $p_1 = 2$ and $2^{\alpha_1} = p^3 - 1$ . In that case $p \neq 2$ and hence $p$ is an odd prime. Since $p^3 - 1 = (p-1)(p^2 + p + 1)$ and $p^2 + p + 1$ is odd for any prime $p$, hence $2^{\alpha_1} \neq p^3 - 1$ .

**Lemma 2.5.** Let $k > 1$ be an integer. The equation $\sigma^*(x) = 2^k$ is always solvable and its solutions are of the form $x$ = Mersenne prime or $x$ = a product of distinct Mersenne primes.

**Proof.** Let $x = p_1^{\alpha_1} p_2^{\alpha_2} .....p_r^{\alpha_r}$, then $(1 + p_1^{\alpha_1})(1 + p_2^{\alpha_2})....(1 + p_r^{\alpha_r}) = 2^k$ , which gives $(1 + p_1^{\alpha_1}) = 2^{k_1}$ , $(1 + p_2^{\alpha_2}) = 2^{k_2}$ , ....$(1 + p_r^{\alpha_r}) = 2^{k_r}$, where $k_1 + k_2 + ... + k_r = k$. Clearly each $p_i$ is odd. Now we consider the equation $p^\alpha = 2^a - 1$ ,$(a > 1)$ .

If $\alpha = 2m$ is an even and $p \geq 3$, then $p$ must be of the form $4h \pm 1$ and $p^2 = 16h \pm 8h + 1 = 8h(2h \pm 1) + 1 = 8j + 1$. Therefore $p^{2m} + 1 = (8j+1)^m + 1 = (8r+1) + 1 = 2(4r+1) \neq 2^a$. If $\alpha = 2m+1$, $(m \geq 0)$, then $p^{2m+1} + 1 = (p+1)(p^{2m} - p^{2m-1} + ... - p + 1)$. Clearly the expression $p^{2m} - p^{2m-1} + ... - p + 1$ is odd. Thus $p^\alpha + 1 \neq 2^a$, when $\alpha = 2m + 1$,$(m > 0)$ . If $m = 0$, then $p = 2^a - 1$, a prime. Any prime of the form $p = 2^a - 1$ is always a Mersenne prime. Thus each $p_i$ is Mersenne prime. Hence the lemma is proved.

**Lemma 2.6.** Let $p$ be a prime and $k > 2$ be an integer. The equation $\sigma^*(x) = p^k$ has solution only for $p = 2$.

**Proof.** Let $x = p_1^{\alpha_1} p_2^{\alpha_2} .....p_r^{\alpha_r}$ .Then proceeding as the lemma 2.5,we have to solve the equation of the form $q^\alpha + 1 = p^k$, where $q$ is a prime. If $q$ is odd , then $q^\alpha = p^k - 1$ must be odd. This is possible only when $p = 2$ and $\alpha = 1$. In that case $\sigma^*(x) = 2^k$ and by the lemma 2.5, this equation is solvable. If $q$ is even, then only possibility is $q = 2$ and $2^\alpha = p^k - 1$ .One can easily show that this equation has no solution for $k > 2$ . It is clear that $p$ is an odd prime. If $k = 2m + 1, (m > 0)$ is an odd, then $p^{2m+1} - 1 = (p-1)(p^{2m} + p^{2m-1} + ... + p + 1)$.Since the expression $p^{2m} + p^{2m-1} + ... + p + 1$ gives an odd number, so in that case $p^{2m+1} - 1 \neq 2^\alpha$ . If $k = 2m + 2, (m > 0)$ is an even (since $k > 2$), then $p^{2m+2} - 1 = 2^\alpha$. This equation gives $p^{m+1} - 1 = 2^a$ and $p^{m+1} + 1 = 2^b$,where $a + b = \alpha$. Solving we obtain $2^b - 2^a = 2$. The last equation has only solution $a = 1$,$b = 2$. Therefore we get $\alpha = 3$. For $\alpha = 3$, the equation $p^{2m+2} - 1 = 2^\alpha$ strictly implies that $m = 0$. But by our assumption $k > 2$. Hence the lemma is proved.

## 3. Results

In this section we discuss our main results.

Following result follows from the definition of $U^*(n)$

**Theorem 3.1.** For all $n \geq 1$, $\sigma^*(U^*(n)) \leq n$.

**Theorem 3.2.** For all $n \geq 2$, $U^*(n) \leq n - 1$

**Proof.** From the lemma 2.1,for all $k \geq 2$ ,we have $\sigma^*(k) \geq k + 1$ . Putting $k = U^*(n)$, one can get $\sigma^*(U^*(n)) \geq U^*(n) + 1$ . Using the theorem 3.1,we obtain $n \geq \sigma^*(U^*(n)) \geq 1 + U^*(n)$, for all $n \geq 2$ .

**Theorem 3.3.** If $p$ is a prime and $\alpha \geq 1$ , then $U^*(p^\alpha + 1) = p^\alpha$

**Proof.** Since for any prime power $p^\alpha$ , we have $\sigma^*(p^\alpha) = p^\alpha + 1$ , so we can write $\sigma^*(p^\alpha)|p^\alpha + 1$. Therefore from the definition of $U^*(n)$, we get $p^\alpha \leq U^*(p^\alpha + 1)$ ,for all $\alpha \geq 1$ .Putting $n = p^\alpha + 1$ in the inequality of the theorem 3.2, we get $U^*(p^\alpha + 1) \leq p^\alpha$ .

**Theorem 3.4.** For $i = 1, 2, ...., r$, let $p_i$ be distinct primes . If $n$ be a positive integer such that $(1 + p_1^{\alpha_1})(1 + p_2^{\alpha_2})....(1 + p_r^{\alpha_r})|n$ , where $\alpha_i \geq 1$ ,then $U^*(n) \geq p_1^{\alpha_1} p_2^{\alpha_2}...p_r^{\alpha_r}$

**Proof.** Since $\sigma^*(p_1^{\alpha_1} p_2^{\alpha_2}...p_r^{\alpha_r}) = (1 + p_1^{\alpha_1})(1 + p_2^{\alpha_2})....(1 + p_r^{\alpha_r})|n$, so from the definition of $U^*(n)$, the result follows.

**Theorem 3.5.**

$$U^*(p) = \begin{cases} 2^m, & \text{if } p = 2^m + 1 \text{ is Fermat prime,} \\ 1, & \text{if } p = 2 \text{ or } p \text{ is not Fermat prime} \end{cases}$$

**Proof.** We have $\sigma^*(k)|p$, when $\sigma^*(k) = p$ or $\sigma^*(k) = 1$ . Thus from the lemma 2.2 and the definition of $U^*(n)$ the result follows.

**Theorem 3.6.**

$$U^*(p^2) = \begin{cases} 3, & \text{if } p = 2, \\ 8, & \text{if } p = 3 \\ 2^m, & \text{if } p = 2^m + 1 > 3 \text{ is Fermat prime,} \\ 1, & \text{if } p \text{ is not Fermat prime} \end{cases}$$

**Proof.** The result follows from the lemma 2.3 and the definition of $U^*(n)$.

**Theorem 3.7.**

$$U^*(p^3) = \begin{cases} 7, & \text{if } p = 2, \\ 8, & \text{if } p = 3 \\ 2^m, & \text{if } p = 2^m + 1 > 3 \text{ is Fermat prime,} \\ 1, & \text{if } p \text{ is not Fermat prime} \end{cases}$$

**Proof.** The result follows from the lemma 2.4 and the definition of $U^*(n)$.

**Theorem 3.8.** $U^*(2^t) = g$, where $g$ is the greatest product $(2^{p_1} - 1)(2^{p_2} - 1)...(2^{p_r} - 1)$ of Mersenne primes, where $p_1 + p_2 + ... + p_r \leq t$ .

**Proof.** Let $\sigma^*(k)|2^t$,then $\sigma^*(k) = 2^a$, where $0 \leq a \leq t$ .From the definition of $U^*(n)$ and the lemma 2.5, the greatest value of such $k$ is $k = g$, where $g = (2^{p_1} - 1)(2^{p_2} - 1)...(2^{p_r} - 1)$, with $p_1 + p_2 + ... + p_r \leq t$

.

**Example 3.1.** For $n = 2^8$, $p_1 + p_2 + ... + p_r = 8$ , so we get $p_1 = 3$, $p_2 = 5$. Therefore $g = (2^{p_1} - 1)(2^{p_2} - 1) = 217$, i.e. $U^*(2^8) = 217$ .

**Theorem 3.9.** For $k > 3$,

$$U^*(p^k) = \begin{cases} g, & \text{if } p = 2, \text{ where } g \text{ is given in the theorem 3.8,} \\ 8, & \text{if } p = 3, \\ 2^m, & \text{if } p = 2^m + 1 > 3 \text{ is Fermat prime,} \\ 1, & \text{if } p \text{ is not Fermat prime} \end{cases}$$

**Proof.** The result follows from the lemma 2.6 and the definition of $U^*(n)$.

**Corollary 3.10.** For any $a \geq 1$, $U^*(7^a) = 1, U^*(11^a) = 1$, $U^*(13^a) = 1, U^*(19^a) = 1$ etc.

**Theorem 3.11.** For $a \geq 1$,any number of the form $n = (2^m + 1)(2^p - 1)^a$, $U^*(n) = 2^l$ for some $l$, where $2^m + 1$ is Fermat prime and $2^p - 1$ is Mersenne prime.

**Proof.** Since 3 is the only prime which is both Mersenne and Fermat prime, so in that case for $a \geq 1$, $n = 3^{a+1}$ from the theorem 3.9, it follows that $U^*(n) = 2^3$. For $n \neq 3^{a+1}$, if $\sigma^*(k)|n = (2^m + 1)(2^p - 1)^a$, then the only possibility is $\sigma^*(k)|2^m + 1$. Therefore the result follows from the lemma 2.2.

**Example 3.2.** $U^*(35) = 2^2$, $U^*(51) = 2^4$, $U^*(7967) = 2^8$.

## 4. Conclusion

We study the maximum function $U^*(n)$ in detail and determine the exact value of $U^*(n)$ if $n$ is prime power. There is also a scope for the study of the function $U^*(n)$ for other values of $n$.

## Acknowledgement

We are grateful to the anonymous referee for reading the manuscript carefully and giving us many insightful comments.

## Conflict of Interest

All authors declare no conflicts of interest in this paper.

## References

1. C. Ashbacker, An introduction to the Smarandache function, Erhus Univ. Press ,Vail, AZ, 1995.
2. David M. Burton, Elementary number theory, Tata McGraw-Hill Sixth Edition, 2007.
3. E. Cohen, *Arithmetical functions associated with the unitary divisors of an integer*. Math. Zeits., **74** (1960), 66-80.
4. P. Moree, H. Roskam, *On an arithmetical function related to Euler's totient and the discriminator*. Fib. Quart., **33** (1995), 332-340.
5. J. Sandor, *On certain generalizations of the Smarandnche function*. Notes Num. Th. Discr. Math., **5** (1999), 41-51.
6. J. Sandor, *A note on two arithmetic functions*. Octogon Math. Mag., **8** (2000), 522-524.
7. J. Sandor, *On a dual of the Pseudo-Smarandache function*. Smarandnche Notition Journal, **13** (2002).
8. J. Sandor, *A note on exponential divisors and related arithmetic functions*. Sci. Magna, **1** (2005), 97-101.

9. J. Sandor, *The unitary totient minimum and maximum functions*. Sci. Studia Univ. "Babes-Bolyai", Mathematica, **2** (2005), 91-100.

10. J. Sandor, *The product of divisors minimum and maximum function*. Scientia Magna, **5** (2009), 13-18.

11. J. Sandor, *On the Euler minimum and maximum functions*. Notes Num. Th. Discr. Math., **15** (2009), 1-8.

# 15

# Approximation of solutions of multi-dimensional linear stochastic differential equations defined by weakly dependent random variables

**Hiroshi Takahashi**[1,*] **and Ken-ichi Yoshihara**[2]

[1] Department of Mathematics, Tokyo Gakugei University, Koganei, Tokyo, 184-8501, Japan
[2] Department of Mathematics, Yokohama National University, Hodogaya, Yokohama, 240-8501, Japan

[*] **Correspondence:** hstaka@u-gakugei.ac.jp

**Abstract:** It is well-known that under suitable conditions there exists a unique solution of a $d$-dimensional linear stochastic differential equation. The explicit expression of the solution, however, is not given in general. Hence, numerical methods to obtain approximate solutions are useful for such stochastic differential equations. In this paper, we consider stochastic difference equations corresponding to linear stochastic differential equations. The difference equations are constructed by weakly dependent random variables, and this formulation is raised by the view points of time series. We show a convergence theorem on the stochastic difference equations.

**Keywords:** Difference equation; weakly dependent random variables; Euler-Maruyama scheme; strong invariance principle; linear stochastic differential equation
**Mathematics Subject Classification:**Primary: 39A50; Secondary: 60H10, 91G60

## 1. Introduction

Stochastic differential equations have been used to describe stochastic models in many areas. Under suitable conditions, a stochastic differential equation has a unique solution, and the representation of the solution is given by a stochastic integral. To express the explicit solutions of stochastic differential equations, numerical methods on stochastic differential equations have been well established. For example, we refer [1] and [3].

In the theory of mathematical finance, Black-Scholes type stochastic differential equation is an important example. Let $\{W(t)\}$ be a standard one-dimensional Wiener process. For fixed constants $\mu \in \mathbb{R}$ and $\sigma > 0$, Black-Scholes model is given by the following stochastic differential equation:

$$dX(t) = X(t)(\nu dt + \sigma dW(t)),\ 0 \leq t \leq T, \tag{1.1}$$

where $T$ is a maturity. Fixing $n \in \mathbb{N}$, we observe the stochastic differential equation at times $\{t_k = kT/n, k = 1, 2, \ldots, n\}$. To simplify the problem, we consider the case where $\nu = 0$. By the Euler-Maruyama scheme, we can consider an approximation of the solution of (1.1) such that

$$X(t_k) - X(t_{k-1}) \simeq \sigma X(t_{k-1})\{W(t_k) - W(t_{k-1})\},\ k = 1, 2, \ldots, n,$$

where $\simeq$ means *nearly equal to*. We rewrite the approximation above such that

$$\sigma\{W(t_k) - W(t_{k-1})\} \simeq \frac{X(t_k) - X(t_{k-1})}{X(t_{k-1})}.$$

In mathematical finance, we regard the difference $W(t_k) - W(t_{k-1})$ as a rate of returns. For the Black-Scholes model, the difference is given by a sequence of independent and identically distributed random variables. If we consider a stochastic difference equation corresponding to the stochastic differential equation (1.1), then time series analysis for the difference can be investigated. From the view of time series analysis, it is natural to consider the model based on dependent data. For the model, the difference should be given by a sequence of dependent random variables.

In [7], Yoshihara studied a stochastic differential equation such that

$$X(t_k) - X(t_{k-1}) = (\nu + \sigma\xi_{k-1})X(t_{k-1}),$$

where $\{\xi_k\}$ is a strictly stationary stochastic sequence. This difference equation corresponds to a Black-Scholes type stochastic differential equation. Under suitable conditions on $\{\xi_k\}$, Yoshihara showed an almost sure convergence theorem for the solution of the difference equation above by using results related to strong Wiener approximations of partial sums of some dependent random variables. Yoshihara's result was extended to multi-dimensional cases in [4], and application to finance models were also considered in [5] and [6]. Following the previous studies, we consider strong approximation of linear stochastic differential equations by weakly dependent random variables in this paper.

Let $\{\xi_k\} = \{(\xi_{k,1}, \ldots, \xi_{k,d})\}$ be a strictly stationary sequence of $d$-dimensional centered random vectors defined on a complete probability space $(\Omega, \mathcal{F}, P)$ and satisfies the strong mixing condition

$$\alpha(n) = \sup_{A\in\mathbf{M}_{-\infty}^{0}, B\in\mathbf{M}_{n}^{\infty}} |P(AB) - P(A)P(B)| \to 0 \quad (n \to \infty), \tag{1.2}$$

where $\mathbf{M}_a^b (a < b)$ denotes the $\sigma$-algebra generated by $\{\xi_a, \ldots, \xi_b\}$. Under some conditions on $\{\xi_k\}$, which shall be mentioned below, there exists a $d \times d$ matrix $\mathbf{\Gamma} = (\gamma_{q,q'})_{q,q'=1,\ldots,d}$ such that

$$\begin{cases} \gamma_{q,q} &= E\xi_{1,q}^2 + 2\sum_{i=2}^{\infty} E\xi_{1,q}\xi_{i,q}, \\ \gamma_{q,q'} &= E\xi_{1,q}\xi_{1,q'} + \sum_{i=2}^{\infty}(E\xi_{1,q}\xi_{i,q'} + \xi_{1,q'}\xi_{i,q}). \end{cases} \tag{1.3}$$

We write

$$\mathbf{R} = (r_{q,q'})_{q,q'=1,\ldots,d} = (E\xi_{1,q}\xi_{1,q'})_{q,q'=1,\ldots,d}. \tag{1.4}$$

We remark that if $\{\xi_k\}$ is a sequence of independent and identically distributed $\mathbb{R}^d$-valued random variables, then $\Gamma$ equals to $\mathbf{R}$. In the paper, we always assume that $d \times d$ matrix $\Gamma$ and $\mathbf{R}$ are always positive definite. Let $\{\mathbf{W}(t), t \geq 0\} = \{(W_1(t), \ldots, W_d(t)), t \geq 0\}$ be a $d$-dimensional Wiener process with covariance matrix $\mathbf{\Gamma}$, i.e.,

$$E\mathbf{W}(t)\mathbf{W}(t)^{\mathsf{T}} = t\mathbf{\Gamma} \quad \text{for all } t \geq 0,$$

where $\mathbf{W}^{\mathsf{T}}$ is the transpose of the matrix $\mathbf{W}$ and set $\{\mathcal{F}_t, t \geq 0\}$ as a filtration generated by the Wiener process.

For a stationary sequence $\{\xi_k\}$, the following strong invariance principle was shown by Liu and Lin [2]. We remark that their result holds under more general conditions:

**Proposition 1.1.** *Let $\{\xi_k\}$ be a stationary strong mixing sequence of centered d-dimensional random vectors. Assume that*

(i) $E[\xi_k] = \mathbf{0}$,
(ii) *$\xi_k$ has the third moment,*
(iii) *for some $\tau > 0$ there exists positive constant c such that $\alpha(n) \leq cn^{-3(1+\tau)}$.*

*Then, on a richer probability space we can redefine the sequence $\{\xi_k\}$ together with a d-dimensional Brownian motion $\{\mathbf{W}(t)\}$ whose covariance matrix is $\mathbf{\Gamma}$ such that*

$$\left|\sum_{k\leq t} \xi_k - \mathbf{W}(t)\right| = O(t^{1/4}) \tag{1.5}$$

*almost surely, where $O(t)$ is the Landau notation, that is, $O(t)$ is any $\mathbb{R}$-valued sequence for which* $\limsup_{t\to\infty}(O(t)/t) < \infty$.

Afterward, for the strong approximation (1.5) of Proposition 1.1 we use a notation such that

$$\sum_{k\leq t} \xi_k \Rightarrow \mathbf{W}(t) \quad \text{a.s.}$$

We denote by $A_n \to A$ a.s. the almost surely convergence of random variables.

## 2. Setting and result

Let $A(t), B_1(t), \ldots, B_d(t)$ be $p \times p$-matrix functions and $a(t), b_1(t), \ldots, b_d(t)$ be $p$-dimensional vector functions. We assume that all components of the above matrix and vector functions have continuous bounded derivatives on $[0, \infty)$. On $(\Omega, \mathcal{F}, P)$, we denote a $d$-dimensional Wiener process with covariance matrix $\mathbf{\Gamma}$ by $\{\tilde{\mathbf{W}}(t), t \geq 0\} = \{(\tilde{W}_1(t), \ldots, \tilde{W}_d(t)), t \geq 0\}$. It is known that the $d$-dimensional linear stochastic differential equation

$$\begin{aligned} dX(t) &= \{A(t)X(t) + a(t)\}dt + \sum_{q=1}^{d}\{B_q(t)X(t) + b_q(t)\}d\tilde{W}_q(t), \\ X(0) &= \boldsymbol{x} \in \mathbb{R}^p \end{aligned} \tag{2.1}$$

has a unique solution $X(t)$, which is given by

$$\Phi(t)\left\{\boldsymbol{x}+\int_0^t \Phi(s)^{-1}\left(a(s)-\sum_{q=1}^{d} B_q(s)b_q(s)\right)ds+\sum_{q=1}^{d}\int_o^t \Phi(s)^{-1}b_q(s)d\tilde{W}_q(s)\right\}, \tag{2.2}$$

where $\Phi(t)$ denotes the solution of the matrix stochastic differential equation such that

$$\begin{aligned} d\Phi(t) &= A(t)\Phi(t)dt+\sum_{q=1}^{d} B_q(t)\Phi(t)d\tilde{W}_q(t), \\ \Phi(0) &= \boldsymbol{I}, \end{aligned} \tag{2.3}$$

where $\boldsymbol{I}$ denote the identity matrix.

We remark that unlike the scalar homogeneous linear equation, we cannot solve (2.3) explicitly for its fundamental solution $\Phi(t)$, even when all of the matrices are constant. If $A, B_1, \ldots, B_d$ are constant matrices and commutative, then we obtain the following explicit expression for the fundamental matrix solution:

$$\tilde{\Phi}(t)=\exp\left\{\left(A-\frac{1}{2}\mathbf{R}\sum_{q=1}^{d} B_q^2\right)t-\sum_{q=1}^{d} B_q\tilde{W}_q(t)\right\}.$$

See Section 2.2 in [1] for more detail. Hence, approximation of solution of stochastic differential equation is a critical tool.

We fix $T>0$ arbitrarily. For an arbitrary positive integer $n$, let $m=[n^{1/6}]$ and $l=[n/m]$, where $[x]$ is a floor function. Using them, we define time point for a sufficiently large $n$ such that

$$\begin{cases} s_{i,j} = \left(\frac{i}{l}+\frac{j}{lm}\right)T, \ i=0,\ldots,l-1, \ j=1,\ldots m, \\ s_{0,0} = 0. \end{cases} \tag{2.4}$$

We identify $s_{k,0}$ with $s_{k-1,m}$ for $k=1,\ldots,l$.

Corresponding to (2.1), we consider the following difference equation:

$$\begin{aligned} \Delta X^{(n)}(s_{i,j}) &= X^{(n)}(s_{i,j})-X^{(n)}(s_{i,j-1}) \\ &:= \left\{A(s_{i,j-1})X^{(n)}(s_{i,j-1})+a(s_{i,j-1})\right\}\frac{T}{lm} \\ &\quad +\sum_{q=1}^{d}\left\{B_q(s_{i,j-1})X^{(n)}(s_{i,j-1})+b_q(s_{i,j-1})\right\}\xi_{im+j,q}\sqrt{\frac{T}{lm}}, \\ X^{(n)}(0) &= \boldsymbol{x}\in\mathbb{R}^p. \end{aligned} \tag{2.5}$$

In addition, corresponding to (2.3), we also consider the following difference equation:

$$\begin{aligned} \Delta\Phi^{(n)}(s_{i,j}) &= \Phi^{(n)}(s_{i,j})-\Phi^{(n)}(s_{i,j-1}) \\ &:= \left\{A(s_{i,j-1})\frac{T}{lm}+\sum_{q=1}^{d} B_q(s_{i,j-1})\xi_{im+j,q}\sqrt{\frac{T}{lm}}\right\}\Phi^{(n)}(s_{i,j-1}), \\ \Phi^{(n)}(0) &= \boldsymbol{I}. \end{aligned} \tag{2.6}$$

For the stochastic difference equation (2.6), we obtain a convergence result by using (i) of Theorem 1 and Theorem 2 in [4].

**Proposition 2.1.** *We assume that a strictly stationary strong mixing sequence* $\{\xi\}$ *satisfies that (i)* $E[|\xi_i|^{6+\delta_1}] < \infty$ *with* $\delta_1 > 0$ *and (ii)* $\alpha(n) \le cn^{-3(1+\tau)}$ *with some* $\tau(\delta) = \tau > 0$. *Then, we have the following:*

*(1) The solution of (2.6) (we denote by* $\Phi^{(n)}(T)$*) is given by*

$$\prod_{i=0}^{l-1}\prod_{j=1}^{m}\exp\left\{I + A(s_{i,j-1})\frac{T}{lm} + \sum_{q=1}^{d} B_q(s_{i,j-1})\xi_{im+j,q}\sqrt{\frac{T}{lm}}\right\}.$$

*(2) On a richer probability space we can redefine the sequence* $\{\xi_k\}$ *together with a d-dimensional Brownian motion* $\{\mathbf{W}(t)\}$ *whose covariance matrix is* $\mathbf{\Gamma}$ *such that* $\Phi^{(n)}(T)$ *converges almost surely to a matrix* $\Phi(T)$ *which corresponds to the solution of the stochastic differential equation (2.3).*

Using the solutions $\Phi^{(n)}(T)$ and $\Phi(T)$, we obtain the solution of the stochastic difference (2.5) as follows:

**Theorem 2.1.** *Under the same assumptions as those in Proposition 2.1, we obtain the following:*

*(1) The solution of (2.5) (we denote by* $X^{(n)}(T)$*) is given by*

$$\Phi^{(n)}(T)\left[\boldsymbol{x} + \sum_{i=0}^{l-1}\sum_{j=1}^{m}\Phi^{(n)}(s_{i,j})^{-1}\left\{a(s_{i,j}) + \sum_{q,q'=1}^{d} B_q(s_{i,j-1})b_{q'}(s_{i,j-1})\xi_{im+j,q}\xi_{im+j,q'}\right\}\frac{T}{lm}\right.$$
$$\left. + \sum_{i=0}^{l-1}\sum_{j=1}^{m}\sum_{q=1}^{d}\Phi^{(n)}(s_{i,j-1})^{-1}b_q(s_{i,j-1})\xi_{im+j,q}\sqrt{\frac{T}{lm}}\right].$$

*(2) On a richer probability space we can redefine the sequence* $\{\xi_k\}$ *together with a p-dimensional Brownian motion* $\{\mathbf{W}(t)\}$ *whose covariance matrix is* $\mathbf{\Gamma}$ *such that* $X^{(n)}(T)$ *converges almost surely to a matrix* $X(T)$ *which corresponds to the solution of the stochastic differential equation (2.2).*

## 3. Proof

Assertion (1) is shown by constructing $X^{(n)}$ in assertion (2). We consider the following difference equation:

$$\begin{aligned}
\Delta Z^{(n)}(s_{i,j}) &= Z^{(n)}(s_{i,j}) - Z^{(n)}(s_{i,j-1}) \\
&:= \Phi^{(n)}(s_{i,j-1})^{-1}\left\{a(s_{i,j-1}) - \sum_{q,q'=1}^{d} B_q(s_{i,j-1})b_{q'}(s_{i,j-1})\xi_{im+j,q}\xi_{im+j,q'}\right\}\frac{T}{lm} \\
&\quad +\Phi^{(n)}(s_{i,j-1})^{-1}\sum_{q=1}^{d} b_q(s_{i,j-1})\xi_{im+j,q}\sqrt{\frac{T}{lm}}, \\
Z(0) &= \boldsymbol{x}.
\end{aligned}$$

Then for each $(s_{i,j})_{0\le i\le l-1, 1\le j\le m}$ we have that

$$\begin{aligned}
\Delta(\Phi^{(n)}Z^{(n)}) &= \Delta\Phi^{(n)}\,Z^{(n)} + \Phi^{(n)}\,\Delta Z^{(n)} + \Delta\Phi^{(n)}\,\Delta Z^{(n)} \\
&=: U_1^{(n)} + U_2^{(n)} + U_3^{(n)}. \qquad (3.1)
\end{aligned}$$

Firstly, we put that $X(t) := \Phi(t)Z(t)$. Then, Theorem 1 in [4] implies that

$$\sum_{i=0}^{l-1}\sum_{j=1}^{m} U_1^{(n)} = \sum_{i=0}^{l-1}\sum_{j=1}^{m}\left\{A(s_{i,j-1})\frac{T}{lm} + \sum_{q=1}^{d} B_q(s_{i,j-1})\xi_{im+j,q}\sqrt{\frac{T}{lm}}\right\}\Phi^{(n)}(s_{i,j-1})Z^{(n)}(s_{i,j-1})$$

$$\Rightarrow \int_0^T A(s)X(s)ds + \sum_{q=1}^{d}\int_0^T B_q(s)X(s)dW_q(s) \quad \text{a.s.} \tag{3.2}$$

Secondly, we consider that

$$\sum_{i=0}^{l-1}\sum_{j=1}^{m} U_2^{(n)} = \sum_{i=0}^{l-1}\sum_{j=1}^{m} a(s_{i,j})\frac{T}{lm} - \sum_{i=0}^{l-1}\sum_{j=1}^{m}\sum_{q,q'=1}^{d} B_q(s_{i,j-1})b_{q'}(s_{i,j-1})\xi_{im+j,q}\xi_{im+j,q'}\frac{T}{lm}$$

$$+\sum_{i=0}^{l-1}\sum_{j=1}^{m}\sum_{q=1}^{d} b_q(s_{i,j-1})\xi_{im+j,q}\sqrt{\frac{T}{lm}}$$

$$=: V_1^{(n)} - V_2^{(n)} + V_3^{(n)}. \tag{3.3}$$

Since $a(t)$ has continuous bounded derivatives on $[0,\infty)$, it is obvious that

$$V_1^{(n)} \to \int_0^T a(s)ds. \tag{3.4}$$

For $V_2^{(n)}$, we use the law of large numbers for $\{\xi_k\}$. Then, we obtain that

$$V_2^{(n)} \to \sum_{q,q'=1}^{d} r_{q,q'}\int_0^T B_q(s)b_{q'}(s)ds \quad \text{a.s.} \tag{3.5}$$

Further, using Theorem 1 in [4] again, we have that

$$V_3^{(n)} \Rightarrow \sum_{q=1}^{d}\int_0^T b_q(s)dW_q(s) \quad \text{a.s.} \tag{3.6}$$

Finally, we consider that

$$\sum_{i=0}^{l-1}\sum_{j=1}^{m} U_3^{(n)} = \sum_{i=0}^{l-1}\sum_{j=1}^{m}\left\{A(s_{i,j-1})\frac{T}{lm} + \sum_{q=1}^{d} B_q(s_{i,j-1})\xi_{im+j,q}\sqrt{\frac{T}{lm}}\right\}$$

$$\left[\left\{a(s_{i,j-1}) - \sum_{q,q'=1}^{d} B_q(s_{i,j-1})b_{q'}(s_{i,j-1})\xi_{im+j,q}\xi_{im+j,q'}\right\}\frac{T}{lm} + \sum_{q=1}^{d} b_q(s_{i,j-1})\xi_{im+j,q}\sqrt{\frac{T}{lm}}\right]. \tag{3.7}$$

Since under the condition of the theorem on $\{\xi_k\}$, there exists a covariant matrix $\mathbf{\Gamma}$. We thus obtain that the right hand side of (3.7) equals almost surely to

$$\frac{T}{lm}\sum_{i=0}^{l-1}\sum_{j=1}^{m}\sum_{q,q'=1}^{d} B_q(s_{i,j-1})b_{q'}(s_{i,j-1})\xi_{m(i-1)+j,q}\xi_{m(i-1)+j,q'} + O((lm)^{-3/2}). \tag{3.8}$$

To estimate the first term of the right hand side of (3.8), we consider the term of random variables. Then we obtain that

$$\frac{1}{lm}\sum_{i=0}^{l-1}\sum_{j=1}^{m}\xi_{m(i-1)+j,q}\xi_{m(i-1)+j,q'} \to r_{q,q'} \quad \text{a.s.} \tag{3.9}$$

In addition, all components of $B_q$ and $b_q (1 \leq q \leq p)$ have continuous bounded derivatives. Thus, there exists some $c > 0$ such that for any $1 \leq j \leq m$

$$\begin{aligned} & \|B_q(s_{i,j-1})b_{q'}(s_{i,j-1}) - B_q(s_{i,0})b_{q'}(s_{i,0})\| \\ & \leq \sum_{k=1}^{j-1} \|B_q(s_{i,k})b_{q'}(s_{i,k}) - B_q(s_{i,0})b_{q'}(s_{i,0})\| \\ & \leq \frac{c(j-1)}{lm} \leq \frac{c}{l}, \end{aligned}$$

which implies that

$$\frac{T}{lm}\sum_{i=0}^{l-1}\sum_{j=1}^{m} B_q(s_{i,j-1})b_{q'}(s_{i,j-1})\xi_{m(i-1)+j,q}\xi_{m(i-1)+j,q'} \to r_{q,q'}\int_0^T B_q(s)b_{q'}(s)ds \quad \text{a.s.} \tag{3.10}$$

Hence, by (3.9) and (3.10) we obtain that

$$\sum_{i=0}^{l-1}\sum_{j=1}^{m} \Delta\Phi^{(n)}(s_{i,j})\Delta Z^{(n)}(s_{i,j}) \to \sum_{q,q'=1}^{p} r_{q,q'}\int_0^T B_q(s)b_{q'}(s)ds \quad \text{a.s.} \tag{3.11}$$

Combining (3.1), (3.2), (3.3)-(3.6), (3.7) and (3.11), we obtain that

$$\begin{aligned} & \sum_{i=0}^{l-1}\sum_{j=1}^{m} \Delta(\Phi^{(n)}(s_{i,j})Z^{(n)}(s_{i,j})) \\ & \Rightarrow \int_0^T A(s)X(s)ds + \sum_{q=1}^{p} B_q(s)X(s)dW_q(s) + \int_0^T a(s)ds + \sum_{q=1}^{p}\int_0^T b_q(s)dW_q(s) \quad \text{a.s.} \\ & = \int_0^T (A(s)X(s) + a(s))ds + \sum_{q=1}^{p}\int_0^T (B_q(s)X(s) + b_q(s))dW_q(s) \quad \text{a.s.} \end{aligned}$$

On the other hand, the solution $X^{(n)}(T)$ of the difference equation

$$\begin{aligned} \Delta X^{(n)}(s_{i,j}) &= X^{(n)}(s_{i,j}) - X^{(n)}(s_{i,j-1}) \\ &:= \Delta(\Phi^{(n)}(s_{i,j})Z^{(n)}(s_{i,j})), \\ X^{(n)}(0) &= \boldsymbol{x} \end{aligned}$$

satisfies that

$$X^{(n)}(T) = \boldsymbol{x} + \sum_{i=0}^{l-1}\sum_{j=1}^{m} \Delta X^{(n)}(s_{i,j}).$$

Thus, we obtain that

$$X^{(n)}(T) \Rightarrow \boldsymbol{x} + \int_0^T (A(s)X(s) + a(s))ds + \sum_{q=1}^{p} \int_0^T (B_q(s)X(s) + b_q)dW_q(s) \quad \text{a.s,}$$

which shows assertion (2) of Theorem 2.1. □

## Acknowledgments

The authors thank Professor R. Naz, Professor M. Torrisi, Professor I. Naeem and Professor C.W. Soh for giving an opportunity to talk in AIMS 2016 Meeting, Orlando, Florida, USA. The research is supported by JSPS KAKENHI Grant number 26800063.

This paper had been almost finished before the second author, Ken-ichi Yoshihara, passed away. I express my gratitude to him for constant encouragement. I will cherish the memory of him forever. I thank the referee for his/her comments which help improve this paper.

## Conflict of Interest

All authors declare no conflicts of interest in this paper.

## References

1. P. E. Kloden and E. Platen, *Numerical solution of stochastic differential equations,* Springer-Verlag, Berlin Heidelberg, 1992.
2. W. Liu and Z. Lin, *Strong approximation for a class of stationary processes*, Stoch. Proc. Appl., **119** (2009), 249-280.
3. X. Mao, *Stochastic differential equations and applications,* second edition. Horwood Publishing Limited, Chichester, 2008.
4. H.Takahashi, S. Kanagawa and K. Yoshihara, *Asymptotic behavior of solutions of some difference equations defined by weakly dependent random vectors*, Stoch. Anal. Appl., **33** (2015), 740-755.
5. H.Takahashi, T. Saigo, S. Kanagawa and K. Yoshihara, *Optimal portfolios based on weakly dependent data*, Dyn. Syst. Differ. Equ. Appl., AIMS Proceedings, (2015), 1041-1049.
6. H.Takahashi, T. Saigo and K. Yoshihara, *Approximation of optimal prices when basic data are weakly dependent*, Dyn. Contin. Discrete Impuls. Syst. Ser. B, **23** (2016), 217-230.
7. K. Yoshihara, *Asymptotic behavior of solutions of Black-Scholes type equations based on weakly dependent random variables*, Yokohama Math. J., **58** (2012), 1-15.

# 16

# Critical blowup in coupled Parity-Time-symmetric nonlinear Schrödinger equations

**Edès Destyl, Silvere Paul Nuiro and Pascal Poullet***

LAMIA, Université des Antilles, Campus de Fouillole, BP 250, Pointe-à-Pitre F-97115 Guadeloupe F.W.I, France

* **Correspondence:** Email: pascal.poullet@univ-antilles.fr

**Abstract:** In this article, we obtain sufficient conditions to obtain finite time blowup in a system of two coupled nonlinear Schrödinger (NLS) equations in the critical case. This system mainly considered here in dimension 2, couples one equation including gain and the other one including losses, constituting a generalization of the model of pulse propagation in birefringent optical fibers. In the spirit of the seminal work of Glassey, the proofs used the virial technique arguments.

**Keywords:** Coupled nonlinear Schrödinger equations; Parity-Time symmetry; finite time blowup; generalized Manakov model

## 1. Introduction

Several applications need to solve and study nonlinear Schrödinger (NLS) equations. Hereafter, a basic model of propagation of weakly dispersive waves is considered by a system of coupled NLS equations which reads as follows:

$$\left\{ \begin{array}{rcl} \imath u_t &=& -\Delta u + \kappa v + \imath \gamma u - (g_{11}|u|^2 + g_{12}|v|^2)u, \\ \imath v_t &=& -\Delta v + \kappa u - \imath \gamma v - (g_{12}|u|^2 + g_{22}|v|^2)v, \end{array} \right. \tag{1}$$

with the coefficients of the nonlinear parts being real. The $\imath$ is the complex such that $\imath^2 = -1$ and the coefficients $\kappa$ and $\gamma$ are positive constants that characterize gain and loss in wave components. This model is known for its pertinence for several applications of nonlinear optics (as birefringent optics fiber) and has been studied by several authors [1, 2, 3, 4, 5, 6, 7, 8]. The parameter $\gamma$ does not influence the model by a damping phenomenon as it appears in each component of the system by opposite sign (see for example some studies of damped NLS [9, 10, 11]).

In addition, one recalls that if the parameter $\gamma$ was equal to zero, the density and the energy would be time invariants, and the model is known as the Hamiltonian version of the generalized Manakov system [5, 6]. In this paper, we focus onto giving sufficient conditions in order to predict that the solution of the Cauchy problem in the critical case blows up in finite time.

Let us recall that the system (1) is considered as a Parity-Time (PT) symmetric system as soon as the coefficients $g_{11} = g_{22}$, which means that if $g_{11} = g_{22}$, the following formal property holds: if $(u(x,t), v(x,t))$ solve the system (1), then the pair $(u_{PT}(x,t), v_{PT}(x,t) := (\bar{v}(x,-t), \bar{u}(x,t))$ also solves the same system (hereafter an overbar stands for the complex conjugation).

The concept of PT-symmetry first emerged from quantum mechanics with the study of pure real spectra of non-Hermitian operators [12]. But for around one decade this PT-symmetry property gained a particular relevance due to its importance in several other branches in Physics: optics, Bose-Einstein Condensates, plasmonic waveguides, electronic circuits, superconductivity,... [13, 7, 14] (and references therein). Indeed, the interplay between this property and the nonlinearity of the system seems to be at the heart of several phenomenon as the behaviour of a single atomic specie in two different ground states of Bose-Einstein Condensates [15], or the modulational instability of the carrier wave between two waveguides [14].

The existence of a unique global solution $(u(t), v(t)) \in C(\mathbf{R}, (H^1(\mathbf{R}))^2)$ of the Cauchy problem for the generalized Manakov system (1) in dimension 1 with $(u(0), v(0)) = (u_0, v_0) \in (H^1(\mathbf{R}))^2$ is known [6]. Also in 1D, we proved recently that if the symmetry is *unbroken* ($\gamma < \kappa$) the $H^1$-norm of the solution cannot blow up in finite time [8]. But in higher dimension, the answer is fully different as the problem is more complex. In the supercritical case, which corresponds to dimension greater or equal to 3, finite time blowup for the system (1) is obtained under suffisient conditions on the parameters [16].

But in the critical case, as far as we know, only partial results are available until now [14]. The existence of finite-time blowup for solutions of a single NLS equation in dimension $n \geq 2$ is well known [17]. Analysis of global existence and blow-up of solutions of the Hamiltonian version of the generalized Manakov system have been studied [18, 19]. And solutions of this Hamiltonian version without linear coupling are known to blow up in finite time in a specific way that is called $L^2$-concentration (mass concentration) [20]. In our focusing case, as soon as the PT-symmetry is unbroken ($\gamma < \kappa$) and the nonlinear coefficients $g_{11}, g_{22}$ and $g_{12}$ equals to 1, a global solution in $(H^1(\mathbf{R}^2))^2$ exists for initial solutions that provide a density remaining limited all over time [6].

Our main contribution is to adapt the sketch of proof that has been developed for the supercritical case by other authors [16]. But to develop it, we had to modify slightly the second time-derivative expression by substituting the first integral Stokes variable using the energy equality. Following that, the proof by contradiction can be led in the spirit of the technique that has been introduced by Glassey [21]. Assuming that the nonlinear coefficients are such that the quadradic form

$$\phi(x,y) = g_{11}x^4 + g_{22}y^4 + 2g_{12}x^2y^2$$

is negative definite, one obtains the non existence of the solution of the Cauchy problem in the interval $[0, T_0]$.

The outline of the paper is the following. The main results consist of gathering in the proposition, the computations of the time-derivative of the density, the time-derivative of the energy, and the first and second time-derivative of the mean square momentum. In a second proposition, we focus onto initial conditions of the Cauchy problem that provide the suffisient blowup conditions of the main theorem that will follow.

## 2. Main results

Let us consider here a local solution in energy space

$$(u(t), v(t)) \in C\left([-t_0, t_0], H^1(\mathbf{R}^2) \times H^1(\mathbf{R}^2)\right)$$

of the Cauchy problem (1)-$(u(0), v(0)) = (u_0, v_0)$, that is obtained by a modified contraction method with Stritcharz estimates [22].

Let us recall that in the Hamiltonian case (for $\gamma = 0$), it is known that for $(u, v) \in C(\mathbf{R}; H^1(\mathbf{R}^n) \times H^1(\mathbf{R}^n))$, the two following quantities are conserved, namely the density, which is defined by,

$$Q(t) = \int (|u|^2 + |v|^2)dx,$$

and the total energy of the system (1)

$$E(t) = \int \left(|\nabla u|^2 + |\nabla v|^2 + \kappa(\bar{u}v + u\bar{v}) - \frac{g_{11}}{2}|u|^4 - \frac{g_{22}}{2}|v|^4 - g_{12}|u|^2|v|^2\right)dx.$$

In the sequel, we also need to introduce the mean square momentum, and its time derivative

$$X(t) = \int |x|^2(|u|^2 + |v|^2)dx, \quad \text{and } Y(t) = \frac{dX(t)}{dt}.$$

In the next proposition some results are recalled for a convenient self-consistent corpus of this paper.

**Proposition 2.1.** *For $(u(t), v(t)) \in C(\mathbf{R}, (H^1(\mathbf{R}^n))^2)$ the solution of the Cauchy problem for the generalized Manakov system (1) with $(u(0), v(0)) = (u_0, v_0) \in (H^1(\mathbf{R}^n))^2$,*

$$\frac{dQ}{dt} = 2\gamma \int (|u|^2 - |v|^2)dx, \tag{2}$$

$$\frac{dE}{dt} = 2\gamma \int (|\nabla u|^2 - |\nabla v|^2 - g_{11}|u|^4 + g_{22}|v|^4)dx, \tag{3}$$

$$Y(t) = 4\mathcal{I}m \int ((\bar{u}x.\nabla u) + (\bar{v}x.\nabla v))dx + 2\gamma \int |x|^2(|u|^2 - |v|^2)dx. \tag{4}$$

$$\frac{dY}{dt} = 4nE(t) + 4(2-n)\int (|\nabla u|^2 + |\nabla v|^2)dx + 4\gamma^2 X(t) +$$
$$8\gamma\mathcal{I}m \int ((\bar{u}x.\nabla u) - (\bar{v}x.\nabla v))dx + 8\kappa\gamma\mathcal{I}m \int |x|^2(v\bar{u})dx - 8\kappa n\mathcal{R}e \int u\bar{v}dx. \tag{5}$$

*Proof.* The equalities (2) and (3) are obvious (see [6, 8]).
To prove equality (4), one needs to take the scalar product of the first equality of the system (1) with $2u$ and the second one with $2v$.

Taking the imaginary part of the first computation, one obtains:

$$\frac{\partial}{\partial t}|u|^2 = -2\mathcal{I}m(\bar{u}\Delta u) + 2\gamma|u|^2.$$

But, one can aslo write:

$$\frac{\partial}{\partial t}|u|^2 = -2\nabla.(\mathcal{I}m(\overline{u}\nabla u)) + 2\gamma|u|^2.$$

Taking the product with $|x|^2$ and after space integration over $\mathbf{R}^n$, one gets:

$$\frac{\partial}{\partial t}\int |x|^2|u|^2 dx = -2\int \nabla.(\mathcal{I}m(\overline{u}\nabla u))|x|^2 dx + 2\gamma \int |x|^2|u|^2 dx.$$

Following an integration by parts of the first integral of the right hand side, the previous equality becomes:

$$\frac{\partial}{\partial t}\int |x|^2|u|^2 dx = 4\int \sum_{k=1}^{n}(\mathcal{I}m(\overline{u}\nabla u))_k x_k dx + 2\gamma \int |x|^2|u|^2 dx.$$

In other words,

$$\frac{\partial}{\partial t}\int |x|^2|u|^2 dx = 4\mathcal{I}m \int (\overline{u}x.\nabla u) dx + 2\gamma \int |x|^2|u|^2 dx. \tag{6}$$

Processing in a similar way with the second equality of the system (1), one gets the following equality.

$$\frac{\partial}{\partial t}\int |x|^2|v|^2 dx = 4\mathcal{I}m \int (\overline{v}x.\nabla v) dx - 2\gamma \int |x|^2|v|^2 dx. \tag{7}$$

Adding the equalities (6) and (7), one recovers the equality (4).

To prove equality (5), one needs to start by taking the scalar product of the first equality of the system (1) with $(2x.\nabla u)$ and the second one with $(2x.\nabla v)$. Following that, after integrating over $\mathbf{R}^n$ each term of the expression that has been obtained, one takes the real part of their sum to obtain:

$$\begin{aligned}2\mathcal{R}e\int \imath\left[(x.\nabla\overline{u})\frac{\partial u}{\partial t} + (x.\nabla\overline{v})\frac{\partial v}{\partial t}\right]dx &= -2\mathcal{R}e\int [(x.\nabla\overline{u})\Delta u + (x.\nabla\overline{v})\Delta v]\,dx \\ &+ 2\kappa\mathcal{R}e\int [(x.\nabla\overline{u})v + (x.\nabla\overline{v})u]\,dx + 2\gamma\mathcal{R}e\int \imath\,[(x.\nabla\overline{u})u + (x.\nabla\overline{v})v]\,dx \\ &- 2\mathcal{R}e\int \left[(x.\nabla\overline{u})(g_{11}|u|^2 + g_{12}|v|^2)u + (x.\nabla\overline{v})(g_{22}|v|^2 + g_{12}|v|^2)v\right]dx.\end{aligned} \tag{8}$$

First, let us denote by $R_1$, $R_2$, $R_3$ and $R_4$, each term of the right hand side of the equality (8) (in the same order of appearance), such that the right hand side equals the sum of each $R_i$.
By several integration by parts, similarly as been done ([22] p. 125, for example), we get

$$R_1 = -2\mathcal{R}e\int [(x.\nabla\overline{u})\Delta u + (x.\nabla\overline{v})\Delta v]\,dx = (2-n)\int (|\nabla u|^2 + |\nabla v|^2)dx.$$

Also, the second term of the right-hand-side of equality (8) can be rewritten as

$$R_2 = 2\kappa\sum_{k=1}^{n}\int \mathcal{R}e\,x_k\frac{\partial}{\partial x_k}(u\overline{v})dx = -2\kappa n\mathcal{R}e\int u\overline{v}dx.$$

By transforming the third term of the right-hand-side of (8), one obtains:

$$R_3 = -2\gamma\mathcal{I}m\int [(x.\nabla\overline{u})u - (x.\nabla\overline{v})v]\,dx.$$

Expressing

$$\frac{\partial \bar{u}}{\partial x_k} = \frac{\partial |u|^2}{\partial x_k} - \bar{u}\frac{\partial u}{\partial x_k},$$

and similarly for the second component, the first part being real, one obtains by taking the imaginary part,

$$\begin{aligned} R_3 &= -2\gamma \mathcal{I}m \sum_{k=1}^{n} \int x_k \left( -\bar{u}\frac{\partial u}{\partial x_k} + \bar{v}\frac{\partial v}{\partial x_k} \right) dx \\ &= 2\gamma \mathcal{I}m \int \left[ (\bar{u}x.\nabla u) - (\bar{v}x.\nabla v) \right] dx. \end{aligned}$$

The last term of the right-hand-side of (8) can be formulated as follows.

$$R_4 = -\sum_{k=1}^{n} \int x_k \left[ g_{11}|u|^2 (2\mathcal{R}e\frac{\partial \bar{u}}{\partial x_k} u) + g_{22}|v|^2 (2\mathcal{R}e\frac{\partial \bar{v}}{\partial x_k} v) \right] dx$$
$$+ g_{12} \sum_{k=1}^{n} \int x_k \left[ |v|^2 (2\mathcal{R}e\frac{\partial \bar{u}}{\partial x_k} u) + |u|^2 (2\mathcal{R}e\frac{\partial \bar{v}}{\partial x_k} v) \right] dx.$$

Hence,

$$R_4 = -\sum_{k=1}^{n} \int x_k \left[ g_{11}|u|^2 \frac{\partial}{\partial x_k}|u|^2 + g_{22}|v|^2 \frac{\partial}{\partial x_k}|v|^2 \right] dx + g_{12} \sum_{k=1}^{n} \int x_k \left[ |v|^2 \frac{\partial}{\partial x_k}|u|^2 + |u|^2 \frac{\partial}{\partial x_k}|v|^2 \right] dx.$$

Finally, one gets:

$$\begin{aligned} R_4 &= -\sum_{k=1}^{n} \int x_k \left[ \frac{1}{2}\frac{\partial}{\partial x_k}(g_{11}|u|^4 + g_{22}|v|^4) + g_{12}\frac{\partial}{\partial x_k}(|u|^2|v|^2) \right] dx \\ &= \frac{n}{2} \int (g_{11}|u|^4 + g_{22}|v|^4) dx + n g_{12} \int |u|^2|v|^2 dx. \end{aligned}$$

The left-hand-side of the equality (8) can be also rewritten as follows:

$$\mathcal{R}e \int \imath \sum_{k=1}^{n} x_k \left[ \left( \frac{\partial \bar{u}}{\partial x_k} u \right)_t - \frac{\partial}{\partial x_k}(u\bar{u}_t) + \left( \frac{\partial \bar{v}}{\partial x_k} v \right)_t - \frac{\partial}{\partial x_k}(u\bar{u}_t) \right] dx$$
$$= -\frac{\partial}{\partial t}\mathcal{I}m \int \left[ (x.\nabla\bar{u})u + (x.\nabla\bar{v})v \right] dx - \mathcal{R}e \int \imath \sum_{k=1}^{n} x_k \frac{\partial}{\partial x_k}(u\bar{u}_t + v\bar{v}_t) dx,$$

and one can still use the system (1) to obtain:

$$\mathcal{R}e \int \imath \sum_{k=1}^{n} x_k \left[ \left( \frac{\partial \bar{u}}{\partial x_k} u \right)_t - \frac{\partial}{\partial x_k}(u\bar{u}_t) + \left( \frac{\partial \bar{v}}{\partial x_k} v \right)_t - \frac{\partial}{\partial x_k}(u\bar{u}_t) \right] dx$$
$$= \frac{\partial}{\partial t}\mathcal{I}m \int \left[ (\bar{u}x.\nabla u) + (\bar{v}x.\nabla v) \right] dx - n\mathcal{R}e \int (|\nabla u|^2 + |\nabla v|^2) dx$$

$$- n\kappa \int (v\bar{u} + u\bar{v})dx + n \int (g_{11}|u|^4 + g_{22}|v|^4 + 2g_{12}|u|^2|v|^2)dx.$$

Consequently, using all the previous computations, equation (8) can be formulated in the following way, by recovering the energy expression,

$$\frac{\partial}{\partial t}\mathcal{I}m \int [(\bar{u}x.\nabla u) + (\bar{v}x.\nabla v)]\, dx = nE(t) + (2-n)\int (|\nabla u|^2 + |\nabla v|^2)dx$$
$$- 2\kappa n\mathcal{R}e \int u\bar{v}dx + \gamma\mathcal{I}m \int [(\bar{u}x.\nabla u) - (\bar{v}x.\nabla v)]\, dx. \quad (9)$$

In other hand, let us take the difference between the scalar product of the first equation of the system (1) with $|x|^2u$ and the second equation with $|x|^2v$. Then, the imaginary part of the computations gives:

$$\frac{1}{2}\frac{\partial}{\partial t}\int |x|^2(|u|^2 - |v|^2)dx = 2\kappa\mathcal{I}m \int |x|^2 v\bar{u}dx + \gamma X(t). \quad (10)$$

By summing side by side 4 times equation (9) with $4\gamma$ times equation (10), we obtain the final formulation of the second time derivative of X. □

In order to state the main result of this paper, let us now introduce several notations. Following the introduction, we will consider the quadratic form $\phi(X, Y)$,

$$\phi(X, Y) := g_{11}X^4 + g_{22}Y^4 + 2g_{12}X^2Y^2, \quad (11)$$

that we assume negative definite. This property is valid as soon as the coefficients $g_{11}, g_{22}, g_{12}$ are all strictly negative, but also if $g_{11}, g_{22} < 0$ and $g_{12} \leq \sqrt{g_{11}g_{22}}$.
Let us define $\rho(t)$, $F(t)$, $M(t)$ and $G(t)$ four real maps by,

$$\begin{aligned}
\rho(t) &= \int_0^t \int_0^\sigma (|\nabla u|^2 + |\nabla v|^2 + |u|^4 + |v|^4)d\tau d\sigma, \\
F(t) &= X(0) + Y(0)t + 6E(0)t^2 + \frac{3\kappa}{\gamma^2}Q(0)(e^{2\gamma t} - 2\gamma t - 1), \\
M(t) &= \sup_{\tau\in[0,t]} F(\tau) + 1, \\
G(t) &= M(t)\Big(\frac{c_1 t}{2} + \exp(\frac{\gamma c_3 t}{c_2}) - 1\Big),
\end{aligned}$$

where the real constants $c_1, c_2$ and $c_3$ being as follows

$$\begin{aligned}
c_1 &= 4\gamma^2(\frac{\kappa}{\gamma} + 3), \\
c_2 &= \begin{cases} \min(2, -2g_{11}, -2g_{22}) & \text{if } g_{12} < 0 \\ \min(2, (-2g_{11} + g_{12}), (-2g_{22} + g_{12})) & \text{if } g_{12} > 0, \end{cases} \\
c_3 &= 24\max(1, |g_{11}|, |g_{22}|).
\end{aligned}$$

**Proposition 2.2.** *Let $C_0$ a real constant defined later according to the initial conditions $(u_0(x), v_0(x))$ of the Cauchy problem for the system (1). Let $\beta = \frac{\gamma c_3}{c_2}$ be another constant, and let us define the real map*

$$\widetilde{M}(t) = 1 + X(0) + C_0(e^{2\gamma t} - 1),$$

*and*

$$\begin{aligned} T_0^+ &= \frac{1}{\beta}\ln\left(1 + \frac{\beta^2}{(1 + X(0))(\beta^2 + c_1)}\right) \\ T_0^- &= \frac{1}{\beta}\ln\left(1 + \frac{\beta^2}{\widetilde{M}(T_0^+)(\beta^2 + c_1)}\right). \end{aligned} \tag{12}$$

*If*

$$E(0) < -\frac{\widetilde{M}(T_0^+)}{6(T_0^-)^2}, \tag{13}$$

*we fix* $C_0 = \frac{|Y(0)|}{2\gamma} + \frac{3\kappa}{\gamma^2}Q(0)$. *And if*

$$Y(0) < \frac{6\kappa}{\gamma}Q(0) - \frac{\widetilde{M}(T_0^+)}{T_0^-}, \tag{14}$$

*we fix* $C_0 = \frac{6|E(0)|}{\gamma^2} + \frac{3\kappa}{\gamma^2}Q(0)$.
*Then, there exists $T_0 \in [T_0^-, T_0^+]$ such that the following conditions are satisfied:*

$$F(T_0) + 1 < 0, \tag{15}$$

$$G(T_0) < 1. \tag{16}$$

The proof of this proposition is similar to that have been done by Dias *et. al.* [16].

*Proof.* Let us introduce

$$\widetilde{G}(t) := M(t)\left(1 + \frac{c_1}{\beta^2}\right)(e^{\beta t} - 1) \quad \text{for } t > 0.$$

It is obvious that for $t > 0$, $\widetilde{G}(t) > G(t)$ and it can be also easily viewed that $\lim_{0^+} \widetilde{G} = 0$, $\lim_{+\infty} \widetilde{G} = +\infty$. By intermediate value theorem, let us then define $T_0$ as the smallest solution of the equation $\widetilde{G}(T_0) = 1$. This automatically implies condition (16), and also,

$$\forall 0 < t < T_0, \qquad T_0 < \frac{1}{\beta}\ln\left(\frac{\beta^2}{M(t)(\beta^2 + c_1)}\right).$$

Using the fact that $M(t) \geq F(0) + 1 = 1 + X(0)$ one obtains that $T_0 \leq T_0^+$.

Moreover, let us first consider that $E(0) \leq 0$, then $F(t) + 1 - 6E(0)t^2 \leq \widetilde{M}(t)$ with $t \geq 0$. Therefore, for all $t \in [0, T_0^+]$ one obtains that

$$M(t) \leq \sup_{\tau\in[0,t]} F(\tau) + 1 - 6E(0)\tau^2 \leq \sup_{\tau\in[0,t]} \widetilde{M}(\tau) = \widetilde{M}(T_0^+),$$

which implies that $T_0 \geq T_0^-$.

Condition (13) implies that $E(0) < -\frac{\widetilde{M}(T_0)}{6T_0^2}$ from which (15) is obtained.

Secondly, let us consider now that condition (14) holds.

Thus, $F(t) - Y(0)t + \frac{6\kappa}{\gamma}Q(0)t \geq F(t)$ for any $Y(0) \leq \frac{6\kappa}{\gamma}Q(0)$. Therefore, for all $t \in [0, T_0^+]$ one obtains also that

$$M(t) \leq \sup_{\tau\in[0,t]} F(\tau) + 1 - Y(0)\tau + \frac{6\kappa}{\gamma}Q(0)\tau \leq \sup_{\tau\in[0,t]} \widetilde{M}(\tau) = \widetilde{M}(T_0^+),$$

which implies that $T_0 \geq T_0^-$.

Condition (14) also implies that $Y(0) \leq \frac{6\kappa}{\gamma}Q(0)$ from which (15) is obtained. □

The Theorem 2.1 will explain how conditions (13) or (14) (which imply (15) and (16) thanks to Proposition 2.2) can be considered as suffisient blowup conditions as soon as the quadratic form $\phi$ defined by (11) is negative definite.

**Theorem 2.1.** *Let $n = 2$ and let us assume that the coefficients $g_{11}$ and $g_{22}$ are strictly negative and either $g_{12} < 0$ or $g_{12} \leq \sqrt{g_{11}g_{22}}$.*
*Assuming moreover that the initial data $(u_0(x), v_0(x))$ of the Cauchy problem for the system (1) are chosen such that there exists $T_0 > 0$ for which the two conditions (15) and (16) hold. Then, the solution of the Cauchy problem for the system (1) does not exist in the interval $[0, T_0]$.*

*Proof.* The proof is done by contradiction. With the notations which have been previously introduced, the first step of the proof is to establish the viriel estimate:

$$X(t) + c_2\rho(t) \leq c_1 \int_0^t \int_0^\sigma X(\tau)d\tau d\sigma + \gamma c_3 \int_0^t \rho(\tau)d\tau + F(t), \tag{17}$$

and secondly, by continuation technique we obtain the non positivity which is unacceptable for a viriel.

Using equality (5) of Proposition 2.1 for $n = 2$, after decomposition of the last term, we get:

$$\frac{d^2X}{dt^2} = 8E(t) + 4\gamma^2X(t) + 8\gamma\mathcal{I}m\int((\overline{u}x.\nabla u) - (\overline{v}x.\nabla v))dx$$
$$+ 8\kappa\gamma\int|x|^2(v\overline{u})dx - 24\kappa\mathcal{R}e\int u\overline{v}dx + 8\kappa\mathcal{R}e\int u\overline{v}dx.$$

In order to substitute the last term of the previous rhs, one can use the total energy definition by this way,

$$4\kappa\int(u\overline{v} + \overline{u}v)dx = 8\kappa\mathcal{R}e\int u\overline{v}dx = 4E(t) - 4\int(|\nabla u|^2 + |\nabla v|^2)dx$$
$$+ 2\int(g_{11}|u|^4 + g_{22}|v|^4 + 2g_{12}|u|^2|v|^2)dx,$$

and one obtains

$$\frac{d^2X}{dt^2} + 4\int(|\nabla u|^2 + |\nabla v|^2)dx - 2\int(g_{11}|u|^4 + g_{22}|v|^4 + 2g_{12}|u|^2|v|^2)dx$$
$$= 12E(t) + 4\gamma^2 X(t) + 8\gamma \mathcal{I}m\int((\bar{u}x.\nabla u) - (\bar{v}x.\nabla v))dx$$
$$+ 8\kappa\gamma \mathcal{I}m\int |x|^2(v\bar{u})dx - 24\kappa \mathcal{R}e\int u\bar{v}dx.$$

According to equality (3) of Proposition 2.1, the following upper bound is straightforward:

$$\frac{dE}{dt} \le 2\gamma \max(1, |g_{11}|, |g_{22}|)(||\nabla u||^2 + ||\nabla v||^2 + ||u||^4_{L^4} + ||v||^4_{L^4}),$$

and after integration,

$$E(t) \le E(0) + \frac{\gamma c_3}{12}\int_0^t (||\nabla u||^2 + ||\nabla v||^2 + ||u||^4_{L^4} + ||v||^4_{L^4})d\tau.$$

Step by step let us formulate an upper bound for the second derivative of the viriel. Using Cauchy-Schwarz inequality and Young inequality with $\varepsilon = 2\gamma$, we obtain

$$8\gamma \mathcal{I}m\int((\bar{u}x.\nabla u) - (\bar{v}x.\nabla v))dx \le 8\gamma\int(|\bar{u}x.\nabla u| + |\bar{v}x.\nabla v|)dx,$$

then,

$$8\gamma \mathcal{I}m\int((\bar{u}x.\nabla u) - (\bar{v}x.\nabla v))dx \le 2(||\nabla u||^2 + ||\nabla v||^2) + 8\gamma^2 X. \tag{18}$$

Also, we get

$$8\kappa\gamma \mathcal{I}m\int |x|^2(v\bar{u})dx \le 4\kappa\gamma X, \tag{19}$$

and using the density upper bound,

$$24\kappa \mathcal{R}e\int u\bar{v}dx = 12\kappa\int(u\bar{v} + \bar{u}v)dx \le 12\kappa Q(0)e^{2\gamma t}. \tag{20}$$

With the notations that have been introduced before, and inequalities (18), (19) and (20) the second derivative of X can be estimated as follows,

$$\frac{d^2X}{dt^2} + c_2(|\nabla u|^2 + |\nabla v|^2 + |u|^4 + |v|^4) \le c_1 X(t)$$
$$+ \gamma c_3\int_0^t (|\nabla u|^2 + |\nabla v|^2 + |u|^4 + |v|^4)d\tau + 12E(0) + 12\kappa Q(0)e^{2\gamma t}.$$

After successively integrate two times in time, the inequality (17) is obtained.

Now, one can continue as Dias *et al.* did in [16] by assuming that the solution of the Cauchy problem for the system (1) exists for all $t \in [0, T_0]$ with $T_0 > 0$ such that conditions (15) and (16) hold. One can then define

$$T_1 = \sup\{t \in [0, T_0] : X(s) \le M(T_0) \text{ for any } s \in [0, t]\}.$$

Due to the viriel estimate (17) and condition (15), one obtains that

$$\begin{aligned} X(t) + c_2\rho(t) &\leq c_1 M(T_0)\frac{T_0^2}{2} + \gamma c_3 \int_0^t \rho(\tau)d\tau + M(T_0) - 1 \\ &< M(T_0) + \gamma c_3 \int_0^t \rho(\tau)d\tau. \end{aligned} \tag{21}$$

Then, the positivity of the viriel and Gronwall's inequality provides the inequality as follows,

$$\rho(t) \leq \frac{M(T_0)}{c_2}\exp(\frac{\gamma c_3}{c_2}t).$$

Using this upper bound back into inequality (17) and due to the definition of $G(t)$, we find that

$$\begin{aligned} X(t) + c_2\rho(t) &\leq M(T_0)\left(c_1\frac{T_0^2}{2} + \exp(\frac{\gamma c_3}{c_2}T_0) - 1\right) + F(t) \\ &\leq G(T_0) + F(t). \end{aligned}$$

As the mapping $\rho(t)$ is positive,

$$X(t) \leq G(T_0) + F(t), \tag{22}$$

then by taking the least upper bound and using the assumption (16), one obtains

$$\sup_{t\in[0,T_0]} X(t) \leq M(T_0) - 1 + G(T_0) < M(T_0).$$

Thus, $T_1 = T_0$, hence using (22) and assumption (16),

$$X(T_0) \leq F(T_0) + 1 < 0.$$

This non positivity being impossible for a viriel, the proof is done. □

## 3. Concluding remarks

This paper aims to study blowup phenomenon in finite time for solution of a coupled system of NLS equations (in critical case) which statisfies a PT-symmetry property. In particular, two sufficient conditions of finite time blowup have been obtained by adapting a method used recently by Dias *et al.* for the problem in the supercritical case. Several complements should be done in the future about this phenomenon, like estimates of the finite blowup times and a numerical procedure suitable to provide some illustrations.

## Acknowledgements

Special thanks to C. Sulem who invited one of the authors to University of Toronto and brought up this problem to our attention.

## Conflict of Interest

All authors declare no conflicts of interest in this paper.

## References


1. Manakov S.V, *On the theory of two-dimensional stationary self-focusing of electromagnetic waves.* Zh. Eksp. Teor. Fiz. **65** (1973), 505-516.
2. Ismail M.S and Taha T.R, *Numerical simulation of coupled nonlinear Schrödinger equation.* Math. Comput. Simulat. **56** (2001), 547-562.
3. Driben R and Malomed B.A, *Stability of solitons in parity-time-symmetric couplers.* Opt. Lett. **36** (2011), 4323-4325.
4. Abdullaev F.K, Konotop V.V, Ögren M, et al. *Zeno effect and switching of solitons in nonlinear couplers.* Opt. Lett. **36** (2011), 4566-4568.
5. Pelinovsky Dmitry E, Localization in periodic potentials, London Mathematical Society -Lecture Note Series 390 Cambridge University Press, (2011).
6. Pelinovsky D.E, Zezyulin D.A and Konotop V.V, *Global existence of solutions to coupled $\mathcal{PT}$-symmetric nonlinear Schrödinger equations.* Int. J. Theor. Phys. **54** (2015), 3920-3931.
7. Suchkov S.V, Sukhorukov A.A, Huang J, et al. *Nonlinear switching and solitons in PT-symmetric photonic systems.* Laser Photonics Rev. **10** (2016), 177-213.
8. Destyl E, Nuiro P.S and Poullet P, *On the global behaviour of solutions of a coupled system of nonlinear Schrödinger equations.* Stud. Appl. Math. in press, (2017).
9. Tsutsumi M, *Nonexistence of global solutions to the Cauchy problem for the damped nonlinear Schrödinger.* SIAM J. Math. Anal. **15** (1984), 317-366.
10. Ohta M and Todorova G, *Remarks on global existence and blowup for damped nonlinear Schrödinger equations.* Discrete Contin. Dyn. Syst. **23** (2009), 1313-1325.
11. Dias J-P and Figueira M, *On the blowup of solutions of a Schrödinger equation with an inhomogeneous damping coefficient.* Commun. Contemp. Math. **16** (2014), 1350036-1350046.
12. Bender C.M and Boettcher S, *Real spectra in non-Hermitian Hamiltonians having $\mathcal{PT}$-symmetry.* Phys. Rev. Lett. **80** (1998), 5243-5246.
13. El-Ganainy R, Makris K.G, Christodoulides D.N, et al. *Theory of coupled optical $\mathcal{PT}$-symmetric structures.* Opt. Lett. **32** (2007), 2632.
14. Konotop V.V, Yang J and Zezyulin D.A, *Nonlinear waves in $\mathcal{PT}$-symmetric systems.* Rev. Mod. Phys. **88** (2016), 035002(59).
15. Kartashov Y.V, Konotop V.V and Zezyulin D.A, *CPT-symmetric spin-orbit-coupled condensate.* EPL. **107** (2014), 50002.
16. Dias J.-P, Figueira M, Konotop V.V et al. *Supercritical blowup in coupled parity-time-symmetric nonlinear Schrödinger equations.* Stud. Appl. Math. **133** (2014), 422-440.
17. Sulem C and Sulem P.-L, The Nonlinear Schrödinger Equation, Springer, New-York, 1999.

18. Lin T.C and Wei J, *Solitary and self-similar solutions of two-component system of nonlinear Schrödinger equations.* Physica D. **220** (2006), 99-115.

19. Jüngel A and Weishäupl R.-M, *Blow-up in two-component nonlinear Schrödinger systems with an external driven field.* Math. Mod. Meth. App. Sci. **23** (2013), 1699-1727.

20. Zhongxue Lü and Zuhan Liu, *$L^2$-concentration of blow-up solutions for two-coupled nonlinear Schrödinger equations.* J. Math. Anal. Appl. **380** (2011), 531-539.

21. Glassey R.T, *On the blowing-up of solutions to the Cauchy problem for the Nonlinear Schrödinger equation.* J. Math. Phys. **18** (1977), 1794-1797.

22. Linares F and Ponce G, Introduction to Nonlinear Dispersive Equations, Universitext, Springer New-York, (2009).

# Permissions

All chapters in this book were first published in MATHEMATICS, by AIMS Press; hereby published with permission under the Creative Commons Attribution License or equivalent. Every chapter published in this book has been scrutinized by our experts. Their significance has been extensively debated. The topics covered herein carry significant findings which will fuel the growth of the discipline. They may even be implemented as practical applications or may be referred to as a beginning point for another development.

The contributors of this book come from diverse backgrounds, making this book a truly international effort. This book will bring forth new frontiers with its revolutionizing research information and detailed analysis of the nascent developments around the world.

We would like to thank all the contributing authors for lending their expertise to make the book truly unique. They have played a crucial role in the development of this book. Without their invaluable contributions this book wouldn't have been possible. They have made vital efforts to compile up to date information on the varied aspects of this subject to make this book a valuable addition to the collection of many professionals and students.

This book was conceptualized with the vision of imparting up-to-date information and advanced data in this field. To ensure the same, a matchless editorial board was set up. Every individual on the board went through rigorous rounds of assessment to prove their worth. After which they invested a large part of their time researching and compiling the most relevant data for our readers.

The editorial board has been involved in producing this book since its inception. They have spent rigorous hours researching and exploring the diverse topics which have resulted in the successful publishing of this book. They have passed on their knowledge of decades through this book. To expedite this challenging task, the publisher supported the team at every step. A small team of assistant editors was also appointed to further simplify the editing procedure and attain best results for the readers.

Apart from the editorial board, the designing team has also invested a significant amount of their time in understanding the subject and creating the most relevant covers. They scrutinized every image to scout for the most suitable representation of the subject and create an appropriate cover for the book.

The publishing team has been an ardent support to the editorial, designing and production team. Their endless efforts to recruit the best for this project, has resulted in the accomplishment of this book. They are a veteran in the field of academics and their pool of knowledge is as vast as their experience in printing. Their expertise and guidance has proved useful at every step. Their uncompromising quality standards have made this book an exceptional effort. Their encouragement from time to time has been an inspiration for everyone.

The publisher and the editorial board hope that this book will prove to be a valuable piece of knowledge for researchers, students, practitioners and scholars across the globe.

# List of Contributors


**Motohiro Sobajima**
Department of Mathematics, Faculty of Science and Technology, Tokyo University of Science, 2641 Yamazaki, Noda-shi, Chiba-ken 278-8510, Japan

**YutaWakasugi**
Graduate School of Mathematics, Nagoya University, Furocho, Chikusaku, Nagoya 464-8602 Japan

**Ahmad Mohammed Alghamdi**
Department of Mathematical Science , Faculty of Applied Science, Umm Alqura University, P. O. Box 14035, Makkah 21955, Saudi Arabia

**Sadek Gala**
Department of Mathematics, University of Mostaganem, Algeria
Dipartimento di Mathematica e Informatica, Università di Catania, Viale Andrea Doria, 6, 95125 Catania, Italy

**Gunduz Caginalp**
Mathematics Department University of Pittsburgh, Pittsburgh, PA 15260, USA

**Khalida Inayat Noor**
Department of Mathematics Comsats Institute of Information Technology Park Road, Islamabad, Pakistan

**Nazar Khan and Qazi Zahoor Ahmad**
Department of Mathematics Abbottabad University of Science and Technology, Abbottabad, Pakistan

**Murugan Suvinthra, Krishnan Balachandran and Rajendran Mabel Lizzy**
Department of Mathematics, Bharathiar University, Coimbatore 641046, India

**Liang Cai and Huan-Huan Zhang**
School of Mathematics and Statistics, Beijing Institute of Technology, Beijing 100081, China

**Li-Yun Pan**
Department of Basic Science, Beijing Institute of Graphic Communication, Beijing 102600, China
Beijing Institute of Education, Beijing 100120, China

**Qitong Ou and Huashui Zhan**
School of Applied Mathematics, Xiamen University of Technology, Xiamen 361024, P. R. China

**Sadek Gala and Mohamed Mechdene**
Department of Mathematics, University of Mostaganem, Box 227, Mostaganem, 27000, Algeria

**Joseph L. Shomberg**
Department of Mathematics and Computer Science, Providence College, 1 Cunningham Square, Providence, Rhode Island 02918, USA

**Nazar Khan, Bilal Khan, Qazi Zahoor Ahmad and Sarfraz Ahmad**
Department of Mathematics Abbottabad University of Science and Technology Abbottabad, Pakistan

**Sadek Gala**
Department of Mathematics, University of Mostaganem, Box 227, Mostaganem, 27000, Algeria
Dipartimento di Matematicae Informatica, Universit`a di Catania Viale Andrea Doria, 6, 95125 Catania, Italy

**Maria Alessandra Ragusa**
Dipartimento di Matematicae Informatica, Universit`a di Catania Viale Andrea Doria, 6, 95125 Catania, Italy

**Huaji Cheng and Yanxia Hu**
School of Mathematics and Physics, North China Electric Power University, Beijing, 102206, China

**Canot Hèlène and Fŕenod Emmanuel**
Universitė de Bretagne-Sud, UMR 6205, LMBA, F-56000 Vannes, France

**Bhabesh Das**
Department of Mathematics, B.P.Chaliha College, Assam-781127, India

**Helen K. Saikia**
Department of Mathematics, Gauhati University, Assam-781014, India

**Hiroshi Takahashi**
Department of Mathematics, Tokyo Gakugei University, Koganei, Tokyo, 184-8501, Japan

**Ken-ichi Yoshihara**
Department of Mathematics, Yokohama National University, Hodogaya, Yokohama, 240-8501,Japan

**Edès Destyl, Silvere Paul Nuiro and Pascal Poullet**
LAMIA, Université des Antilles, Campus de Fouillole, BP 250, Pointe-à-Pitre F-97115 Guadeloupe F.W.I, France

# Index

www.ingramcontent.com/pod-product-compliance
Lightning Source LLC
Chambersburg PA
CBHW082034190326
41458CB00010B/3368
*9781632386311*